WHY SEX

MATTERS

Why Sex Matters

A Darwinian Look

at Human Behavior

Bobbi S. Low

PRINCETON UNIVERSITY PRESS

PRINCETON, NEW JERSEY

Library of Congress Cataloging-in-Publication Data

Low, Bobbi S.
Why sex matters : a Darwinian look at human behavior /
Bobbi S. Low.
p. cm.
Includes bibliographical references.
ISBN 0-691-02895-8 (alk. paper)
1. Human evolution. 2. Sex role. 3. Nature and nurture.
4. Sociobiology. I. Title.
GN281.4.L68 1999
305.3—dc21   99-24612

FOR MY SON MICHAEL,
and for my academic lineage: parental,
sibling, and offspring generations

Contents

Preface

SEX DIFFERENCES are central to our lives, wherever and whenever—however—we live. And we all think about them, from Professor Henry Higgins in *My Fair Lady* ("Why can't a woman be more like a man?") to Sigmund Freud ("What do women want?") to actor Charles Boyer ("Vive la différence!"). Are these differences genetically programmed: snakes and snails and puppy-dog tails for boys versus sugar and spice and everything nice for girls? Or are we trapped by our societies into roles that may be uncongenial to us simply because we do, or do not, have a Y chromosome? This is a fascinating tangle: what do the widely acclaimed (and equally widely denied) differences between men and women mean in terms of the ways in which men and women use resources, take risks, make war, and raise children? Which differences are lasting, which are ephemeral? If we follow the real differences through time, across space, and into different environments, what might they mean in today's societies?

We are asking these questions at an exciting time. New research in evolutionary theory, combined with findings from anthropology, psychology, sociology, and economics, supports the perhaps unsettling view that men and women have indeed evolved to behave differently—that, although environmental conditions can exaggerate or minimize these differences in male and female behaviors, under most conditions each sex has been successful as a result of very different behaviors. I will argue that many apparently complex behaviors and sex differences in fact arise from simple conditions that are conducive to analysis.

I begin with the fundamental principle of evolutionary biology, that all living organisms have evolved to seek and use resources to enhance their reproductive success. They strive for matings, invest in children or help other genetic relatives, and build genetically profitable relationships. In biology, this is not a controversial proposition, and it follows that all organisms will act as though they are able to calculate costs and benefits. Futhermore, in biological terms the currencies are, in the end, reproductive: that is, who survives and who reproduces best? This principle seems so simple that it is hard to imagine that diverse and complicated behaviors could arise from it. Yet they do, because the ecological conditions that shape success vary so widely.

There is growing evidence that humans are not immune from this principle, for in order to survive and persist, we humans must solve the same ecological problems as all other species. Evolutionists argue, therefore, that people have evolved to behave in ways that do, or did, contribute to their reproductive success. This approach can help us answer apparently diverse, unconnected questions such as the following: Why are there so few women warriors? Why were chastity belts designed for women, not men? Why aren't old women seen as sexy, but old men often are? Why are practices such as infanticide routine in some cultures and forbidden in others? Many of these questions can be posed only by using an evolutionary approach; in other approaches they have represented problems, or "noise."

I present three themes in this work. First, resources are useful in human survival and reproduction; like other living things, we have evolved to wrest resources from the environment for our benefit. Second, the two sexes tend to differ in how they can use resources most effectively to accomplish survival and reproduction. Third, how each sex accomplishes these ends relies not only (and not obviously) on differences in genes, but on differences in environment—there are no identified genes specific for polygyny, for example, but in many environments the trends for male mammals to profit from trying to be polygynous are strong.

These intertwined motifs of resource utility, sex differences, and environmental constraints soon lead us to consider other problems—for example, status striving and risk taking. Why is homicide largely a male enterprise? Why are men and women jealous about different things? Differences such as these give rise to the grander issues of population numbers, resource consumption, and sustainability. As human populations have grown and technologies have become more efficient, the utility of resources in simple survival and reproduction leads us to a series of dilemmas: Why did family sizes fall in nineteenth-century Europe and North America (the "demographic transition")? Why is today's demographic transition in the developing world so different? What is the impact of the "global village"—the evolutionary novelty that our actions here and now affect others' lives far away? What can we do about the fact that many of today's problems have relatively straightforward technical solutions—which will work only if we can see the interests of strangers in strange lands as equal to our own, something it never paid our ancestors to do?

To follow these threads, I will begin with the basic arguments and assumptions of behavioral and evolutionary ecology: selfish genes, conflicts of interest, and why two (and not more or fewer) sexes have specialized to reproduce through different behaviors (chapters 1–3). Then I ask: How do these basic sex differences, whose theory we understand, actually play out in other primates, as well as humans (chapters 4–6)? Next, I take an empirical glance at the diverse ways in which both traditional and transitional societies make a living, how men's and women's roles and lives diverge, and how even marriage is affected by ecology and resources (chapters 7, 8). The complexity of these patterns leads us back to basic theory to explore how conflicts of interest are mediated, literally from the level of genes in genomes to whole societies (chapters 9, 10). Sex differences and conflicts of interest help us predict why there are so few women warriors or high-roller politicians in most societies—and the kinds of societies in which they are likely to occur (chapters 11–14). And finally I ask: How does our evolutionary past interact with current global population and resource consumption problems (chapter 15)? Have we, in creating novel environments, changed the rules so that now it may even be detrimental to "strive" to our utmost abilities? Have we gotten ourselves into a bind in which the behaviors we have evolved to do, and do ever more efficiently, are now the behaviors that threaten our very existence?

It is a messy business to try to sort out the intricacies of sex, power, and resources, both in humans and in other animals. I will try to avoid some popular but diverting issues debated within the fields I draw on: whether natural selection or historical accident is more important in evolution; whether one must know mechanisms to understand evolution; whether our environments today are so new

that we can deduce nothing.[1] As we wend our way through these issues, I hope that my own positions and reasoning will become clear.

I will be drawing cross-cultural examples from traditional, historical, and modern societies; and from physical, physiological, demographic, and behavioral data. I'll share my interest and my concern,
but not offer cut-and-dried solutions. My purpose is to say: Here is
a puzzle, a conundrum—what ideas, old and new, can we use to
solve it? I would like to reach scholars in the traditional human disciplines with concepts that may be new and tantalizing to them. It is
my hope that experts in other fields will find themselves saying: "I
know a way to tackle that problem; my field can contribute something here although it's not the sort of problem I usually analyze."
My sense is that many crucial problems haven't been solved because
we stick to our own disciplinary approaches, and that no one will be
able to solve them alone in an attempt to use interdisciplinary approaches. But a number of us, reaching across boundaries with tolerance and patience, might make some progress in getting answers.

Acknowledgments

I T SEEMS so easy to start out—but writing a book, like childbirth, turns out to be something most of us will repeat only if we forget many of the details of the process. Writing is a complex process, and many people must get credit for the accomplishment.

Thanks go to some long-suffering souls who read and reviewed various chapters at various stages of writing (some having suffered through more than one draft): Robert Axelrod, Mary Brinton, Cameron Campbell, Helena Cronin, Lee Cronk, Martin Daly, Josh Epstein, Steven Frank, W. D. Hamilton, Kristen Hawkes, Henry Horn, Laura Howard, Bill Irons, Misty McPhee, John Mitani, Randolph Nesse, Karen Parker, Jen Parody, Carl Simon, Barbara Smuts, George Williams, and Margo Wilson.

Thanks also to the following people for discussions, joint work on problems leading to parts of this manuscript, or for providing data: Richard Alexander, Scott Atran, Monique Borgerhoff Mulder, Elizabeth Cashdan, Napoleon Chagnon, Jae Choe, Alice Clarke, Klaus Dietz, Derek Dimcheff, Elizabeth Hill, Kim Hill, Magdalena Hurtado, Deborah Judge, Hillard Kaplan, John Knodel, Jeffrey Kurland, Richard Nisbett, Elinor Ostrom, Andrew Richards, Matt Ridley, Mark Siddall, B. Holly Smith, Rachel Smolker, Eric Smith, and Joao Sousa.

And thanks to these people for the ever-underappreciated but always necessary logistic support: Lisa DeBruin, Rebecca Howell, Jamie Kryscynski, Kristen LeBlond, and Melissa Slotnik. Carole Shadley suffered many partial drafts and valiantly strove to make sense of them, and to format the manuscript clearly. Acquisitions editor Jack Repcheck of Princeton University Press valiantly weaned me (or tried) from overuse of academic jargon. Kristen Gager, Linda

Chang, and Alice Calaprice of Princeton University Press actually made the manuscript come together in publishable form.

Along the way, I profited enormously from discussions with seminar participants in Natural Resources and Environment 505, Human Resource Ecology, at the University of Michigan, and with participants in the German-American Postdoctoral Workshop in Bielefeld, Germany (1996) and Ann Arbor, Michigan (1997).

Carl Simon has provided humor and a sense of perspective, and my son, Michael, provided a reason to expend ridiculous amounts of parental effort.

All remaining errors and blunders are, sadly, mine alone.

WHY SEX

MATTERS

1.

Introduction

Probably a crab would be filled with a sense of personal outrage if it could hear us class it without ado or apology as a crustacean, and thus dispose of it. "I am no such thing," it would say: "I am MYSELF, MYSELF *alone."*
—William James, *Varieties of Religious Experience* (1902)

To the same natural effects we must, as far as possible, assign the same causes.
—Sir Isaac Newton

WHY CAN'T A WOMAN be more like a man?" wailed Professor Henry Higgins in *My Fair Lady*, the musical derived from George Bernard Shaw's *Pygmalion*. Certainly in many societies, across time, there have been women who were "more like a man." Think of Joan of Arc, who was burned at the stake (on the minor charge of wearing men's clothing); or of George Sand, of whom Elizabeth Browning said, "You are such a large-brained woman and a large-hearted man." Yet in part, we remember such women *because* they are singular, whether we envy their ability to break free or imagine that they missed a lot. What really contributes to the patterns we see, and to their exceptions?

In this book I want to explore sex differences from a relatively unusual perspective, one that is often misunderstood. Understanding and explaining human behavior is a central concern for all of us. But doing so—especially when sex differences are the issue—presents a real dilemma. We are complicated, highly social beings. We live in a staggering array of environments, both ecological and social. Our families, lovers, and friends are not exemplars or prototypes, but unique, particular individuals. None of us wants to be "reduced" to some formula.

For, like William James's crab, we *know* we are above crude analysis. Even the name we give ourselves, *Homo sapiens*, reflects both the value that we give to understanding, and the fact that we feel ourselves to be special. Like that crab, many people may be appalled at the approach I will use here, that is, to assume that we humans are as predictable as other animals in our behavior, and are governed by the same rules. And I want to begin with *simple* rules, no less.

Many of us assume that humans operate under rules that are different from those of other species, that our rules are culturally based rather than biological. I will ask: What can we learn if we begin without assuming that this were true? I want to explore what a biologist would predict if he or she knew only that here was a smart, upright-walking, highly social primate and *nothing more.* I will explore the ecology of being male and female, beginning with simple rules and with what I can discern about the environments with which the evolutionary rules interact. The approach I use, behavioral ecology, is an evolutionary approach with roots in Charles Darwin's work. It focuses on the question, How do environmental conditions influence our behavior and our lifetimes?, and has proved profitable in exploring other realms of human behaviors.[1] Behavioral ecology and its intellectual relatives seek to understand how relatively simple operating rules interact with historical accidents, and with temporal and spatial specifics, to yield a rich diversity of patterns. There is no doubt that genes influence not only our physical structure and physiology, but our behavior; there is no doubt that historical accident often plays a role; nor is there any doubt that cultural and social pressures can influence behavior. But where lies the balance? Perhaps by beginning with very simple rules and assumptions, we can gain some insight.[2]

Vampire Stories and Beyond

Humans have always sought to explain the patterns they see. In fact, one of the strongest selective forces on human behavior has been to understand pattern, not only in order to deal with environmental variation, but to be the first in one's tribe who is able to predict events (imagine being the first human to predict a solar eclipse). But creating stories that more or less match our observations is not science but folklore. Consider vampires. From Bram Stoker through Anne Rice, from Bela Lugosi to Tom Cruise and Leslie Nielsen, vampires have always fascinated us: aristocratic, sexy, dangerous, and invincible. Vampire folklore provides a wonderful example of how our need to explain something can drive us to spin stories that *seem* to explain what we see, can be hard to refute, but nonetheless do not reflect what actually happens.

The folklorist Paul Barber, in a delightful examination of vampire myths around the world, notes that the ways people in preindustrial societies interpreted phenomena associated with death and the decay of corpses are "from our perspective, quite wrong. What makes them interesting, however, is that they are also usually coherent, cover all the data, and provide the rationale for some common practices that seem, at first, to be inexplicable."[3]

The variety of myths and legends about vampires all begin simply: death—especially unexpected or unusual death—brings more death. If someone died, "why" was likely to be unknown, and epidemics leading to death and more death were once far more common than today. Once they were buried (often without coffins), not all corpses had the decency to stay below ground. In a prebacteriology culture, people weren't likely to see a "flailing" corpse as the natural by-product of bacterial decay, but rather as the will of the dead person, or as the rejection of the corpse by Mother Earth. Since death brings death, those first to die (as in an epidemic) were dangerous and somehow had to be disarmed so that they could not continue to bring death. Only when all "changing" ceased, and ashes or bones alone remained, was the corpse neutral, inactive, and no longer dangerous. People thus began with a repeatable observation: that death brings death. This applies not only to vampires, but to the general idea that dead people call to their relatives and friends and must be propitiated to protect those still alive. Because they had no knowledge of disease transmission, people imbued the corpse with dangerous properties. Not a bad idea, particularly in times of plague, when unexpected deaths were frequent and vampire fears were heightened—but an idea that led to a misinterpretation of the normal signs of decomposition.

In folklore, in a variety of societies around the world, vampires are described as undecomposed; they have a ruddy or dark complexion, do not suffer from rigor mortis, are swollen or plump, have blood at the mouth and/or in internal cavities, and grow new skin or nails after burial. In an interesting twist, people suspected of being vampires were frequently buried differently, in ways making a diagnosis of vampirism *more*, not less, likely. Suspected vampires were often buried face down, so that if they tried to claw their way out of the earth to torment others, they would dig themselves in deeper. Because blood settles into the lowest capillaries after death, face-down burial meant that the body's face (rather than the back) would be dark and ruddy. And a ruddy face was believed to be a sign of a vampire. Some putative vampires were buried with lime to hasten their decomposition—but lime in fact retards it. Thus, someone who died in an unusual way and was feared to become a vampire was likely to be buried face down, with lime—and thus to have a ruddy face, to decompose more slowly, and, on exhumation, to be confirmed as a vampire. Such practices thus reinforced mistaken beliefs.

Folklore about vampires arises from an entirely sensible and consistent desire to explain something in the absence of complete information. But what people *say* about what they see and do can be a rotten path to explanation. Although observers called the corpses undecomposed, they described unmistakable signs of rot (e.g., a stench). Descriptions such as "ruddy" or "swollen" that were used to assert failure to decompose are in fact signs of the ordinary (but variable) process of decomposition.

It is important to separate carefully what people *describe*, as they see, hear, and smell what happens, from the *causes they attribute*. The observation that a corpse stinks is, in fact, consistent with the contention that decay is occurring. It is important to avoid this kind of muddle at all times, not only when we are no longer likely to believe in something like vampirism. Being led astray by "vampire myths" that sound reasonable but are untested is most likely to occur when a behavior is complicated and we want to believe the stories we tell.

EXPLAINING BEHAVIOR WITHOUT FOLKLORE

Other species' behavior can be more complex than we realize. An excellent example of such complexity generated by the interaction of operating rules (genes), environment, and historical accident is biologist Bernd Heinrich's work on food-sharing in ravens,

Corvus corax. Heinrich saw a curious behavior on a hike in Maine late one October: a group of ravens, feasting on a dead moose, were giving a distinctive, loud, and high-pitched "yell."[4] Because other ravens were attracted to the yelling, the result was that there were more competitors at the kill and thus less food for each of the yellers. Why didn't the raven who first found the kill just keep quiet? And why weren't the ravens now fighting over the kill?

If we saw humans share like this, or if we extended our often untested social perceptions about humans to ravens, we would probably think how kind all this indiscriminate sharing is. In fact, that would constitute a vampire story; behavioral ecologists find true genetically costly altruism to be so rare, as I will explain below, as to be a fluke in nonhumans. The ravens who shared seemed to be doing so at a cost. Heinrich's first question was whether the "called" birds were related to the callers, for sharing with individuals who have at least some genes in common can help copies of one's own genes. After much work marking and recapturing ravens in the field, Heinrich was able to eliminate the possibility that ravens were summoning their kin. Where next?

Heinrich made a series of careful observations, comparing the behavior of different ravens under different circumstances. His summary begins with eleven clues, and proceeds from simpler to more complex deductions.[5] Without giving away the whole plot, I can say that one of his major findings was that *adult* ravens are territorial, controlling access to any carcasses in their territory and driving off any juveniles found on a carcass. When a *juvenile* found a carcass, it was likely to "yell," attracting other juveniles (the largest group at a carcass was about 1500). When enough juveniles were present, the resident territorial adults could not drive all of them off. So the cost of additional juvenile competitors could be offset by the benefit of attracting a group large enough to stay on the carcass even if adults were nearby. Clearly the costs and benefits of yelling would differ under different circumstances. Heinrich did not simply create a plausible vampire or just-so story about the juvenile behavior he observed. Instead, he observed, made hypotheses, and tested them to discover the most likely functional reason for the ravens' behavior.

Heinrich could not know what role any gene plays in this behavior, so he used a technique called the "phenotypic gambit" to make testable predictions: starting from what he saw—the phenotype—

he made certain assumptions. He assumed that, whatever the relationship between genes and the set of behaviors he saw, enough time had passed for the system to come to equilibrium, and thus what he saw represented the outcome of competing strategies.[6] The phenotypic gambit is a powerful tool (although controversial for some), and I'll return to it below.

For ravens as for humans, both ecological and social conditions can change the costs and benefits of any action. Heinrich, of course, had to observe the ravens without their cooperation; he had to concentrate on what the ravens actually did. Sociologists, psychologists, and anthropologists find it useful to interview people, and this tells us something about what people themselves imagine they are doing. Because we can make decisions consciously, often we assume that unless a behavior is consciously considered, it is of no interest. Yet many other species routinely learn, and behave in complicated ways, without (so far as we can tell, at least) consciousness—or at least without the ability to share abstract sentiment through speech. Furthermore, as I noted above, it is presumptuous to assume that people's conscious attributions of their behavior is analytically helpful, and can cause real trouble.

Behavioral ecologists cannot interview ravens about why they call to other ravens when they find a carcass (who knows what reasons a raven might give, anyway?), and they don't know the genetics of the situation (is such calling the result of a single gene's action?). For these reasons, behavioral ecologists concentrate on what happens—on what behaviors show up under what conditions. If we take the same approach in looking at human behavior, we will lose some information about people's intentions, but we won't get distracted by our human reports of conscious reasons. And such lack of distraction may prove useful, for what people say is often not consistent with what we observe them doing.[7]

Perhaps new connections will appear as we look past what we imagine behavioral causes to be and as we look beyond what people say about why they act certain ways and examine carefully what sorts of behaviors we see in particular environments. Without requiring consciousness or rationality (or even speculating on their existence), we can ask what behaviors will be profitable under what environmental conditions. Then we can ask explicitly how conscious, cultural influences can influence the costs and benefits of these behaviors.

Kinds of "Why" Questions

Heinrich's analysis of raven behavior highlights an important distinction. To understand "why" we do things, to explain both the behaviors that seem almost universal or unvarying and those that vary greatly, we can seek answers in different ways. "Why" questions have two principal complementary forms in biology: "proximate" and "ultimate" explanations.[8] Why do birds migrate? One answer might be "changing day length causes hormonal changes, triggering migration." Both changing day length and changing hormones are *proximate* triggers, or cues. If we could interview birds, we might have another set of proximate causes, the equivalent of our reasons: "I really hate the cold," "it makes me feel good," "that way I get to see my relatives." However, proximate answers are no help in explaining why one species migrates while others don't, why not all individuals in this species migrate (costs and benefits may differ for older, younger, weak, or healthy individuals), or why day length rather than some other cue, or a combination of cues, has become the trigger.

The *ultimate* cause of migration always concerns reproductive success. Seasonal better-versus-worse geographic shifts in foraging and nesting areas mean that individuals who seek the better areas, shifting seasonally, leave more descendants than those who remain in one area. When day length is the most reliable predictor of these seasonal shifts, individuals who use it as a cue will fare better than those that use some other proximate cue or fail to migrate. Thus, we would predict migration patterns triggered by day length for birds that are (for example) insectivores or nectar-eaters in northern temperate regions; their food disappears seasonally. We expect variation in which individuals migrate when the benefits and costs of migration in terms of survival and reproduction differ for older, prime-age birds, compared with yearlings, for example. Proximate cues and ultimate (selective) causes tell us very different things.

It is useful to ask questions about both "proximate" triggers and "ultimate" selective cause, and it is important to understand that these two approaches are not alternatives but complement each other. Proximate triggers, the *mechanisms that release behaviors,* are sometimes also called "causes." They tell us what kinds of environmental factors are important. The ultimate cause of a behavior's existence, in evolutionary terms, is always its impact on family (or genetic line) persistence through time. We seldom think of such matters, perhaps because few of us can now trace our ancestry in the same way a certain schoolteacher from Cheddar, England, could: a preserved "bog man" from ancient times was found, through DNA analyses, to be a clear direct ancestor of his. Nonetheless, the persistence of genes through time, and the clustering of genes in family lines, is real.

Sometimes it is important to ask "Why?" at the proximate level. Suppose we wish to ask about variation in human fertility. Lowered fertility could have the proximate "cause" of later marriage age, and an ultimate selective cause of greater lineage success through fewer, better-invested children than through more numerous, but less able, children. We humans would naturally think about the first of these, but seldom about the second.

We could interview people about their conscious reproductive decisions, the proximate causes. If you were interested in manipulating what people will do, this would be the appropriate level at which to ask the question. Behavioral ecology, in contrast, seeks to discover which behaviors, in particular environments, result in greater success (more about definitions of success in a moment). It starts with a bias toward "ultimate" questions, although it seldom can ignore proximate correlates. We can profit from disaggregating—teasing apart—behavior patterns in a population: who does what, under what circumstances. Consider: we might discover that, in a particular society, men who marry younger women have more children over their lifetimes than men who marry older women; we would not then be surprised to find a "proximate" social preference for youth in wives, nor would we be terribly surprised to find that older wealthy men in this society marry younger women more often than do poor men. The behavioral ecologist is more interested in the first question: Does marrying a young wife affect a man's lifetime reproduction? A cultural anthropologist, on the other hand, would be more interested in the ways women and men make marital decisions in this society.

Answers to both kinds of "why" questions are informative. And certainly any human society can make decisions to foster behaviors that are counterproductive in terms of ordinary natural selection, though no proximate "cause" is likely to remain common for more than a few generations if it does not serve an ultimate selective cause.

For example, the Shakers are a religious group that imposes celibacy on all its members. That's certainly a cultural rule with biological implications: the few Shakers remaining today are not being replaced.

Proximate mechanisms can enrich our understanding. The usefulness of specific mechanisms depends on (1) what is most predictable in the external environment, and (2) what internal devices already exist in the organism. When our primary concern is intervention (as in medicine or family planning), these particulars of proximate mechanisms become important. To understand the ultimate evolutionary purpose, we are more likely to study the correlations between organismal traits and environmental conditions. For example, when predictably timed periods of very cold weather alternate with food-rich moderate-temperature times, we expect trees to lose leaves (to conserve water), nectar-eating birds to migrate, and so forth.

SIMPLE RULES, COMPLEX OUTCOMES

Although humans are more complicated than other species in many ways, the exercise of asking questions in the same way about ourselves as we do about other animals may be instructive for two reasons. First, other species, like ravens, are often more complex than we realize—and we learn much from studying their behavior. Second, even as we tout our human complexity, we sometimes offer remarkably simplistic explanations about human culture and behavior. If we apply the same standards of repeatability and hypothesis testing to our own behavior as we do to that of other species, perhaps we can gain new insights.

My explorations here assume that humans are indeed animals, even if elegantly complex ones, and that they are therefore subject without special exemption to the general rules of natural selection, the rules that govern behavior and life history among living things. Though we don't know much about genetic specifics yet, it is clear that genes are a "currency" to be maximized in various behavioral equations. We can explore to see what we can learn about human behavioral patterns by considering genes alongside more standard currencies like status or money.[9] The philosophy of keeping underlying assumptions as simple as possible is sometimes called Occam's razor in the sciences. To paraphrase Einstein, "Keep things as simple as possible, but no simpler." That is, seek the simplest model that still explains what we see. If we start with the simplest model, whether verbal or formal, we can see where it fails—where we have ignored complexity that we must now consider.

HUMANS AS CRITTERS

Despite our cultural complexity, we humans must solve the same ecological problems as all other organisms in order to survive and to reproduce. That is, in any environment, individuals must extract sufficient resources to survive and to reproduce in competition (sometimes cooperative) with others, both among our own and different species. Perhaps because of the scope of our actions, we seldom think, except in the most personal terms, about the impact of our behavior on our genetic lineage. Yet when one family lineage dies out, it is replaced by other competing lineages. Remember the old story about a farmer, thinking of buying a bull, who asks the seller about the bull's potency? "Well," drawls the seller, "he comes from a long line of fertile ancestors."[10] The same is true for each of us.

I will apply to humans, at least for the purposes of generating hypotheses, the central paradigm in biology: What would it mean if humans, like other living organisms, have evolved to maximize their genetic contribution to future generations through producing offspring and assisting nondescendant relatives such as nephews, nieces, and siblings? How will the particular strategies that accomplish such maximization differ in specific ways in different environments? And, just as for other mammals, how will these strategies typically differ between the sexes?

This is a complicated endeavor, at best. We change our own environments probably more often and more completely than any other organism. Further, history contains not only "selective" events, but events that are random with regard to fitness: when Mount Vesuvius erupted, the evolutionarily fittest Pompeian died as well as the least fit. Such histories complicate our problem: most of the time, we can expect the emergence of strategies that produce, compared to other strategies, the largest increase in genetic contribution—but sometimes sheer historical chance can alter what we see.[11]

Since we know so little about how much effect any particular genetic locus has on any particular behavior, behavioral ecologists must assume that behaviors are the product of the *interaction* of genes and environment—not the result purely of genetics *or* environment. By using the phenotypic gambit, we assume that when we look at behavior, we are seeing the result of gene-environment interactions over time, and that the most common behaviors in an environment are working well compared to available alternatives. We ask when and how environmental conditions (including social conditions) change individuals' genetic costs and benefits. If we understand how particular conditions are likely to affect behavior, and if we are cautious, we can predict the kinds of behavior we are likely to see.[12]

It is important to note that predictions are not absolute, but statistical. We do not predict that genetically costly behaviors never arise, only that they will not become and remain common. Of course, there is a catch. Although we probably know more about the genes of *Homo sapiens* than any other species except perhaps fruit flies, some yeasts, and some prokaryotes, we know the specific genetics of only very few behaviors.[13] Historical accident can present problems as well. Our inability to have predicted the Pompeian tragedy probably changed human population genetics at least locally. Thus we can get unexpected and interesting results from simple rules and historical accident. Despite such complications, the phenotypic gambit is a good place to start—it works in many cases, helping us simplify and clarify what we see, as well as highlighting those behaviors that are more complicated than we had thought. When it doesn't work, we have learned something valuable; when we do not find what we expect, we look for alternative explanations, usually more complex ones. In animal behavior, this has proved to be an extremely powerful technique, as in the example of Heinrich's ravens.

The rules may be simple, but rules never operate in a vacuum, and environments can be varied and complex. Both physical and social/cultural environments are major determinants of what strategies will succeed. Humans are remarkably complicated and flexible organisms, and human environments, with their elaboration of social and cultural rules, are multifaceted. There is little that tells us, in most cases, how important the various possible influences are, or what the relative role of genes versus individual experience is. But if we ask about human behavior without assuming that humans are qualitatively different from other animals, perhaps we can get rid of the false dichotomy that has persisted between "biological" and "social" causes (both defined narrowly, with only proximate mechanisms considered as hypotheses).

Earlier "biological" hypotheses were typically concerned with, for example, whether sex differences could clearly be related to hormonal or brain lateralization differences. This is of interest, but not related to (and not contradicting) questions about the selective importance of sex differences. "Biological determinism" has often been inferred from such observations that, for example, a behavior occurs in all cultures, and/or a behavior occurs at a typical age. But to do this ignores the possible differences in the ecology of succeeding as a male or female mammal, and simply makes assumptions based on analogy. It is surely misleading to assume a dichotomy between some sort of "biological/genetic determinism," assumed to be fixed and immutable, and "social" causes of behavior, assumed to have no correlation with genotype. Most biologists now think that *all* behavior is likely to be the result of interactions between genes and environment, and that experience is important for many species, not just humans.[14]

Both verbal and quantitative behavioral ecology differ from older approaches to behavior in two crucial regards: (1) currencies to be maximized are not simply economic or social, but also genetic; and (2) following from this, an individual will treat others differently on the basis of what those others can do for that individual's genetic representation, for example, treating kin and reciprocators better than others.[15] We predict some widespread biases, and we can test for them. Thus, for example, "society" is not our primary concern as we dispense social and economic largesse;[16] we typically leave our wealth to our children and nondescendant relatives unless we have none, or unless we have so much wealth that we can take care of our kin as well as endow foundations and chairs in universities (with, as a colleague noted, our family's name attached).

With all these complexities, what then does behavioral ecology suggest about a view of "human nature"? It suggests that some traditional approaches have previously ignored an important currency: genes. It suggests that we do indeed look a lot like calculators, though that we are not necessarily more conscious in calculation than other species, which may forage as optimally as if they carried

Hewlett-Packard calculators in their cheek pouches, for example.[17] It suggests that ecological constraints are important in setting limits to the strategies that will, and will not, "work," and that human social complexity cannot be ignored.[18]

Although genes are a "currency," we seldom know the actual genetic influence on any particular trait, as I already noted. Using the approach I outline here, this lack of information need not keep us from testing hypotheses. What we *are* able to measure, both in modeled systems like genetic algorithms and in empirical behavioral ecological studies of many species, including humans, is this: *in any particular environment, what is the success of variants with different traits in reproducing, and how strong is the parent-offspring correlation in traits?* Genes are more important in this view as a currency to be conserved and multiplied than as behavioral dictators, because external environment, development, and genes interact in a complex way. As the geneticist Theodosius Dobzhansky once put it: "Inheritance is particulate, but development is unitary. Everything in the organism is the result of the interactions of all genes, subject to the environment to which they are exposed."[19] That is, though we conceptualize the effects of "a" gene as though it were separable, no gene acts alone; it is embedded along with other genes in a particular organism, which develops in a particular environment—and all this affects how the genetic influence plays out.

Many different particular internal mechanisms may be called upon to create complex behavioral responses. If we search too hard for the mechanism in each particular case, or if we ignore development and ecology, we may miss the forest as we stumble about in the trees. Consider this metaphor. The link between genes and bits of body or behavior is rather like the link between a cake recipe and the resulting cake. There is little one-to-one mapping. One cannot pick up this crumb and match it to that word or phrase in the recipe. Rather, the words of the recipe, like the genes in the chromosomes, together comprise a set of instructions for carrying out a process: development. In most cases, changing a word or phrase in the recipe will not change a particular crumb; more likely, it will subtly change the cake's characteristics. Changing "baking powder" to "yeast" will change the cake considerably, but not in a particulate way. So will a recipe "mutation" to a sharply different oven temperature.[20]

To extend the metaphor a little further, perhaps the reason we initially find this complexity confusing is that there are some well-known one-gene–one-trait correlations that seem particulate—like changing "walnuts" to "pecans" in a recipe—and these have become quite famous. For example, the disease phenylketonuria arises from an individual getting two alleles for the disease at a single locus; this results in disturbed metabolism of phenylalanine (an important amino acid; the condition is diagnosed by a peculiar odor in the urine as phenylpyruvic acid is excreted), and leads to mental retardation.[21] So here is a case of a "one word" (single locus) change that is particulate.

We are discovering other examples, and they make the front page of the news about once a month. But these dramatic single-gene effects can distract us from the ordinary, more subtle paths, and it is among these more subtle and complex interactive paths that I think we must look for the important links among environmental conditions, gene persistence, and observed behavior. Hence, my focus is not on allelic specifics or precise models, but on more general problems that we've not been able to model precisely. As a classic text on behavioral genetics concludes, for the majority of behaviors studied so far, there is clear evidence of substantial genetic influence, though seldom any evidence of really particulate single-gene–single-trait relationship.[22]

Here, I focus on questions about the "current utility," in selective terms, of different behaviors: What advantage, or disadvantage, accrues to an individual by virtue of having this trait in this environment?[23] This isn't always simple. The *process of optimization* (in each particular population, better strategies displace and replace inferior ones, and the best *available* strategies prevail) is different from the *state of optimality* (the best imaginable fit between strategy and environmental conditions). Because selection acts only on existing variants, optimization is always local (these variants in this environment—some prevail, some disappear) and often incomplete. As a result, we will see variety, perhaps a lot of rather similar, pretty-good varieties, not necessarily a settled, singular strategy.

Another difficulty in asking about utility or optimality is that one is asking about trade-offs, and the "phenotypic correlation" can hide them. For example, suppose I hypothesize that, for an individual, what is spent on housing cannot be spent on transportation. So I would expect a negative relationship between housing and transportation expenses. But, when I measure, I find a positive relationship. Does that mean that no trade-off exists? No, it probably means that I have compared quite noncomparable individuals. If I compare only graduate students, or hunter-gatherers in a particular society, rather than lump them together with professors and millionaires (who have enough money to own both a mansion and a Porsche), the range of variation obscures the trade-offs.[24]

As we look at species, including our own, in a variety of environments—some quite new—we will find variety. We might find an excellent fit between trait and environment because of a long evolutionary history of unchanging selection. For example, the fact that Arctic fish can die of heat prostration at temperatures cold enough to freeze humans to death reflects a long selective history of constant cold. Or, we could see a trait that is currently advantageous, but one we are certain is not a specific evolved adaptation, like running away from a fast-approaching truck. Since there were no trucks in the Pleistocene, the evolved rule was probably something like "run away from large, fast-moving things," and trucks, though relatively new, fit the same general category as dangerous fast predators of the Pleistocene. Because the process of optimization is complex and few traits can be easily isolated from developmental and historical constraints, we may rarely see a really fine-tuned "fit" between any single trait and environmental conditions.[25] Finally, humans seem to me to be at least as likely as any other species to show interpopulation differences not only as a result of natural selection and adaptation (perhaps sickle-cell anemia allele frequencies), but because of historical—and cultural-historical—events (e.g., lactase distribution in humans; see chapter 10). There are very real difficulties, but I hope defining the problem as one of current utility may help avoid some of the less useful controversy.

This approach, I hope, creates natural linkages: to empirical fields of human behavior such as anthropology, psychology, and sociology on the one hand; and to genetics, behavioral genetics, and population theory on the other. Scholars in each of these (and other) fields have information and perspective on constraints, and on how to consider human behavior. The "current utility" approach links us to what we know about other species as well.

We know something about the ecological and genetic components of behavior, and about what behaviors become common under what circumstances, but our knowledge is still unconnected across disciplines; my work daily leads me to conversations with colleagues with whom I have more shared interests than shared knowledge. Now we need to reach across disciplines, and I hope experts in other fields will read this not as a postulated expert disquisition in their field, but as an invitation to contribute what they know to solving the questions I raise.

2.

Racing the Red Queen:
Selfish Genes and Their Strategies

Now here, you see, it takes all the running you can get to do, to keep in the same place. If you want to get somewhere else, you must run at least twice as fast as that!
—The Red Queen in Lewis Carroll's *Through the Looking-Glass*

ALICE HAD some trouble following the Red Queen's logic, that one has to run as fast as one can just to stay in place because everything else in the landscape is running as well. Biologists, however, find the image an apt one. Consider Matt Ridley's engaging book on the origins of sexual selection, which he chose to call *The Red Queen* in recognition of the problem that the sexes continually change each other's costs and benefits. In a way, much of biology is a record of such selective arms races.[1] Ecology is rife with examples: if faster rabbits escape coyotes, tomorrow's rabbits are faster than today's—but once this is true, fast rabbits put pressure on coyotes, so that faster or sneakier coyotes become the only successful ones. For us humans, our families, friends, and rivals are forces to be reckoned with. Such "social selection" (chapter 10) is surely a good example of the Red Queen's problem: the goal you seek is situated in a moving landscape, and it may always be moving away from you.

At the core of behavioral ecology rests the notion Richard Dawkins aptly called the "selfish gene," the idea that genes that get themselves copied into more and more individuals will be the genes that prevail and persist through time. This measure of success is a modern version of the simple logic first employed explicitly by Darwin.[2] Genes compete for locations on the chromosome, and groups of genes make what biologist Leigh Van Valen called a "parliament" (they interact to produce complex effects); individuals housing effective parliaments survive and reproduce relatively better than those who don't.

At some level, this competition among genes is like the old adage, "I don't have to outrun the bear; I only have to outrun you."[3] That is, no guarantees exist that the chance events of mutation, recombination, and drift, combined with the filtering of natural selection, have generated the best possible combinations; it is only true that, at any moment, relatively more effective combinations do better than others. Individuals thus never represent the "best conceivable" combinations of genes plus environment, only the "currently most effective"—which may be superb or less than wonderful, and likely to disappear if a better alternative appears.

Although the idea of selfish genes is simple, a great complexity results. On the one hand, genes can affect more than one trait; on the other, groups of genes can cooperatively affect a single trait. Genes are carried about by individuals, yet genes in one individual may affect the success of genes in other individuals. Some information goes from generation to generation through the cytoplasm, not the genes, and in humans and in some other species, some information is transferred across individuals through culture. Individuals differ genetically, and they live in varied physical and social environments. Thus, while individual strategies for survival and reproduction are all-important, their analysis may be complicated.

We know a great deal about the evolution and ecology of resource use in other species: the costs and benefits; the impact of various environmental conditions; the evolution of sex differences.[4] Genes, history, and environment interact, but the basic patterns are clear. Recent empirical tests suggest that the relative power of chance, selection, and history can differ under various circumstances. Nonetheless, for traits strongly correlated with fitness, even when chance is great in the environment, natural selection is still powerful.[5]

If we humans, like other species, evolved simply to get resources and to survive long enough to get duplicates of our genes into the next generation, why are we so complicated about the process? Even in our life history (chapter 6), we humans are unusual in the pace of our maturation, growth, and reproduction. Socially, we not only live in families (common in many species), but cluster together in villages, cities, nations. We ally ourselves with one another in more complicated ways than most species, and our groups are based on more than simple reciprocity. We have formal trade agreements and schooling, art galleries, transportation networks, and so on.

This complexity and diversity seem a far cry from any simple set of strategies. Yet despite our complexity, this simple observation is true: those of us alive today are the descendants of those that successfully survived and reproduced in past environments. Historical accidents can happen, but the rules still are true. This means that genetically selfish behaviors, those that enhance an individual's total genetic representation, are always favored by ordinary natural selection.

Why doesn't the favoring of genetically selfish behavior always result in bloody outright battles? The short answer is: (1) sometimes it does, (2) when it does not, it is because of the costs of attempting bloody battles. Rules operate under environmental pressures. While all living things have evolved to acquire and use resources to survive and reproduce, the ways they do so are constrained by ecological conditions. Individuals that use the most effective and efficient resource strategies in any particular environment are those that tend to survive and reproduce; but there is no reason to suspect that what works in the desert will work in the river, or that what works among small kin-based societies will work in nation-states. Further, fertility is complex. Although a simplistic interpretation might imply that the best strategy is to produce as many offspring as possible as soon as possible, this is seldom, in fact, a winning strategy even for relatively nonsocial animals. In some environments, only "superkids" survive and reproduce at all; the result is that fertility responds to the cost of parental investment to make offspring successful.[6] Life history theory is, in fact, largely the study of trade-offs: size against number of offspring, for example. Finally, in social animals, other individuals create some of the most important environmental pressures, and rampant short-term self-interest will often fail (more on this in later chapters). Clearly, rich diversity is likely.

The starting point is that, other things being equal, individuals that use efficient strategies produce more offspring for the next generation than their competitors. But other things are often *not* equal. Are we sure that we are measuring costs and benefits correctly? At first glance, many behaviors appear counterproductive. For example, infanticide occurs commonly in many species, including lions, ground squirrels, and a number of primates.[7] To determine whether infanticide is an evolutionary "mistake" or an effective strategy, we must ask: Who commits infanticide, and under what circumstances? Who profits from it? Most infanticide is committed by reproductive competitors; its evolutionary logic seems clear. Sometimes, however, parents kill their own children. How shall we view this? Surely parental infanticide is an evolutionary mistake that decreases, rather than increases, reproductive success? Indeed, evolutionary mistakes are possible. But across species, except for rare pathologies, infanticide is found under specific ecological and social circumstances, and its impact in these circumstances is an increased lineage success for the killer—even when the killer is a parent.[8]

How can this be? In species in which a successful offspring requires considerable parental effort, there are circumstances in which terminating a particular investment pays off—for example, a deformed offspring or a mother's poor health. In these circumstances, parents win who discriminate by investing more in offspring that are more likely to be successful, and investing less in weak or deformed offspring—even in the extreme case of infanticide. Other similar puzzles of apparently maladaptive, yet common, behaviors include lethal conflict (when does it pay to risk getting killed?), delayed reproduction (how can waiting to reproduce increase one's reproductive success?), and sterility (the ultimate in nonreproduction).[9] Consider honeybees, in which all females except the queen are sterile. How can a (female) worker bee's genetic representation possibly be increased by remaining sterile and devoting her life to caring for the queen's, her sister's, eggs?[10] Darwin worried about how a trait like sterility could be inherited. He understood that honeybee colonies were somehow special, but left the solution for future researchers.

Perhaps the most blatant examples of hard-to-explain phenomena fall under the heading Darwin called "sexual selection." He understood that anything that helped you survive would be "favored," but unless you also reproduced, that trait would disappear. The reason Darwin treated sexual selection separately from "ordinary" natural selection was that successful sexual strategies in so many species were also dangerous, life-threatening strategies, usually associated with male-male competition. Darwin struggled to understand how such behaviors could be favored by selection. It took much observation to determine that (1) sometimes the most effective thing you can do is take a huge risk, and (2) in a sexual species, reproducing means that you must face both the competitors of your own sex and the preferences of the opposite sex you wish to win. So, Darwin argued, others of your species could "select" just as effectively as the horse or cattle breeders who change the genetics of future race horses and milk cows by selecting who gets to breed. Sex differences, how they arose, and how and when they are maintained are at the heart of my exploration, and I will return to these themes repeatedly.

WHOSE GENES COUNT, AND WHY? KIN SELECTION

Though the concept of selfish genes is a simple one, it has been repeatedly misinterpreted, just like Darwin's original formulation. How horrible, "Nature red in tooth and claw"! How could we possibly believe such a noxious idea, when we can see generosity all about us in many species. It remained for biologist W. D. Hamilton to quantify and formalize some of the most important genetic costs and benefits of behavior—and his formula, though simple, is to much of biology what $E = mc^2$ is to physics.

The first general rule is that a behavior will become common only if its genetic benefits outweigh its costs—if $b > c$. Hamilton pointed out that, since "nondescendant" relatives of any individual ("ego") such as nieces, nephews, and siblings share genes with ego, helping these relatives (even if it has a direct cost to ego) can help ego's genes. To be favored, there must be a net genetic benefit. Relatives shares only some genes with the helper, and this varies with r, the degree of relatedness; your sister has more genes (higher r) in common with you than your fifth cousin. So the benefit must be discounted, and only the help that goes to identical genes counts. Hamilton pointed out that "giving" behavior should evolve whenever $rb > c$, that is, whenever the benefit to the recipient b times the degree of relatedness r is greater than the cost to the doer—for example, whenever the benefit to one's sibling, who shares on average one-half of one's genes, exceeds twice the cost of one's act, or when the benefit to each of two siblings exceeds the cost of the act.[11] Notice that this also means that not all help to kin will be favored: if the cost to self exceeds the (benefit * relatedness of kin), the behavior should disappear. And sometimes being positively mean to one's kin pays— when $b_{self} > rc_{rel}$ (simply reorganizing the above).

This concept had been recognized informally for some time. The mathematical biologist Haldane had noted that, while he would not give his life to save his brother (who would, on average, share half his genes), he would die for two brothers, or eight cousins.[12] Hamilton applied this concept of *inclusive fitness maximization* or *kin selection* to Darwin's puzzling problem of sterile castes in social insects (surely it is a great sacrifice not to reproduce and spend one's life caring for a relative's offspring). He could show why Darwin was correct in maintaining that the self-sacrificial behavior persisted only because it occurred in the familial context, and that it would not persist in other sorts of groups.[13]

The point is that a behavior can be genetically profitable even if, to the casual observer, it appears to be costly and of benefit to others (table 2.1). This distinction between what seems to be true and what a behavior does for relative genetic representation is a thorny one, but one we cannot ignore. Measuring both costs and benefits correctly is crucial.

How many genes we are likely to share with another individual (r, the degree of relatedness) is one key to how much we will profit

TABLE 2.1.
Categorization of the Impacts of Behaviors on Phenotypic and Genetic Condition.

Behavior	Apparent Effect	Genetic Effect
Overt competition	Profitable ("selfish")	Profitable ("selfish")
Parenting, nepotism, reciprocity	Costly ("altruistic")	Profitable ("selfish")
BECAUSE NATURAL SELECTION FAVORS ONLY GENETICALLY PROFITABLE BEHAVIORS, UNDER NATURAL SELECTION BEHAVIORS ABOVE THIS BOX SHOULD BE COMMON; BELOW THE BOX, RARE.		
??	Profitable ("selfish")	Costly ("altruistic")
Mother Teresa?	Costly ("altruistic")	Costly ("altruistic")

Source: Modified from Alexander 1974.

Notes: If we look only at superficial, apparent ("phenotypic") impact, we miss crucial differences. ?? = this category is so very rare that I have trouble imagining a non-controversial example: Perhaps a rich miser (phenotypically selfish, since he is a miser) who disinherits his family, leaving an anonymous gift to a home for unwed mothers (genetically altruistic, since he hurts his relatives in order to help genes, not IBD). See chapters 9 and 10 for further exploration.

by helping. As humans, we are likely to have, on average, half of our genes in common with a sister or brother. All adult humans have two alleles at each locus, or location, on every chromosome (with some special conditions for the X and Y sex chromosomes); these two alleles can be duplicates (homozygous; perhaps both are the allele we call *a*), or different (heterozygous, perhaps *a* and *A*). We received one set from our father, one set from our mother. Mother's egg and father's sperm are produced in a lotterylike process called *meiosis;* each resultant egg or sperm has only one set of chromosomes. An egg or sperm has some chance of getting either allele at any location. Suppose my mother is heterozygous, and has *aA* at the locus I am interested in. My sister and I might both receive an *a* or *A* from Mom, or one of us might have *a* and the other *A* (and be totally different in what we inherited from Mom). This pattern is true for each genetic location, and true for what we received from our mother and our father. As a result, we share about one-half of our genes with our full sister, and one-fourth of our genes with her child (fig. 2.1). Your own child, of course, shares exactly half your genes.

Hamilton suggested that striking phenomena follow from the fact that social insects have the peculiar genetic arrangement in which mothers produce sons by laying unfertilized eggs. Because males have no father and get only one set of alleles from their mother, they have and pass on exactly the same genetic material in each sperm. So in social insects like honeybees, full sisters share identical genes through their father. They can never be less than one-half alike if they have the same father, and they are more closely related to each other on average (3/4) than mothers are to daughters (1/2). Sterile female workers, Hamilton suggested, were not paying, but gaining, genetically by raising their three-fourths-alike sisters rather than half-alike daughters. While this hypothesis may not fully explain eusociality, Hamilton's statement of the theory of kin selection was more general and is an important part of the general theory of natural selection. Hamilton's summary hypothesis makes a strong, testable prediction: "The social behavior of a species evolves in such a way that in each distinct behavior-evoking situation the individual will seem to value his neighbor's fitness against his own according to the coefficients of relationship appropriate to that situation."[14]

Thus, other things being equal, we expect individuals to treat their kin more gently than strangers, and to treat close kin more gently than distant kin. Even though we expect no organism to be able to

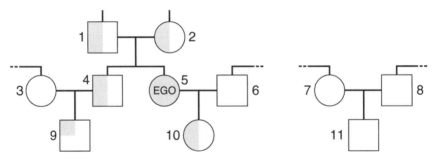

Figure 2.1. Hamilton's (1964) theory of inclusive fitness maximization suggests that behavior will evolve as a result of its costs and benefits not only directly to the doer, but on those to whom the behavior is directed, depending on the degree of relatedness. Helping behavior will be favored as it helps more genes that are identical by descent, and less favored as its effects are diluted by genes that are not identical; thus the proportion of shared genes is important to calculate. Here, Ego (5) receives half her genes from her father (1) and half from her mother (2). Her daughter (10) receives half her genes from Ego. Because her brother (4) is a full sibling, with the same mother and father, he will, on average, be genetically half identical to Ego; his daughter (9) will be exactly one-half like him, and on average, one-quarter like Ego. Individuals 3, 8, and 11 (relatives of Ego's mate) share no genes that are identical by descent. (From Williams 1992a)

calculate its relatedness to others, we expect them to *act* as though they could. In fact, of course, what we see is that organisms treat those with whom they grow up differently from others, because under most circumstances in most species, individuals grow up among relatives, not strangers. Hamilton describes an individual's strategy for making effective use of proximate cues of relatedness.[15]

In many species this proximate cue of nearness works because the individuals with whom one grows up typically *are* one's relatives. This is why, for example, a researcher can fool ground squirrel mothers into adopting unrelated babies before her own children emerge from the nest to forage above ground, but not after, when there is opportunity for mothers to make costly mistakes, as youngsters occasionally blunder down the wrong burrow. Research on a wide variety of species has so far supported Hamilton's prediction.[16] Each individual has reproductive interests, but these interests are shared to a predictable degree by others who also share common genes—genes identical by descent. Overtly selfish behavior, "nature red in tooth and claw," is thus only one route to enhanced genetic representation (table 2.1, fig. 2.1). Individually risky or energetically costly behaviors and helping one's kin can also be genetically profitable. We expect organisms, including humans, frequently to engage in activities that benefit relatives. Further, highly selfish behaviors will not be genetically profitable in the longer run if they harm relatives too much.[17] Human history has a wealth of examples: the impact of family on death rates in crisis situations, alliances and internecine warfare in Icelandic and English history, and others.[18]

How much relatives profit from helping or harming one another will depend on the degree of relatedness as well as on the costs and benefits of the act. Helping relatives, even at some cost to oneself—and helping friends who will reciprocate, for example—can be genetically profitable, and many behaviors that *appear* to cost the performer are actually profitable in terms of genetic representation. Here, it is the actual effect on reproduction and genes that we care about. Reciprocal coalitions can be complex and quite elaborate in humans (think of politics, and of warfare), and they affect men and women differently; I will return to them in chapters 10–14.

SUMMING UP THE BASICS: ASSUMPTIONS AND OBJECTIONS

The forces of selection, including kin selection, explain much of the variation in the behavior of individuals in other species. To ignore them completely when it comes to humans would be absurd. Imagine how much credence you would give to me if I told you that gravity exists, making all other animals behave in certain ways (walking on the ground, being able to glide only a short distance, or expending considerable effort to fly), and then went on to explain that humans, on the other hand, are exempt because we have cultural phenomena such as airplanes. We have more complexity and more variation in *how* we can travel on the ground or in the water, and because of our cultural transmission and elaboration (including inventing and sharing technology), we are not bound to the ground; we can clearly fly in ways unknown among other species. And when we fly, it is because of our cultural innovations and personal desires, not because we "need" to. Nonetheless, you would probably still bet that gravity influences all of our travel. While we can circumvent gravity in sophisticated ways, it never disappears as a force to be reckoned with.

This is not the place for a lengthy review of behavioral basics; some excellent reviews already exist.[19] But because the behavioral outcomes I explore here involve not only rules but environments and history, I want to be explicit about what I am, and am not, suggesting about our human sex differences. Behavioral ecology asks whether, if we know about environmental conditions (including social conditions) and how those conditions change actors' (genetic) costs and benefits, we can predict the kinds of behavior we are likely to see. To explore this possibility requires assumptions. These include the following:

1. Organisms are generally well suited to the environments in which they live; they achieve success in any environment by getting resources that enhance their survival and reproduction. Strategies we see have had time to compete against other strategies, and "what you see is what you get"—it is what has worked best (the phenotypic gambit).

2. Only heritable variation is appropriately considered in testing predictions about changes in gene frequencies over time. An individual can assist its genes to spread
   a. by reproduction, the most direct method; or
   b. by assisting individuals carrying copies, identical by descent, of its genes (kin selection); or by helping individuals that do not carry identical copies of its genes, if such assistance is returned in genetically effective ways (reciprocity).

   This implies that individuals who help reproductive competitors without any reciprocation will lose descendant representation in existing lineages. In short, people like them will become rare or cease to occur.

3. Organisms that are more efficient in getting resources in any environment will survive and reproduce better than others. In the evolutionary history of all species, there have been important proximate correlates of reproductive success, including resources (food, territory), rank (status or power), wealth, and, in highly social species such as humans, social "reputation."

4. Further, no organism, including humans, has evolved to perceive or assess directly the spread of genes; rather, organisms behave as though these proximate correlates were their goal. Thus, species

may find themselves in novel evolutionary environments, and individuals' behavior may be currently maladaptive; when this is true, the assumption of the phenotypic gambit is violated.

5. In their "deep" objectives—in what they evolved to do—humans are not qualitatively different from other living organisms. Like other living things, they evolved to get and use resources to survive and enhance the spread of their genes. To evolutionary biologists, this is parallel to arguing that humans, while they can make airplanes and fly, for example, are still subject to the laws of gravity. Yet because they are highly intelligent and highly social, humans are the likeliest of species to be in novel environments, making it a complex task to make assumption 4, and to distinguish evolutionary history from current utility.[20]

Perhaps because the study of human social and sexual behavior has in the past "belonged" to fields in which only humans were studied, this is a novel approach. A behavioral ecologist would answer the query, "How do I love thee? Let me count the ways," far less poetically than Elizabeth Barrett Browning. For several reasons, this approach may not be palatable to all. It may also be controversial; certain assumptions are simply not accepted by everyone. Many think, for example, that it is inappropriate to use the same general principles to examine human behavior as to study the behavior of other species, even to test hypotheses; and many feel that not all behaviors can be reasonably analyzed.

A widely held discomfort with any evolutionary approach to human behavior is the one reflected in the following (perhaps apocryphal) story, which I first heard attributed to Bishop Wilberforce's wife. When Darwin presented his theory of natural selection, hers was a typical response as she leaned over to say to her friend, "My dear, have you heard Mr. Darwin's theory that we are all descended from apes? Let us hope that this not be true; and, if true, let us hope that it not become generally known."[21] It is easy to agree with the bishop's wife, or to fear that knowledge of behavioral ecology will lead us to make bad or cynical policy. This confusion of "is" and "ought" is sometimes called the "naturalistic fallacy." Because evolution is simply genetic change over time and thus amoral, its analysis is analytic rather than normative, and it has no bearing on human moral decisions. To observe that something is true does not constitute a moral endorsement.[22]

Human complexity and flexibility raise special analytic concerns. Simply because a trait looks useful, we cannot assume that it evolved through the action of selection. It is important to articulate the proposition one is using to generate hypotheses, and to state one's assumptions.[23]

In relatively stable environments, we expect selective pressures to remain similar over time. Thus, for example, constant cold leads to stumpy limbs and cold-tolerant physiology. This does not mean that any particular observation can offer strong support; the camel's hump, for example, might have come about for nonselective reasons, because of natural selection, or due to some combination; and humans are a particularly difficult entity to study. But if we accumulate a series of a priori predictions, they can grow into a significant number of directional results. In other species, special insulating hair and feathers, found in numerous species—but only in Arctic and Antarctic environments—suggest selection. For complex phenomena in complex species, such as social and sexual behavior in humans, this teasing apart is an issue to which we must pay careful attention.

What about exceptions, what a colleague of mine calls the "Bongo Bongo" argument? "This is how the world works, and you can see I'm right, for among the Bongo Bongo they . . . ." Or the reverse: "Your view of how the world works is obviously wrong, for among the Bongo Bongo, they are absolutely altruistic to everyone." Or: "Gravity cannot be a natural force, for birds and humans can fly." Hardly persuasive. Behavioral ecological predictions are statistical, arguing that at any moment, behaviors that, in a particular environment, get genes passed on will increase relative to other behaviors, and that (employing the phenotypic gambit, and statistics) we are likely to be able to detect this trend. Thus, if we find that the Bongo Bongos are truly genetically altruistic, this means only that the Bongo Bongo are likely to decline over time, to be replaced by competing peoples—and this is a testable prediction.

All of the arguments in this book, in the context of natural selection, are statistical propositions that, other things being equal, individuals with certain traits will be, or will become, more common than competitors in particular environments. None is a statement of absolutes. Consider: Many people would consider Mother Theresa a genetic altruist, helping nonrelatives for no genetic payback. If I argue that genetic selfishness is favored, I am not suggesting that we will never have a Mother Theresa—only that, over time, her genetic lineage will likely decrease compared to others, and she and others who are *genetically* altruistic will remain a rarity. We are making predictions about what a statistician would call "central tendencies" rather than rare exceptions.

NOVEL EVOLUTIONARY ENVIRONMENTS: CAN THE PRINCIPLES STILL HOLD?

A very large question remains: How far can we usefully explore our current behavior? Environments that are new and novel in an evolutionary sense introduce significant complexity. After all, it might be easy to see how reciprocity and discrimination are favored in people living in small bands, interacting daily with the same few people, and protecting some resource against outsiders. But today our societies are large and complex. We may interact with literally hundreds of people; we do business daily with people we have never seen before; we have information about what happened today across the world, to strangers we may never see. History also complicates our problem: we know of many nonselective accidents in our history that are equal to the destruction of Pompeii.

Novelty poses a great difficulty for studying human behavior; through our cleverness, we constantly create environments that are novel in selectively important ways for ourselves and other species. The behaviors that helped hunter-gatherers in the savanna may not be useful in the suburbs of Chicago.[24]

Novel evolutionary events influence the behavior and demography of other animals as well as humans. For example, a male chimpanzee in a well-studied group gained dominance status by banging together empty metal containers instead of the more traditional branches. In Great Britain, Great Tits began to feed out of milk bottles that were evolutionarily novel, although the bird's probing behaviors had evolved to forage on bark and twigs and their digestive systems certainly did not evolve to deal with milk. On the East Coast of the United States, gulls, evolved as generalist feeders, showed marked increases in population density as a result of an increase in garbage dumps, while other seabirds declined in abundance due to gull predation.[25] Consider the Arctic fish discussed in chapter 1: because the water has, for millennia, been just above freezing, there are several Arctic fish species in which one can kill an individual—

through heat stress—just by warming it up to temperatures we would still find frigid. However, if global warming continues, the water in some areas will become warmer, and whether any fish in a particular species still retain the genes for warmth tolerance is a toss-up. For the Arctic fish, over many fish generations, cold tolerance was profitable, and ability to tolerate warmth was not. But now we are changing the environment.

The impact of novel environments is heightened by the fact that no organism, including humans, has evolved to be aware of ultimate selective effects, but only of proximate cues. Selection acts in a way that what enhances our survivorship or reproduction—forming friendships, having sex—tends to be perceived as pleasurable; and acts that typically detract from our survivorship or reproduction—for example, getting burned—are unpleasant or painful. But this relationship can change when the environment changes. Consider a simple example of novelty. In nature, sweet foods are seldom harmful, and sour and bitter tastes are often correlated with the presence of harmful alkaloids. Thus a preference for sweet tastes (a good proximate cue to nutrient-rich, safe food) became widespread in omnivores, including humans. In most past environments, it was difficult to obtain enough sugar to create problems of obesity. Once we humans invented technologies for refining and concentrating sugar, we created foods that had enormous concentrations of sugar, breaking the selective link between sweet taste, the proximate cue, and good food source that had previously led to enhanced survivorship and reproduction. But proximate cues drive the system, and selection acts as a passive sieve. So we retain a preference for the sweet taste that can make us fat and fill our teeth with cavities.

Because we humans can modify our environments so extensively, and because our cultural transmission can respond more quickly than genetic intergenerational transmission, we are frequently in novel environments. It is surely fair to ask how far we can really expect to see selectively advantageous behavior in our current environments. As an example, consider the following: In other species, and in preindustrial human societies for which we have data, males who have more resources typically have more offspring (usually because they have more mates) than others (chapters 4, 7, 8). Now, in much of the world, effective contraception has broken the link between resource accumulation and fertility (the Pill is so extraordinary that *The Economist* recently included it in a list of the Seven

Modern Wonders of the World). For the first time in all of history, men and women could be on equal sexual footing; sex could be without parental consequences for either. Certainly this is a novel environment. But we humans, like other species, operate on proximate cues, not any awareness of selection; the existence of the Pill, breaking the link between sexual pleasure and parenthood, doesn't mean that sex is no longer fun. So do wealthier men today, as in the past, have more children than others? As I explain in chapter 15, possibly not, because of the novel environment. But they do have more sex, if they want it.[26]

MORE THAN ANTS OR PEACOCKS: LIFETIMES, CULTURE, ECOLOGY, AND VARIATION

My explanation of the basics of selfish genes does not yet come close to being useful in looking at human behavior in all its diversity and complexity. Three important ingredients remain to join in the interplay: the social impacts of sexual reproduction (chapters 3–5) and how these play out in human lifetimes (chapter 6); the influence of external environmental influences (chapters 7, 8); and the intense pressures of group living (social influences), which elaborate reciprocity as a social force beyond anything we see in other species (chapters 9 and 10).

Three phenomena—kin selection, reciprocity, and sexual selection, or how we interact with family, friends, and mates—lie at the heart of why we behave as we do in many circumstances. The basics give us a perspective on complex phenomena, and I will try to weave together the themes of the ecology of resource consumption and sex differences in different ecological situations. My central task here is to ask why we behave as we do, especially about resource issues, and why the sexes differ so consistently in some areas, and not at all in others.

I began by assuming that we humans share some constraints with other animals (and plants, too, for that matter): we must get resources to survive and reproduce; parents and offspring are more like each other than like strangers; and what is effective in one environment won't necessarily work in another. Although only heritable variation is important, in complex social animals like ourselves, cultural transmission is one kind of heritability, and interaction be- ·

tween cultural and genetic transmission can certainly complicate analysis (chapter 10).

The advantage to this approach is that we can frame questions in ways that have some rigor and repeatability. That is, we can go beyond convenient "Just So" stories such as "How the Camel Got Its Hump" or what causes vampires. It's not easy; in its crudest formulation, natural selection theory sounds like a circular argument (what works, works, so if you see it, it must be working).[27] Of course, so do the principles of physics. We also are tackling complex and interrelated phenomena. The trick is to figure out for *what* traits we wish to predict the direction, along with *what* environmental forces we predict will favor this versus that version of the trait. Then we can predict the direction before we go out to measure it: "If the environment is A, then version X of our trait should increase over time; if the environment is B, version Y should increase." Then, if we see what we expect, we have support (but not proof) that we are likely to be at least partially right; and we know that if we see something else (e.g., version Y runs riot in environment A), we must be wrong or else there is a factor missing from our analysis.

The philosopher Helena Cronin developed the apt imagery of the "ant and the peacock" to suggest how kin selection (ants) and sexual selection (peacocks) influence the lives of living things.[28] I wish to explore here not only the similar ways in which selection has acted on us (to what extent we are all ants and peacocks), but also the specific ways in which conditions in the external environment are good predictors of differences in the ways men and women approach resources. Finally, I will discuss the reproductive impacts of today's evolutionarily novel environment.

3.

The Ecology of Sex Differences

Of those who were born as men, all that were cowardly and spent
their life in wrongdoing were transformed at the second birth into
women . . . . Such is the origin of women, and of all that is female.
—Plato

We acknowledge a biological difference between men and women,
but in and of itself this difference does not imply an oppressive
relation between the sexes. The battle of the sexes is not biological.
—Editorial Collective, *Questions féministes* (1977)

It is theory that determines what we can see.
—Albert Einstein

A FAVORITE CARTOON of mine shows two deer, a buck
and a pretty annoyed-looking doe standing on a hillside. The buck
is tilting his head, saying, "So I like rutting—so sue me." In the genetic gambling casino, success depends not only on individual
strengths and weaknesses, but on environmental conditions, and
whether or not there are groups to contend with. The buck, however,
highlights an influence on all sexual species: from fish to flying
squirrels, from Hanuman langurs to humans, males and females of
most species experience different costs and benefits in reproducing,
and these differences influence both lifetimes and social behavior.
Professor Higgins's plaintive cry could be universal among males of
sexual species (as could the converse: why can't males be more like
females?).

Sexual reproduction means a loss in genetic representation, since
half your offspring's genes are identical to someone else, not to you.
So "why is there sex?" is an important question.[1] Sexual reproduc-

tion is in fact common, certainly among vertebrates, suggesting that the loss of genetic representation is compensated—but how? Most current hypotheses argue in some form that unpredictably changing conditions may make it so valuable for an individual to produce variable offspring that the genetic cost becomes worthwhile. For example, W. D. Hamilton has argued that, in the face of rapidly evolving parasites, producing offspring exactly like yourself, as in asexual or some parthenogenetic cases (even if you are maximally fit and resistant), is futile and costly, for your parasites, having many generations during your lifetime, can always evolve new strategies faster than you can respond.

Sexual reproduction is far more diverse than you would think from looking at humans: there are multisex species like the lowly slime mold, female-only species like a number of fish and lizards, bisexual species (many plants, snails), environmentally induced sexual or asexual species (aphids), and environmentally induced sex changers (some fish). Consider the Blue-Headed Wrasse (*Thalassoma bifasciatum*), a coral reef fish. Most individuals begin life as a female. Large females make more eggs than small females, but even small females can make some eggs and have no trouble getting them fertilized. Male-male competition is severe and risky; only very large males get mates, but they are highly successful—so size has a very different impact on likely reproductive success for males versus females.[2] If one could choose, and Blue-Headed Wrasses in some sense can, it would be reproductively more profitable to be a female when one is small, and become a male only when one is very large and highly competitive.

How do big males come to exist? Within any group, if an experimenter acts like a predator and removes the large male, the largest female switches sex, becoming a bright, yellow-and-violet-colored male, in a process that takes about a week. If that male dies or if an experimenter again removes it, the next-largest female switches to become a male. And if two females are close in size, they can jockey back and forth, changing and rechanging sex until one of them wins the battle to become the supermale. Sex is not genetically but socially or ecologically determined in this species. A very few individuals follow an alternate strategy: they are born male and remain male, though they are forever small and inconspicuous and must follow a sneaker's strategy, darting in to deposit sperm when a large male courts a female. These sneaker males are rare, for their success is low.[3]

Here I want to explore, for sexual species, the interplay between sexual strategies and environmental conditions, to begin asking when and how the optimal strategies of males and females differ—the ecology of sex differences, if you will.

Sex and Strategies

At the heart of all life history strategies lies a single problem: How can effort (calories and time spent, risks taken) be spent optimally? An organism faces a central constraint: what is spent for one purpose cannot be retrieved and spent for something else. An organism can spend effort on maintaining its body, or it may spend effort in reproducing. The relative payoffs for these competing activities differ with age, competitors, mildness of external conditions—with many things. The central problem facing an individual is how best to spend effort to replicate genes: allocating effort well, surviving and reproducing relatively better than one's competitors. This is the key to being relatively "fit." Surviving and reproducing was the original meaning of "fitness" as used by Darwin, though there is a plethora of additional uses today.[4] To analyze the contribution to fitness of particular strategies, such as age at maturity in particular environments, we test specific predictions about what behaviors should succeed best in specific environments.

First we must back up and ask a series of questions: Why is there sexual reproduction in the first place?[5] Why does the number of sexes almost always reduce to two, not three or more? The answer is not immediately obvious. Some slime molds, for example, have about a dozen "sexes." There are also single-sex, all-female (parthenogenetic) species. In some of these, like whiptail lizards (*Cnemidophorous*), some females behave like males in mating, though no sperm or eggs are exchanged.[6] In other species like *Poeciliopsis* fish, females mate with males of other species. They use the sperm to start the physical process of egg development, and then throw out the male genetic material. Some species reproduce parthenogenetically so long as the environment is stable; females produce diploid daughter eggs without mating. When conditions begin to deteriorate, the females, again parthenogenetically, produce diploid daughters and haploid males (from unfertilized eggs). These mate, and thus sexually produce daughters—who are physically different from their parthenogenetically produced mothers and are ready to overwinter.[7]

There are also plants and animals in which each individual has male and female parts: they are the hermaphrodites. Most of these do not "self"—they mate with another hermaphrodite, producing more variable offspring than they could by selfing. These observations lend some support to the general argument that part of the point of sexual reproduction is the creation of variable offspring in unpredictably changing environments.[8] In some hermaphroditic species, individuals are first one sex and then the other (small maple trees are male; pollen is cheaper than seeds); sometimes the sequence of sexes is not fixed but triggered by changes in the social environment, like the Blue-headed Wrasses. In other species (usually low-density, relatively sessile species), individuals are both sexes simultaneously.

Even though Blue-Headed Wrasses are unusual in changing sex, they are typical in another way. Like most sexual species, there are only males and females—two sexes.[9] One clue lies in the observation that reproducing in sexual species requires two quite different sorts of effort: getting a mate (mating effort: striving to gain resources or status, getting mates), and raising healthy offspring (parental effort such as feeding, protecting, and teaching offspring).

Imagine a population of something like jellyfish floating in the ocean and reproducing by releasing into the sea haploid gametes, each carrying half the adult number of chromosomes. These can combine with gametes from other individuals to make zygotes— new offspring. This is a basic and simple form of sexual reproduction. Since in this example any gamete can recombine with any other gamete, even "sibling" gametes from the same individual, there is nothing so specialized as eggs or sperm. But the scene is set for a conflict of interest that becomes important: a gamete requires, to make a zygote, genes from another gamete. Genes are in the nucleus, but cells also have some genetic instructions in the cytoplasm (the rest of the cell). And while the gamete needs the genes (in the nucleus) from another gamete, it is better off without the extra instructions (in the cytoplasm) from the other gamete. I'll return to this in a moment.

First, imagine that there is variation in the size of gametes released: from very small to extremely large. Remember the two tasks to be accomplished in contributing to a successful zygote: hitching up with another gamete, and making a sturdy, well-endowed zygote. Simple physical laws lead us to conclude that the tiny gametes, those cheapest to produce and likely to move farthest in currents, will have the greatest chance of bumping into other gametes. The largest gametes have resources to live longer, and also have the most to contribute, producing the best-endowed zygotes. Over time, because the smallest and largest gametes are favored for the two distinct tasks and middle-sized gametes lose out to both, most systems reduce to individuals that specialize in making either large or small gametes (fig. 3.1).

One individual could be hermaphroditic, producing both sorts of gametes. At first this seems like a genetically profitable strategy, but it is rare—for reasons that lie at the heart of male-female differences and suggest why, once there are only two sexes, they tend to behave in predictable ways. The only advantages to a small gamete are that it gets there faster and is energetically cheap; the interests of small gametes are promoted by behaviors like traveling far, searching, seeking mates. The only advantage to a large gamete is its contribution to a healthy, well-endowed, competitive zygote; but any risks, such as those of roaming far, may be counter to its interests. So typically it is more profitable for a single individual to make—and promote the success of—only one of the two gamete types. This pattern, *anisogamy* (unlike gametes), is so ubiquitous that, without thinking about it, we tend to call small gametes "sperm" and small-gamete-makers "males," and to call large gametes "eggs," and large-gamete-makers "females."

Any behavior by the gamete carrier that enhances the advantage of either the small gamete or the large gamete is likely to be favored. If one carries small gametes (whose only advantage is meeting many other gametes), then traveling far and wide, spreading those sperm about, will be a better strategy than sitting alone in a safe place, meeting only those who do travel. And so we associate sets of behavioral characteristics with each of the sexes: the risk-taking travelers we call males; the risk-averse nurturers, females.[10] And we're almost always right—the number of sex-role-reversed species is minuscule.

This observation may help us puzzle out why males and females diverge so strikingly in so many species like deer, seals, and sea lions,[11] and what conditions lead to the sexes behaving similarly in species like Canada Geese. Within the single species *Homo sapiens*, the variety of behaviors between the sexes is extraordinary. Perhaps stepping back and looking first at the general rules of sexual repro-

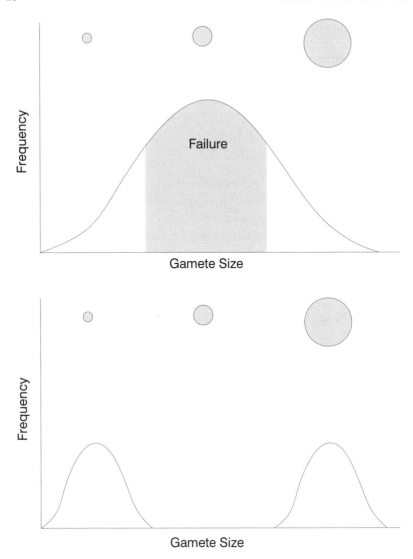

Figure 3.1. Two feats are required for a gamete to contribute to a successful zygote: it must "find" another gamete, and the resulting zygote must be well endowed and healthy. Small gametes do the first task well; large gametes accomplish the second. Middle-sized gametes lose out to both extremes. This disruptive selection leads to anisogamy (unlike gametes) and a bimodal size distribution: small gametes (sperm) and large gametes (eggs). (See text for further discussion and for the role of cytoplasmic conflict in the development of anisogamy.)

duction in other, simpler species will be useful to tease out why the general pattern exists, and what conditions make exceptions profitable.[12]

Once specialization into small- or large-gamete making exists, any subsequent changes in *either* gamete *or* carrier that enhances these specific advantages will be favored by natural selection. That is, if ever a proto-sperm is produced which is not round but ovoid (hydrodynamically superior) and has a tail, it will be favored in the race to the egg. Proto-sperm-making individuals who travel farther, thus finding other gametes over a larger area, will be favored. Proto-egg makers who invest less and less in other functions, and more in big healthy future offspring, will be favored. Thus, starting from the simple physics of gametes bumping into other gametes, a strong bias toward differentiation is built up. Both physical and behavioral characteristics are affected, for the benefits and costs of searching and endowing differ. Perhaps the rarity of simultaneous hermaphroditism in animals is related to this phenomenon, for it's hard to search and to endow, both maximally, at the same time.[13]

The nuclear genes in the gametes had to fight for their spot—in meiosis, the competition is among genes within the individual. But in sexual species like humans that produce eggs and sperm, the cytoplasm in the gamete also brings along information and "interests," and there is still the problem that each gamete would "rather" not have to deal with the other's cytoplasmic material. Much of our understanding of the species that do have more than two sexes arises from biologist Lawrence Hurst's examination of this cytoplasmic "war of all against all."[14] If a zygote is formed by the fusion of two gametes, there can exist a strong conflict of interest (mother's versus father's genes and cytoplasmic material).

Typically, the egg (or egg producer) has mechanisms to keep out the cytoplasmic material from the sperm.[15] In contrast, when sex consists not of fusion between two gametes but of conjugation—transferring just a nucleus across a "pipe" between two exchanging cells—there is no conflict between the cytoplasmic materials, and there can be any number of sexes.[16] There is even a "hypotrich" ciliate, a microbe, that has both sorts of reproduction; in fusion sex, it behaves as if it had only two sexes, but in conjugation sex, it has many. The slime molds with thirteen sexes have a complicated hierarchical sort of fusion sex. Sex 13 always contributes the organelles,

A.

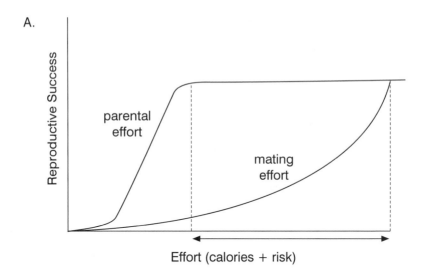

B.

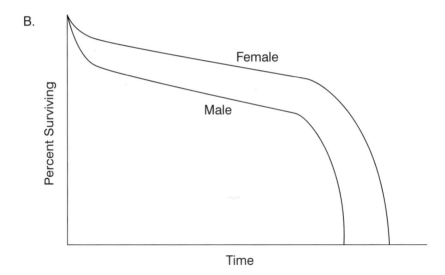

Figure 3.2. (A) Mating effort has a large "fixed cost" compared to parental effort. The arrow highlights the region of additional expenditure required to match the payoff for mating effort compared to parental effort. (B) As a result, risk taking is more profitable for mating-effort specialists (usually male), who show lower survivorship than parental-effort specialists.

whomever it mates with. Sex 12 contributes the organelles if it mates with 11, 10, etc., but not if it mates with 13. And so on.

While both conjugation and gamete sex are options for very simple species, complex multicellular species (including humans, of course), are stuck with sex from the fusion of gametes, and cytoplasmic warfare. Separate sexes, numbering two, are an effective way to get genes repackaged and to resolve some of the conflict between parents' genes. One answer, then, to Professor Higgins's sad cry, "Why can't a woman be more like a man?" is that once there are two sexes, with different paths to success in reproduction, the strategies that work for each are likely to be very different, just as in the Blue-Headed Wrasse.

These behavioral specializations, into mating (seeking) versus parental (nurturing) effort, have profound consequences for trends in behavioral differences between the sexes.[17] Mating and parental effort have very different patterns in the ways individuals profit from them ("reproductive payoff curves"; fig. 3.2A). Mating effort, typical of males in most species, has what economists call a large "fixed cost"; that is, much effort must be invested to get any return whatsoever, but after some level, great additional gains come from just a little more investment. So specializing in pure mating effort has great impact on a male's life. A male must grow large if physical combat is any part of competition; he may have to range far; perhaps he must grow weapons or decorations, like a moose or a peacock; he may have to fight—all these have costs. A red deer male, even to attempt a first mating, must grow large, involving both energy and an opportunity cost of delayed maturation, for what is invested into fighting for status cannot be put into growth, and if size matters for success, a male who switches too soon from growth to conflict will remain forever small and unsuccessful. He must grow antlers, and he must fight for dominance and control of good feeding grounds. All of this is required just to break into the mating game. After this great initial investment, though, a mating-specialist male's payoff curve rises steeply, for additional matings cost little.

For parental investors (usually the sex we call female—the large-gamete makers), the starting condition is that each offspring costs approximately as much parental effort as any other. Getting a mate is almost never a problem, though selecting a good one may be. What is invested in one offspring can seldom be recycled and reinvested in another; for example, though nests may represent generalizable parental effort (reused for several clutches), feeding is off-spring-specific—what this baby eats, that one can't.[18] In many species, the maximum possible number of successful offspring is likely to be lower for females than for males, but males will vary more in their success, and more males will fail entirely to reproduce.

THE ECOLOGY OF BEING MALE AND FEMALE

The "male-female phenomenon" is pervasive. As we have seen, beginning with simple rules of physics (those jellyfish gametes floating in the sea), consistent differences arise in the costs and benefits of mating versus parental effort. This means that there are likely to be consistent differences between males and females: in behavior, and in how much success varies.

For both sexes in many mammals, five traits seem to contribute to reproductive success: age, body size, dominance rank, early development patterns (high early growth rate), and quality of mates chosen. For females, these typically contribute to getting good nutrition and converting it for offspring. When males take risks in direct mating competition, they die sooner than females.[19] Thus, for males, lifetime breeding success is most variable in species with direct conflict over mating access, whether by prolonged defense of territories as in elephant seals or in competition over single females, as in some butterflies.[20] Male lifetime breeding success tends to be least variable in species in which males compete indirectly, as do men in many human societies.

The roles of intrasexual conflict and resource striving have been thoroughly studied in red deer and elephant seals, and these species reflect the general mammalian picture well.[21] In elephant seals, males compete physically; they grow to be much larger than females, but this means that they suffer the attendant cost of maturing later, since maturing late means more exposure to possible death before any chance of reproducing. Males compete for the control of sandy beaches on which females give birth. Over 80 percent of all male elephant seals ever born die without reproducing, but a highly successful male may have over ninety offspring in his life. Reproductive success varies greatly among males, whether it is measured seasonally or over lifetimes. Females, too, suffer great variation in success; approximately 60 percent of females ever born die without giving birth. Most of these are killed as youngsters before they mature; unlike males, almost no adult females fail to reproduce. The most successful females have about ten offspring in their lifetimes. The reproductive stakes are much greater for males than females because so many males fail, and variance is so high. A male who does not compete fiercely gets nothing: so males take more risks, and fight more. Fights can be severe. Seal pups are often injured or killed as males fight. Because females return to the same or nearby beaches every year, but male owners change, the probability that an infant killed is the offspring of the territorial male is difficult to determine.

In red deer, a few males control harems of from one to more than twenty females, or hinds, while most remain "bachelors," lurking nearby with no females. Harem holders do virtually all of the breeding with the hinds in their harems. However, because controlling a harem is so expensive, the tenure of harem holders is shorter than the breeding season, and male reproductive success does not vary so much as harem size. Variation in reproductive success among stags is a function of harem size, duration of harem-holding tenure, rutting area, fighting ability, and (less closely) life span. Stags fight to get harems and matings. Big, long-lived stags who are good fighters have the most offspring. Since most females have one calf each year, their reproductive variation is small; it depends on female life span and calf mortality.

Thus, for stags, size and ability to gain dominance are crucial; for hinds, keeping oneself and one's calf alive is important. The reproductive return for resources expended by the two sexes is quite different (fig. 3.2a) and follows the general pattern: male red deer reproductive success varies more than female reproductive success, and male competition is more direct and physically riskier than female competition. Both poorly- and well-invested hinds can be successful in producing offspring, although their condition does matter subtly. Male calves that are born early in the season (with a long time to grow before their first winter), and at a high birth weight, are more successful than those born "late and light." Not surprisingly, the sons of dominant hinds, in good condition, are more likely to be born heavy and early.

Sons of dominant hinds have greater reproductive success than daughters of dominant hinds; daughters of lower-status hinds have greater reproductive success than sons of low-dominance hinds. (This pattern has interesting parallels in some human societies, in

which degree of polygyny and wealth are related to relative investment in sons versus daughters; see chapters 6 and 7). Further, dominant hinds produce more sons than low-status hinds. Dominance in
hinds is related to their own birth weight and their weight as adults.
Resources still matter for hinds' reproductive success, even though
the impact is less than that on stags and may be seen more strongly
after a generation's lag, in the sons' success. Not only genetic en-
dowment, but also social and historical factors, influence what happens to any individual.

Roamers and Homers

The relatively simple physics of "getting there" versus "investing"
at the gamete stage means that the most profitable investment strategies of males and females are likely to differ. It shouldn't surprise us
that the bearers of little gametes—whose only advantage, remember, was getting gametes into warm, safe places—might be predicted
to roam about more. In polygynous species with no male parental
care, males do simply spend most of their reproductive effort in
searching for mates.

   A variety of interesting consequences follow, highlighted by comparing males and females in two different but closely related species
of voles: small, blunt-nosed, small-eared meadow mice. *Microtus
pennsylvanicus,* the meadow vole, is a typical polygynous species.
Male *M. pennsylvanicus,* who roam more widely than females as they
seek mates, have much better spatial abilities than females, who are
risk-averse and stay much closer to home. If you look at their brain
structure, the brain parts devoted to spatial ability and memory are
larger in male than female *pennsylvanicus.* On the other hand, both
male and female *M. pinetorum,* a monogamous close relative, have
brain structure (and spatial abilities) similar to female meadow
voles.[22] The sex differences arise out of the ecology of reproducing
successfully in monogamous (more parental males) versus polygynous (more "searching" males) systems.

   Given what we know about the polygynous background of humans, you'd suspect sex differences in human spatial ability as well,
so long as men and women have differed in their use of space. And
that is what we see.[23] Men and women seem to be particularly good
at different kinds of spatial tasks: men outscore women by about 67
percent on mental rotation tasks, while women outscore men by 89

percent on remembering the location of objects. Perhaps men's mental-rotation, map-reading skills are related to their history of hunting (more on sexual division of labor in chapter 7); women, as gatherers, might profit more from noticing and tracking the location of small things—landmarks—in the environment. In experiments involving looking at pictures or sitting in rooms and being required to recall items they had seen, women outscored men by 60 to 70 percent.

The basic trends toward sex differences lead to a central and pervasive life-history trade-off. How is it most profitable to allocate reproductive effort: as high-risk, high-gain mating effort, or as offspring-specific, true parental investment? There is a real conflict between what an individual does to be successful in each. Although ecological conditions can change the relative benefits of specialization versus combining mating and parental effort, the "default" condition is a specialization by each sex into either mating or parental effort.

In mammals, these trade-offs are set into sharp relief, for females, having mammae, are specialized to nurse offspring. This specialization of females for nutritional investment means that males typically can profit by specializing in mating effort rather than parental effort. Within each specialization, however, there is still diversity in how an individual can spend either mating or parental effort.

Mating Effort

Mating-specialist males have really a limited number of possible strategies to secure mates: they can try to control females, control or gather resources useful to females, or display for females, independent of resources. Females can expend mating effort, but because females produce the larger gametes with higher offspring-specific costs, males are more likely, in most environments, to profit from mating effort than females. This is especially true for mammals (see above). Each of these mating strategies depends on the abundance, predictability, and defensibility of resources and gives rise to a different mating system.[24]

Resource Control

In many species, males that can gather, commandeer, or sequester resources that are useful to females have a distinct reproductive advantage. When such resources are reliably—predictably—abundant, and economically defensible, males will attempt to control territories and exclude other males. Thus, male elephant seals come to shore before females, and compete to control the few sandy patches along the rocky coast where females must come ashore to give birth; red deer stags control good feeding grounds, excluding other males.[25] In non-human species, territoriality is likely only when resources useful to females are both reliable and economically defensible.

Human systems are far more complex. Even considering only "territory," for example, humans have lineage-held lands that are inherited. We also have recognized property rights. That is, we humans have third-party interests and interventions, so that we see not only simple individual or coalitional territory defense, but situations in which intruders are punished by uninvolved third parties (judges, juries) for invading territories accepted by those same third parties as reasonable.

Land and its contents are only one sort of resource. Status (dominance), in other species, and status and wealth in humans are very real resources (chapters 4, 7, and 8). Finally, there are non-resource-based mating criteria: "sense of humor," and "considerateness" (chapter 5).[26] Underlying this diversity, it is nonetheless true that men who control resources in most societies have a reproductive advantage.

Harems and Mate Guarding

Harems, of great interest to most of the men I know, are fairly common in other mammals and their form is influenced by ecological conditions. When females tend to remain in groups (usually to avoid predators), but useful resources are not reliable or not economically defensible, males may attempt to control females' movements, and thus the access of other males to females. Unsuccessful males cruise near the harem of a successful male and may try to entice females away.

Men are polygynous in many human societies, with more than one mate (whether a marital partner or not). In these societies, guarding and controlling access to women is a large problem for men. In some famous large societies, harem control is elaborated, with eunuchs to protect the harem and severe punishments for transgressing males. In other species, a male controls a harem as large as he can alone or (rarely) with a reciprocator male. In human history, there have been caliphs and emperors with literally thousands of concubines, maintained with the assistance of men who have been castrated well before they reached an age to make such a choice themselves.[27] Human harems can therefore be larger and their maintenance more complex than those in other species.

By now it should be no surprise that reproductive matters in many societies are treated as absolutely core, and that legal measures to discourage sexual transgressions are among the harshest of punishments. Here's an example from the Ashanti about the punishment for a man who sought favors from a king's wife (the woman was beheaded; more about the Ashanti in chapter 12):

> The culprit, through whose cheeks a sepow knife has already been thrust, is taken. . . . The nasal septum is now pierced, and through the aperture is threaded a thorny creeper . . . by which he is led about. Four other sepow knives are now thrust through various parts of his body, care being taken not to press them so deeply as to wound any vital spot. He is now led by the rope creeper . . . to Akyeremade, where the chief of that stool would scrape his leg, facetiously remarking as he did so . . . "I am scraping perfume for my wives" next to the house of the chief of Asafo, where his left ear is cut off; thence to Bantama . . . where the Ashanti generalissimo . . . scrapes bare the right shin bone.[28]

Then he was made to dance all day. After dark, his arms were cut off at the elbows, and his legs at the knee. He was ordered to continue dancing, but since he couldn't, his buttock flesh was cut off and he was set on a pile of gunpowder, which was then set alight. Eventually, the chief gave permission to cut off the offender's head. Thus, although my male colleagues who dream of harems always imagine being the harem master, a little consideration to the individual impacts of reproductive variance is in order!

Tending Bonds

Tending bonds are a form of short-term mate guarding: a male identifies a receptive female and remains with her during her fertile period, chasing rival males away in a temporary "consort" relationship. Then he leaves.[29] The pattern we see, of timing and length of the consortship, suggests that the reproductive point is for the male to deny other males any access to that female until his sperm have successfully fertilized her.

In nonhuman vertebrates, tending bonds occur in specific ecological conditions. When females can raise offspring successfully alone, and there is safe heavy cover, females tend to be solitary; males can seek out females with relatively little harassment from other males. When the terrain is very open, large groups are common, and tending males usually lead the female away from the group for mating during her fertile period (typically a male's dominance status is crucial in his success at this).

Humans, of course, are extraordinary in that females can maintain continual receptivity, and ovulation is quite difficult to detect. Indeed, some scholars suggest that the evolutionary point of this peculiar situation is to make it difficult for males to get away with such temporary alliances. So, for a variety of reasons, human males may profit from rather steady and constant association with one or more females (longer-term mate guarding) rather than brief "tending bond" periods.[30]

Leks

Leks are typically systems in which males defend reproductively useless resources, often with considerable display, and at great risk.[31] Leks may look like territorial systems, but nothing of value is defended. The profitability of such a strategy is not immediately obvious. Generally, if males can control females or resources females need, they have a stronger bid in the mating game; if neither resources nor females are controllable but appropriate staging areas exist, they may display in leks. By lekking, they simply advertise their "quality"—how well they have survived, what great risks they can take by displaying, how strong their sons might be.

It is important that the males are making unbluffably costly advertisements. Since males have nothing to offer but their genes, females assess male ability to perform costly feats, as a result of males' genes and history. Males may advertise costly looks (bright colors are expensive to make and maintain, and visible to predators) or dangerous calling behavior, depending on the environment. For example, among grouse, Prairie Chickens display visually, while forest-dwelling grouse "drum" on a perch and perform a short flight in which they make sharp snapping sounds with their wings.

Females visit the lek, assess male quality, and choose a male. Male reproductive success can be extremely skewed, for females typically choose to mate only with those males able to control the central hardest-to-defend territories. There is an unsolved problem with the maintenance of such systems. Imagine an extreme situation, in which all the females in a generation chose a single male. The sons born to these females vary in their abilities; where does this variation arise? Since all sons carry genes from the same father, even if Dad is heterozygous little variation comes from this source; more comes from differences among genes from their various mothers, or Mom's nutritional condition, or environmental conditions while growing. It would seem that what females can use to choose (current male ability) is imperfectly related to male genetic quality; the formal arguments make this clear. The question remains: How are such systems maintained?[32] One likely explanation, introduced above, is the parasite version of the Red Queen hypothesis: parasites with particular characteristics exert pressures that mediate selection *and* polymorphism in the host population.[33]

Scramble Competition

When neither resources nor females can be controlled economically and no good display grounds exist, males may display to females independent of either real or symbolic resources. This is most likely when females gather in predictable, somewhat centralized places (e.g., mayflies over appropriate water for laying their eggs); then males also congregate, and display. This strategy is, in some sense, the weakest from a male's point of view.

Human mating effort can have elements of all of these kinds of competition, but, of course, some are more common than others, and what's effective may differ during a man's lifetime. Our variation of "tending bonds," in which males temporarily sequester females away from the attention of other males, is common, usually among males who have little but genes to offer.[34] Perhaps resource-based polygyny is the most common pattern cross-culturally; as we will see in chapter 7, in quite diverse societies males with higher status or wealth manage to have more mates.[35] Some strategies are likelier than others at different times in one's life; teenage males often behave in what looks like scramble competition to a biologist, and sometimes I can't help visualizing gang territories—and perhaps discos and soirees among the elite—as kinds of leks.

Parental Effort

After mating, both sexes face the problem of whether and how best to invest in their offspring. In other species, specialization by each sex into either mating or parental effort is most common, but human males, by and large, do invest parentally in one form or another. Parental expenditure or effort has obvious costs to mating effort: what is spent irretrievably on offspring is unavailable for mate-seeking (or for other offspring). As Darwin noted: "The only check to a continued augmentation of fertility in each organism seems to be either the expenditure of more power and the greater risks run by the parents that produce a more numerous progeny, or the contingency of very numerous eggs and young being produced of smaller size, or less vigorous, or subsequently not so well nurtured."[36]

Thus, in addition to the evolutionary pressure to optimize investment in getting mates versus raising offspring, organisms have to solve the problem of how best to raise their offspring. *Parental investment* is the amount of effort invested in any offspring that would otherwise be available for other uses. From the point of view of the particular offspring, this is a most appropriate concept, although it may be difficult to measure. From the parents' point of view, problems of investment must be considered over more than one clutch or litter—in fact, over a lifetime. *Parental effort* is the sum of parental investment over any defined period.[37] It can include not only offspring-specific true parental investment (like nursing), but also "reusable" effort, such as building a large nest, or digging a den—parental care that can be used for many offspring.

Spending parental effort effectively is not a simple problem. Parental effort can be invested once (*semelparity*, single birth) or more than once (*iteroparity*). Whether "once" or "more than once" is most effective depends on the parent's extrinsic chances of survival; what are its chances, independent of reproducing, of surviving for another attempt? If those chances are very low, or quite uncertain, an individual may do best to devote everything, even dying in the attempt, to reproducing this time.[38] Across species there is great variety. Atlantic salmon return to their spawning grounds from the ocean, breed, go back to the sea, and return next year. A really successful adult can return many times. Pacific salmon, whose chances of living another year are lower anyway, return once, and exhaust themselves in breeding.

Perhaps the most curious cases are the species in which one sex lives to reproduce several times, while the other sex of the same species dies after reproducing once. In one octopus, females use themselves up protecting offspring, while males mate as many times as they can. In contrast, there is a marsupial mouse in Australia in which the male dies after copulation, while the female is iteroparous. And, of course, there are a number of insects (praying mantises, some crickets) in which the male dies or is killed by the female, and his body is used to feed either the female or the offspring.

Obviously, parental effort can be expended in male or female offspring, and the profits from, and costs of, male versus female offspring can differ, so it is not a trivial life history decision.[39] Parental effort can be expended as physical biomass, or as behavioral investment; but that represents only the first strategic layer. Within effort spent as biomass, there is still the problem of optimizing the size of each offspring versus number of offspring. Parental effort can be spent as behavioral care (feeding, teaching, guarding), or in other ways (e.g., biochemical defense). In sexual organisms, because of the evolution of large eggs and small hydrodynamic sperm, there is an overall trend toward greater parental expenditure by females.

Variance in Reproductive Success: Mating versus Parental Strategists

This bias, then, is clear: the carriers of small, "finder" gametes will profit by behaving in ways that enhance the finding capabilities of the small gametes; the carriers of big "nurturing" gametes will profit by enhancing nurturance.[40] Males can be parental, however, and in nonhuman species there are important ecological correlates to great male parental effort. One common form of male parental care is feeding, as in many songbirds. Another is offspring protection, either with the female (as in geese) or providing care while the mother recovers (as in some sandpipers).[41] In poison-arrow frogs, the female lays eggs, and the male carries them about until metamorphosis in his vocal pouch; since his skin is full of curare, both he and the eggs are safe from predation.

But male parental care can be a very mixed bag, for all involved. Obviously we do not expect expensive male parental care if a male's paternity is uncertain; so mate guarding typically accompanies male parental care. Further, some forms of male parental care may really function as mating effort. Particularly in some primates, for example, a male may be much more likely to care for a juvenile when he is likely to reap matings from the mother. And the relative worth of guarding, versus caring, versus mating effort, are often frequency dependent and not obvious. The bottom line is that male mating competition may have much stronger influence in shaping male strategies than parenting payoffs, and even when males appear to be exerting parental effort, they well may be spending mating effort.[42]

The story doesn't end there, however; if that were the whole story, all species would be polygynous, and mating systems would not vary. In fact, there are further ecological correlations of polygynous, monogamous, and polyandrous mating systems. An important aspect of mating systems for biologists is the relative *variance in reproductive success* (fig. 3.3).[43] In mammals, especially, since females are equipped to feed dependent offspring, there is a bias toward polygyny. In polygynous systems, because only a few males ever get a mate, great expenditure and risk taking may be worthwhile. As a result, in polygynous systems more males die than females at most ages (fig. 3.2B). Being a male is a high-risk, high-gain strategy. For a female, making a son is thus also a costly high-risk, high-gain strategy.

Rarely can males gain sufficient reproductive success (RS) from parental care to compensate for the lost RS of forfeited matings, so we see many polygynous systems either with no male parental care or with "generalizable" male parental effort. For instance, male Redwinged Blackbirds keep watch for predators, and males in a number of primate species (including humans) act as mentors and protectors for their offspring (chapter 4). But both groups are polygynous. Monogamy is defined by biologists as a system in which variance in male and female reproductive success is equal. Monogamy is typically associated with phenomena such as complex learned behavior, as in many vertebrate predators, or highly competitive environments faced by offspring (so that only superbly invested, costly offspring can succeed). At issue is whether offspring can profit enough from male parental care (of the offspring-specific sort, like feeding) that it "pays" a male genetically to give that care. And whenever a male can either spend paternal investment in generalizable ways (so he is free to seek more matings) or disguise mating effort as paternal effort, he is likely to do so. The reproductive efforts of males and females may converge in monogamous systems, compared to, say, harem polygynous systems, but those interests are still identical only in restricted circumstances.

Although we call many human societies "monogamous," this is

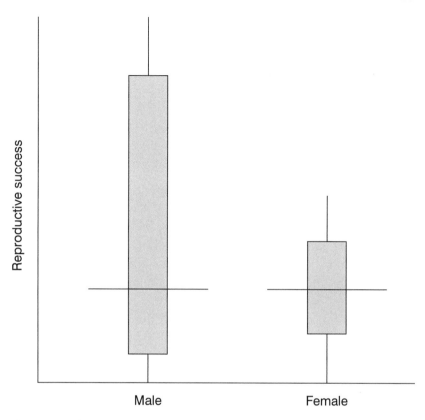

Figure 3.3. The average reproductive success (RS) of all males in a generation must equal the average reproductive success of all females, but the variance in reproductive success may differ greatly between the sexes. In polygynous species in which males specialize in mating effort and suffer high failure and mortality rates (fig. 3.2), the RS variance will be much greater in males than in females.

clearly a biological misnomer. Most societies with one-spouse-at-a-time rules would be called polygynous in a biological definition: more men than women fail to marry, and more men than women re-marry after death or divorce, producing families in these later unions. The most reproductive men have many more children than the most fertile women. All of these phenomena increase the variability of men's reproductive success compared to women's, making us polygynous by a biologist's definition.

Polyandrous mating systems are not only the rarest, but perhaps the most curious ecologically. In these systems, females are likely to spend more mating effort than males, while males specialize in parental care. Polyandrous systems are relatively more common among birds compared to mammals, perhaps because in bird species with high parental care the males, in contrast to most mammals, can feed offspring as well as females. Because egg biomass is true parental investment that cannot be reused for another offspring or for mating, and because there is some size threshold for success in eggs, females in polyandrous systems never achieve the number of offspring (and thus the skewed variance) of males in many polygynous systems. But we do see the expected trend: the eggs of monogamous bird species tend to be larger and less numerous than the eggs of closely related polyandrous species.

Defining mating systems by the relative variance in male versus female reproductive success can be useful. For one thing, it allows us to calculate the relative intensity of sexual selection on the two sexes in any system. It's important to remember that it is the *relative* variance in reproductive success of males and females that is important, rather than the absolute level of either variance. Monogamous systems can show high variance, for example if many offspring die—but variance is still similar for the two sexes.[44]

In the next three chapters, I will look first at how the interaction of ecological constraints and these trends toward differences play out in primates, and then at how resource conflicts of interest develop and persist through human lifetimes.

THE FAR SIDE by Gary Larson

4.

Sex, Status, and Reproduction among the Apes

Power is the ultimate aphrodisiac.
—Henry Kissinger, *New York Times,* January 19, 1971

I don't know a single head of state who hasn't yielded to some kind of carnal temptation, small or large. That in itself is reason to govern.
—François Mitterand, quoted by William Styron,
New Yorker, October 12, 1998

WE MUST REMEMBER that humans are primates, not deer or seals. We are complicated and diverse; ecological, social, and historical conditions can all contribute to the patterns we observe. Yet there are real regularities to how these forces interact, and sometimes we can make better sense of complex, apparently eclectic happenings through the selection lens. Consider the Salem witchcraft trials (more in chapter 10), during which Katherine Harrison was first immune from accusations, then targeted, then became safe once again through a powerful male accuser's turnabout to protector; self-interest and power differentials clearly had influence. But the basic relationship between resources and reproduction is complicated by ecological influences on how striving can occur. Let us explore further the impacts of male and female resource and reproductive value in different environments.

Most primates, like elephant seals, red deer, and other mammals, are polygynous; males specialize in mating effort, and females specialize in parental effort. This has important effects on male versus female payoffs for striving and risk taking. In red deer and elephant seals, male striving for dominance is central to succeeding; strategy matters, but pure size and strength are still very important. Primate species show a diversity of social arrangements, and males' and females' costs and benefits vary. While strength isn't irrelevant to a male's success in most primates, intelligence and social skills appear to be important as well in gaining status and mates. Here I want to focus on two main questions: How does the (spatial and temporal) distribution of important resources affect male and female reproductive options? How do the resulting mating systems vary?

THE ECOLOGY OF DOMINANCE AND RS IN PRIMATES

Henry Kissinger, in the above epigraph, fairly described the mating market among most primates, including, arguably, humans. In nonhuman primates, the relationship of dominance to reproductive success is complex. External environmental conditions (e.g., cover, predation risk) influence the sort of living arrangements that succeed, and the living arrangements influence male and female strategies, similarities, and differences. The strategies of the two sexes further interact, and for both sexes there can be trade-offs. Under all the variation, in most species there is a simple bottom line: reliably, higher-status males get more copulations and/or more offspring than other males.[1]

Simple individual dominance in male-male competition doesn't explain what we see.[2] Sometimes "outsider" males join harems during the females' receptive periods and appear to mate without constraint. It may pay males to seek attention and reciprocity from females, even when females are not sexually receptive, for in many primates females can influence male success through mate choice, "special relationships," and support of particular males in social interactions.[3] In some species, such as Japanese and Barbary macaques, high-ranking males clearly get more matings than others. But the number of copulations is not necessarily the issue; for in many primates, females (like human females, but less dramatically so) may mate when they are not actually fertile.[4] Male success comes not simply from maximizing copulations, but from mating with females during their very short periods of actual fertility; this access appears to be commandeered by the dominant male in most studied species.

In most species, the intensity of competition is related to the number of males in a group: the more male competitors are present, the more intensely males must compete. This intensity is further influenced by the payoff: the number of fecundable females in the group. The competition works as we would guess: genetic markers allowing accurate paternity assessment typically (though not always) show that rank and sexual access are positively correlated for male primates.[5]

The relationship between rank and reproduction went unrecognized for years, perhaps because interactions in many primates are subtle.[6] Consider chimpanzees: status is complicated. There is a formal dominance hierarchy, established through aggressive encounters; once the matter is settled, subordinates display submissive greetings to dominants, and dominants show friendly behavior toward subordinates. Acceptance behaviors lower tension in the group, and cooperative relationships exist. "Real" dominance in any particular situation does not always reflect formal rank, and individuals may achieve power in particular situations without challenging the formal hierarchy. So, for example, female chimps sometimes calmly appropriate food or resting spots from males. Males sometimes negotiate access to estrous females, avoiding overt aggression (but remember that females may be receptive but not fecundable).[7]

Patterns differ among populations of chimpanzees, for example, in the degree of "possessiveness" by the alpha male, the proportion of matings that appear to be opportunistic, and the role of females in initiating sexual encounters; however, in most studies, "rank hath its privileges" for male chimpanzees.[8] Among bonobos, or pygmy chimpanzees, females form alliances, and high-ranking females, at least, appear to be able to make choices among males without restriction. Male aggression appears milder and more quickly reconciled than in chimpanzees, possibly because females are gregarious, and receptive longer. Male dominance rank still affects mating chances, though, and male-male competition is more intense than was previously thought.[9]

In polygynous New World primates, as among the great apes, most studies find that high-ranking males have an advantage in sexual access: golden lion tamarins, some capuchins, red howler monkeys, Costa Rican squirrel monkeys, and spider monkeys.[10]

What about females? Female primates are physiologically limited in the number of offspring they can produce, and the relationship between females' rank and their reproduction is perhaps less clear than male patterns.[11] High-ranking females in several macaque and baboon species reach maturity earlier than others, giving them a reproductive advantage. In several other species, the infants of dominant mothers survive better than other infants. Three studies on macaque species reported relatively "even," low-variance, female reproductive success: 34 percent of females produced 38 percent of offspring, 40 percent of females produced 38 percent of offspring, etc. In gelada baboons and chimpanzees, female rank and reproductive success appear to be positively related. In all, female dominance explains some of the variation we see in female primate reproductive success, but variance is less, and dominance may explain less of reproductive success, than for male primates.

Female striving carries costs in at least some primates: very high status female savannah baboons, who compete to retain their status, suffer some fertility costs.[12] Compared to low-ranking females, they have several reproductive advantages: they mate more often, have shorter interbirth intervals, achieve better infant survival, and their daughters mature more quickly. But the overall relationship between female rank and lifetime reproduction is not significant. Why? High-ranking females also show characteristics we would associate with stress in humans; they suffer more miscarriages and fertility problems than other females. These are unevenly distributed; some high-ranking females are highly successful, others have no offspring. If one excludes the no-offspring females, there *is* a relationship between rank and lifetime reproduction. In analyzing rank and fertility of female primates, we have a lot to learn.

Ecological Aspects of Mating Systems

Whenever females can raise offspring successfully alone, the likely outcome is polygyny, with female specialization in parental effort and male specialization in mating effort. However, ecological conditions set the stage for the particulars of solitary-versus-group living, the kind of group, and the mating system. About half of primate species live in multifemale groups. Avoiding predation through "selfish herd" groups, and defense of food by multifemale groups are two major hypotheses for the formation of primate groups.[13]

While females tend to distribute themselves in response to predation and food pressures, males tend to distribute themselves in response to female distributions. Male-male rivalry can be a deadly
business. But it is also variable. Among baboons, there is a famous
example, in which three relatively closely related baboon species live
quite differently, depending on the combined ecological pressures of
food richness and predation pressure: one lives in multimale groups
in an open, dangerous habitat with predators (several males, but no
single male, can protect against the predators); the second lives in a
resource-poor but safer habitat in single-male groups; the third,
which moves between both sorts of habitats, switches social structure with habitat. There is an ecological chain of influences that
makes sense of the variation.

The rare monogamous single-pair primates tend to live in heavy
cover. There are three hypotheses about the evolution of monogamy
in primates. One simply argues that when two-parent care is
markedly more effective than maternal care alone, monogamy will
evolve. The second suggests that monogamy arises when mated
males are able to protect their infants from infanticide by other
males. The third suggests that for primates, as for many other
species, ecological factors influence a male's ability to monopolize
more than one female.[14] There are few tests, but I suspect the last is
the strongest hypothesis.

Among the New World primates, there are good examples of the
relationships among ecological factors and male and female strategies. When female reproduction is strongly constrained by seasonality, males are more likely to stay in their natal groups, forfeit exclusive mating opportunities, and tolerate infants. Males are more
likely to disperse and compete for exclusivity when the reproductive
payoffs are higher. Saddleback tamarins (*Saguinus fuscicollis*) show
an interesting ecological response. They live in heavy cover, and one
might expect monogamous pairing. However, they give birth to
twins, who weigh half their mother's weight by the time they are
weaned; not surprisingly, mothers do relatively little infant carrying.
A commonly observed group is a pair plus an extra male, in a
polyandrous trio.[15] Here, male and female constraints obviously interact.

Primates also vary in the degree of male-female size differences.
Gorilla males are huge compared to females; gibbon males are barely
larger than females. Here, too, ecological influences are important.
Pair-living, single-male, and multimale group species differ in ex-

traordinary ways. Males in single-male group species face high-risk, high-gain challenges from other males to fight for control of the females; they are physically much larger than females and have very large canines, as in gorillas. Males in multimale group species have a somewhat more subtle problem to deal with as well: other males are always about, and sometimes in shifting coalitions. Sneak copulations by other males in the group are, like overt fights, a serious problem. Males in such species are bigger than females (though not so strikingly as the single-male species), have large canines, but also have extremely large testes; here competition between males exists not only between individuals, but carries down to the level of sperm competition. Males in pair-living species are not much larger than females and have relatively the smallest canines, the smallest testes. The ecology of group living influences male-male competition—and thus body size, canine size, and testis size.[16]

SEX, RESOURCES, AND THE ECOLOGY
OF HUMAN REPRODUCTION

Differences in the reproductive ecology of the two sexes in humans create opportunities for men and women to use resources quite differently in reproductive competition. These conditions foster important sex differences.[17] We expect to find some general trends in behavioral differences between the sexes, and we do, beginning at birth. Newborn boys cry more, respond less to parental comforting, and require more holding. Newborn girls are more "cuddly" than boys; they respond more strongly to adult faces, and to being held. Boys are somewhat more interested than girls in inanimate, nonsocial objects. Boys seem to begin technical problem solving sooner, and wander farther from home earlier. These differences are seen very early and occur across cultures.[18]

While these sex differences are likely to have genetic components, none are "genetically determined" in any straightforward way. Very few sex-related differences have a clear chromosome-trait correlation; of those that do, most are disabling medical conditions.[19] Here is an exceptional example: babies missing an X chromosome (XO rather than XX; Turner's syndrome) are 98 percent likely to die before birth; those surviving show mental deficiency. Consider such a girl: her single X chromosome comes either from her mother or her father. If her X comes from her father, she will have better social skills than if the X comes from her mother.[20] Now, in normal infants, sons have XY, and the X always comes from the mother, while daughters have an X from their mother and an X from their father. It is far too early to make much of this, but there is surely a possibility that the "sugar and spice" versus "snakes and snails" folk wisdom reflects some genetic influence.

A well-known geneticist was quoted recently as having said that "we already have a genetic marker for violent behavior: it's called the Y chromosome." Indeed, in many species, the Y chromosome is a good "proximate" marker for competitive behaviors, for all the ecological reasons I have reviewed. Genetic patterns reflect the ultimate evolutionary causes: most differences between normal "XX" and "XY" individuals arise not because an X or a Y chromosome "dictates" anything, but because the *ecology of achieving reproductive success* differs for males and females. Even if the Y chromosome alone dictates nothing, its bearers live in environments that influence what strategies will be successful in getting genes passed into offspring. The return curves of figure 3.2A set many parameters of success for mating-effort specialists (usually male, in mammals XY)— who must compete, sometimes violently; and parental specialists (usually female, XX), who typically reproduce successfully by being risk averse.

Some ecological conditions favor great male expenditure of offspring-specific effort; in these environments we expect behavioral and even physical convergence between the sexes. The less each sex specializes and the more they do the same things, the more alike they will be. We thus expect sex differences to vary among environments, among mating systems, and among cultures as a result of more subtle cultural influences. And indeed, humans are extraordinarily variable, as we will see.

Resource Value and Men's Reproductive Success
in Human Societies

Just as the above discussion implies, men in most societies that have been studied use resources—wealth or status—to gain reproductively, typically through polygyny: additional wives. In such polygynous societies men's ability to marry and to reproduce successfully varies, sometimes enormously, so great expenditure and great risk taking may be profitable. I will elaborate on this point later, but here I want to suggest the general patterns we expect, and in fact see.

In more than one hundred well-studied societies, there are clear formal reproductive rewards for men associated with status: high-ranking men have the right to more wives, and they have significantly more children than others.[21] In many other societies, there are no formal societal rules (such as "men of status X may have two wives," etc.), but wealthy men simply can afford to marry more often than poorer men. Among the Iranian Turkmen, richer men have more wives and more children than poorer men (chapter 7). Among the African Kipsigis, richer men marry younger (higher reproductive value) wives and produce more children than poorer men. On the Pacific island of Ifaluk, men who hold political power have more wives and more children than others. The status-reproductive success pattern holds not only in these societies, but in others as diverse as the Meru of Kenya, the east African pastoralist Mukogodo, the agricultural Hausa, the Trinidadians, and the Micronesian islanders.

Even in societies such as the Yanomamö and Ache of South America and the !Kung of the Kalahari in southern Africa, in which few physical resources are owned, male striving results in male status, effective in marital negotiations. Among the Yanomamö, coalitions of related men are important (chapters 7, 13). So male kin for coalitions represent a resource, and men manipulate kinship terms to maximize their affiliations with powerful men. Further, men can only marry women in lineages that have a particular relationship to their own, so men try to "redefine" their standing in ways that make more women available for mates. Among the Ache, good hunters have more children than other men (chapter 7). In the Kalahari Desert, the !Kung, living in a resource-limited environment, are almost entirely monogamous—but 5 percent of the men manage to have two wives. Thus in quite varied societies, wealth or status and reproductive success are positively correlated for men. Have we changed today? I explore this problem in chapters 7, 8, and 15.

Reproductive Value and Women's Reproductive Success

What about women? "Reproductive value" is the probable number of daughters a female will have in the rest of her lifetime; it is a function of the probability of living to any given age and the likely number of children born at that age. It has been used to make predictions about migration, contraception, and population growth.[22] It is also useful in understanding trends in marriage age and remarriage rates. Thus, in societies with bridewealth (74 percent of the 862 societies in the Ethnographic Atlas) or some other exchange of goods at marriage, younger women might be expected to command a higher bride "price" than older women, for they will likely have more children in the course of the marriage—and they do.[23]

Men typically set great store on high reproductive value, though it may not be explicitly identified as such—men may simply note that they think old women aren't very interesting. Under such conditions, men with greater economic resources may be able to command, or be chosen by, women with higher reproductive value in the marriage market.[24] And indeed, in empirical work such as that on the agricultural and pastoral Kipsigis, researchers found that the bridewealth required for a woman was directly related to her reproductive value. With the introduction of Western technology and medicine (novel evolutionary events), differentials have been reduced (see chapter 7).

Poor men might choose to court older (lower reproductive value) women who have accumulated their own resources, explicitly trading reproductive value for resource value. In eighteenth- and nineteenth-century Scandinavia, daughters of upper-middle-class men (who would marry richer men) were considered women (marriageable) at eighteen years, while daughters of poorer men, who would marry poorer men, were not considered marriageable until years later, in their mid-to-late twenties (see chapter 8). Meanwhile, richer men provided resources themselves, married younger women, and gained high reproductive value. Similar patterns with men's wealth and women's reproductive value existed in eighteenth-century England.[25]

Resources and status also affect women's reproduction, but, as in other primates, apparently not in the same way or at the same levels as for males. In traditional societies, resources strongly affect women's reproduction when they are limiting (e.g., malnutrition) and result in fewer children, but women can almost never use resources to gain the extraordinary reproductive success of highly polygynous males. There is a possible exception: societies in which the descent system allows highly successful women to concentrate resources in their sons who (in contrast to their mothers and sisters) can use those resources to become successful polygynists (see chapter 11).[26]

Male resources increase male fertility at the "high end" of reproductive variation—rich resources make the most successful males very successful, typically because they can acquire more than one wife.[27] Women's resources avert failure at the "low end" of the variation—women need enough resources to raise healthy children. However, through most of our evolutionary history it has not profited women to strive for great amounts, since they typically could not convert such excess resources into reproductive gain.[28] And, as I noted above, the fact that true parental investment is offspring-specific means that women face a conflict between what they can gain from getting resources versus investing in offspring. For men, since status and resources are so often currencies in mating effort, what's spent on getting resources typically enhances reproductive success; for women (indeed, for female primates in general) such striving may lower their reproduction.[29]

THE ECOLOGY OF HUMAN MATING SYSTEMS

There are ecological influences on this general trend for men and women, as among other primate males and females, to succeed through different strategies. As in other mammals, there are no clear dominant ecological influences leading to polygyny in humans; rather, it is the "default" strategy. Among other primates, polygyny is associated with a female and her offspring being "economically" independent: females can feed their offspring without male assistance. Such polygyny would represent a fairly clear sexual specialization into the mating and parental effort curves of figure 3.2A.

Human polygyny is a social institution involving not just mating but social rules about marriage. It has great biological impact, and to make matters more confusing, it is defined differently by social scientists and biologists,[30] so it is important to clarify here what I mean. As I noted in earlier chapters, behavioral ecologists use the terms "monogamous" and "polygynous" to focus on the impact of sexual selection; it is the relative variance in reproductive success (fig. 3.3) between the two sexes that is of interest. Societies such as Western industrial nations today that impose a one-spouse-at-a-time rule would be called "monogamous" or "serially monogamous" if one were interested in social rules. Most such societies, however, have, as a result of sex differences in remarriage (more men remarry than women) and fertility in second and subsequent marriages (men have children in second and subsequent marriages more than women), much greater variance in male than female reproductive success. A behavioral ecologist would call them functionally polygynous.[31]

The degree of polygyny does have some ecological correlates similar to those of polygyny in nonhuman species.[32] Patterns of parasite risk, rainfall seasonality, irrigation, and hunting explain 46 percent of the observed patterns in human polygyny.[33] The most powerful ecological correlate of the degree of polygyny found so far is perhaps a surprising one: pathogen stress (fig. 4.1). There are good, though perhaps not obvious, ecological reasons. Environmental unpredictability may make it difficult to "track" best phenotypes for an environment; in this case, the most successful parent will be one who produces offspring with genotypes likely, in turn, to produce new genetic combinations.[34]

Polygynous men, of course, have not only more variable offspring, but more offspring than monogamous men, so we must look further before claiming that offspring variability might have a functional role. Powerful men will promote polygyny whenever they can, whether or not pathogen stress is present. But monogamy is absent in high-pathogen areas. The degree of polygyny is really a threshold pattern, as in other species, rather than a linear relationship.

A clue that genetic variability in children is important comes from marriage preference: polygynous men in pathogen-laden parts of the world are more likely to marry exogamously (outside of their group), especially through capture of women from other societies. Such marital outreach results in more variable children for men. Sororal polygyny, in which a man marries sisters (and his children would be less variable genetically), is rare in areas of pathogen stress. Thus, there is a difference in the kind of polygyny if we compare high- and low-pathogen areas: co-wives are more genetically different in high-pathogen areas. From a woman's (or her family's) point of view, being the second wife of a healthy man may be preferable to being the sole wife of a parasitized man; thus, women may prefer polygyny in highly parasitized or disease-ridden areas, and men's and women's interests (typically more divergent in polygyny than in monogamy) may converge.[35]

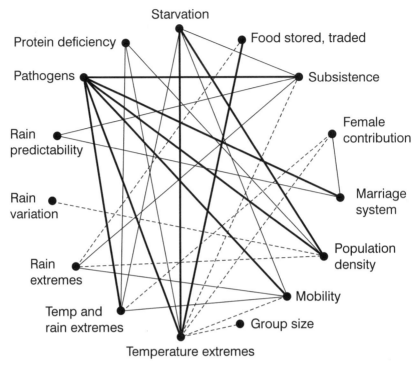

Figure 4.1. Ecological conditions and human social arrangements show a number of associations, summarized here. Bold lines indicate at least one relationship of significance, $p < .001$, or more than one relationship of significance, $p < .01$. Light solid lines indicate at least one relationship of $p < .01$ or more than one of $p < .05$. Dotted lines indicate at least one relationship of significance $p < .05$. Perhaps the most striking (and possibly underappreciated) relationship is that exposure to serious pathogens correlates with a number of social phenomena. (See text for further details.)

Polygyny is much more common in Africa and South America than, for example, in Europe, so it is important to ask if there is simply some sort of covariance of pathogens and socially determined patterns of polygyny. We can examine patterns within each region; if the covariation around the world is a side effect, then we do not expect pathogen stress and polygyny to covary within regions. Within the high-pathogen but socially diverse tropics, within Africa, Eurasia, South and North America, and the Mediterranean, pathogen stress and polygyny covary. Thus neither simple geography nor cultural diffusion of polygynous practices within high-pathogen regions is likely to be the source of the patterns we see. In the Pacific, there have been no societies in the "high" pathogen stress category, and because the relationship is a threshold relationship, no statistical relationship is apparent there.

Pathogen stress alone accounts for 28 percent of the variation in the degree of polygyny around the world, independent of geographic region or any other factor. Polygyny, no matter how it is measured, increases with pathogen stress, and the *type* of polygyny (nonsororal versus sororal) is further correlated with pathogen stress. Thus, pathogen stress may be a real environmental uncertainty—a stress—that both renders fewer men acceptable as mates and favors the production of offspring destined to produce radical new parasite resistance in descendant generations.

Resources and the Kind of Polygyny

As we saw in the last chapter, the ways in which males compete, and females choose, varies ecologically even among polygynous species. A major determinant is whether males can control resources useful to females; if they can, resource-based polygyny will exist. When controllable resources are lacking, males are reduced to trying to control females or to scramble competition (chapter 3).[36] When resources can be accumulated and defended, they are seldom distributed equally among individuals, and when resources are unequally distributed, so typically is reproductive success.

In one way, humans are highly unusual polygynists. In other species, as we have seen, typically males compete and females choose. But "third party" patterns in humans extend to mate choice. In many societies, others, not the bride-to-be, make the choice.

When societies lack rules of inheritance, suggesting that there is little to inherit, men typically do not exchange goods for women, but exchange women;[37] when there are no societal rules about wealth or hereditary class stratification, men are similarly more likely to exchange women than goods. But even in such societies, resources are not irrelevant to the pattern of exchange.

As already noted, when men purchase wives (bridewealth societies), younger (higher reproductive value) women are worth a higher bride price.[38] The currency of choice varies: sometimes women are purchased with cattle, as among the Kanuri people; sometimes with sheep, as among the Yomut Turkmen; sometimes with pigs, as among the Tsembaga-Maring; or a combination.[39] When large surpluses can be stored and men control important resources, wealthy men can negotiate for more wives than others.[40] If variance in men's number of wives arises from choice of men (by women or their families) based on the men's resource control, then men's use of resources in reproductive effort should correlate with the degree of polygyny. Indeed, the use of some payment by men (e.g., bridewealth or bride service), rather than exchange of women or payment of dowry, is strongly associated with the degree of polygyny.[41] When men can accumulate resources, the variability in how many wives they can afford increases; when men cannot, and they exchange women, they have less variance.

In some societies, a bride's family pays a dowry, reflecting an interesting twist on the sexual utility of resources. Dowry is fifty times more common in monogamous, stratified societies than in polygynous or nonstratified ones; in these societies males vary greatly in their status and wealth, and women married to wealthy, high-status men will benefit reproductively. So it may pay fathers of brides to compete, bargaining for wealthier men as mates. In some of these societies, it appears that poorer women's families must pay more dowry than wealthy women's families.[42] Insofar as they fail to be able to do this, the stratification is intensified. One example of dowry as female competition is that in modern rural India. Since about 1950, demographic shifts have resulted in a decline in potential grooms for potential brides of marriageable ages—and dowries have risen steadily. By 1990, a dowry was likely to be over 50 percent of a household's assets. Wives from poor families, able to pay less in dowry, may be less likely to marry; if they marry, they have a high risk of spousal abuse.[43]

Perhaps surprising to those of us in Western societies (in which close relatives are typically forbidden to marry), in many societies first cousins are the preferred marriage partners. Four kinds of first-cousin marriages are possible. A man could marry (1) his father's brother's daughter (FBD), (2) his mother's brother's daughter (MBD), (3) his father's sister's daughter (FZD), or (4) his mother's sister's daughter (MZD). Anthropologists also distinguish "parallel" (FBD, MZD) and "cross" (FZD, MBD) cousin marriage patterns. These combinations have very different implications for resource control and coalitions, and despite a great deal of complexity, some patterns emerge in the way resources and kin coalitions influence the choice of mates among cousins. FBD strengthens reciprocity and nepotism among male paternal kin; it is associated with patrilocal residence, and men have relatively great social power and resource control in such societies.[44] MBD, which strengthens reciprocity and nepotism among male maternal kin, is associated with matrilocal and avunculocal residence.[45] FZD marriage allows high-status men to keep control of resources even in matrilocal societies, and is often practiced by such men.[46] MZD could enhance reciprocity and nepotism among female kin; perhaps because women so seldom control significant heritable resources, it is virtually unknown.

When men's sources of power are unpredictable, and women have sufficient resources to be independent, men cannot always control women. In such societies, "serial monogamy," really a sort of temporal polygyny, can result (just as in other polygynous systems) in high variance in men's reproductive success. This is the case among the Ache (see chapter 7) and the Cuna Indians.[47] Ache men and women have perhaps ten spouses in a lifetime; the Cuna four or five. Resource distribution, coalitions for mate competition, inbreeding avoidance, and nepotism (familial coalitions) are all important in marriage patterns.

Stratification and Striving in Polygynous Societies

The patterns we see in many species—a few males who are highly sucessful in reproducing, and many males who fail—suggest that for males, at least, variance in reproductive success may become very high. That is what lies behind the high degree of risk taking in males of many species. It is true that variance in reproductive success can be high for both sexes, as in monogamous systems with high death rates of offspring. However, in general, the degree of polygyny is thought to be roughly correlated with the intensity of sexual selection, and measures of the intensity of sexual selection typically involve calculating variance in reproductive success.[48]

Social stratification, however, can complicate our attempts to measure this correlation. All calculations of variance are by definition based on a sample from a single population. Such measures are unambiguous, for example, in comparing all adult males in a population with scramble competition: there will be a mean success, and a range of variation in success. However, in many species, all individuals of one sex are not in the "same" population with regard to reproductive competition. In some fig wasps, for example, there are two kinds of males, winged and wingless. They follow very different strategies and have access to different populations of females; wingless males mate with their sisters before the sisters emerge from the fig wasp, while winged males are larger and emerge to seek other females and fight other males for them.[49] The two kinds of males are not really in the same "population" of competitors. In fact, in any such species with clear "alternate male strategies," a *reproductive stratification* exists that makes it inappropriate to combine individuals from different groups or strata.

Stratification can be permanent, as in the above examples; or ontogenetic, as, for example, in species such as bullfrogs in which males move through age or size categories in which they have considerably different means and variance in their success. In species in which there are true "conditional" strategies (all males may switch among strategies),[50] stratification is not an issue except as it interacts with survivorship; but when genotype or age influence possible strategies, stratification can have a powerful influence. In humans, cultural stratification (e.g., heritable class) can have the same reproductive impact as permanent stratification.[51]

One important ramification of sexual selection theory is predicting how hard males should "strive." For this reason, it is important to incorporate any stratification in analyses, for it influences success of striving.[52] How hard one should strive and how many risks one should take depend greatly on the costs and benefits of striving, and these depend on whether payoffs are constrained by one's stratum.

It has often been observed that intensity of male fights depends on how closely matched the males are in size and other similar attributes of power.[53] In bluegill sunfish, and in Ruffs, a European lek-breeding bird,[54] there are clearly two sorts of males who behave very differently. In sunfish, large males pursuing a territorial strategy exert more effort and take more risks than small males, who become female mimics. Both the mean success and the variation in success are greater for large males than for small ones, and any attempt to understand the intensity of sexual selection would fail if these males were lumped in the same analyses.

To begin thinking about such systems effectively, it might be useful to partition the variance. For striving behavior by an individual in any stratum, what are the costs? What are the opportunities for gain—from striving, and from being in a particular stratum? This question of "What produces variation in success?" is not simply a male question; for example, in rhesus macaques, female status has a matrilineally inherited component,[55] suggesting that female reproductive success associated with status can also sometimes be partitioned into an inherited rank component and a behavioral striving component.

Our ability to study behavioral dimorphisms in the context of sexual selection would be more precise if we first partitioned variances in male and female reproductive successes into a nonbehavioral component, including morphological characters and inherited resources, and a second component that would predict behavior.[56] Informally, biologists Steve Frank and James Crow have suggested a simple method for quantifying the *opportunity for reproductive gain* through striving and risky competitive behavior, in which competitive success has both a stratum-related component independent of competitive behavior, and a variable component in which success depends on the intensity of striving.[57] Their method suggests that:

1. When opportunities for gain through striving differ among classes or strata, different behavioral patterns are expected: *high-variance strata will contain the most competitive and risk-taking individuals.*
2. Within-class variance is most important in strata that are on average most successful (i.e., when R_i^2 is high in the model).
3. Among species in which status explains a similar proportion of variance, if two species differ in the amount of total variance explained by heritable rank, then the two species are expected to differ correspondingly in the levels of aggression over status.

Using this model, an example can be placed in the wider context of partitioning variance into behavioral and nonbehavioral components. Some interesting insights follow from the general model. Among human societies, variance in male reproductive success can differ greatly.[58] The extent to which resources (status, wealth, etc.) are inherited also varies widely. For a society in which the variance in resource control among males is high and resources or status are heritable (e.g., strong patrilines), resource control can create stratification, and thus may well influence the utility of striving and achievement—and risk taking.

Heritability of resources should be inversely related to striving behavior. When heritability of resources is high, then the opportunity for gain within strata is likely to be low and competitive behavior muted. Data on striving behavior are far more difficult to obtain than data showing that resources influence reproductive success, but some patterns are clear. The more polygynous the society (the higher the potential rewards), the more sons are taught to strive using several measures—but only in nonstratified societies, in which a man's striving can make a difference in his ability to marry (chapter 7). When a man's reproductive success is largely set by his heritable resources or social/class position, and unlikely to be changed much by potentially expensive and dangerous striving, parents are unlikely to teach the value of striving.

Women's Gains and Losses in Polygyny

Successfully polygynous men are always reproductively better off than their nonpolygynous competitors; that's why it is worth all the cost and all the risk. But the situation is more complicated for women. Above, it appeared that when serious pathogens made some men poor choices, women might prefer polygyny. And sometimes polygynous marriage with a high-status man appears to be preferred by women or their families, even when there are no apparent reproductive benefits.[59]

Women often suffer costs in polygynous systems: in a number of societies, second and subsequent polygynous wives have lower fertility than monogamous wives, or than first wives in polygynous households.[60] Children are likely to survive less well in polygynous households, and a major cause of divorce in polygynous societies is conflict among co-wives.[61]

A variety of proximate factors undoubtedly interact: for example, men may be older when they marry their "later" wives; women who are not considered desirable are likely both to marry late (and thus have low reproductive value) and to be a later wife. Nonetheless, the net result is that within a polygynous society, a woman's (or her family's) choice between an already married man and a not-yet-married man may be complicated.

THE ECOLOGY OF MONOGAMY AND POLYANDRY

Women seldom fully share men's reproductive interests. Males will strive for polygyny when resources are sufficient; when females can be an independent unit with their offspring, polygyny predominates. We expect to see monogamy and polyandry when resources are more limited—in harsh and unproductive habitats, when men may do better reproductively by helping to raise a child with true parental investment, rather than continuing their mating effort.[62] We expect social groups in such environments to be small, and to have relatively little variation in the resources controlled by individuals.[63]

At a crude level, this pattern holds: in the Standard Cross-Cultural Sample, highly polygynous societies are found in areas of the world with significantly higher plant productivity (a measure of environmental richness) than others; polyandrous societies are found in areas of significantly lower plant productivity; and there is no difference in the plant productivity of "monogamous" and mildly polygynous societies. Here, more than ever, it is important to distinguish between the anthropological and the ecological definitions of "monogamy."[64] In this case, when the definitions of marriage systems of the societies in the Standard Cross-Cultural Sample come from ethnographies, the term "monogamous" means monogamous-to-mildly polygynous, and thus it is not surprising that "monogamous" societies show no difference from mildly polygynous ones. Similarly, reading the ethnographic descriptions of societies deemed "monogamous" leads one to conclude that even when a few men manage (through skill in hunting, or getting novel sources of income) to be polygynous when most men remain monogamous, it is in habitats with a poor resource base—insufficient for many men to manage to gain more than a single wife.

Polyandry is extremely rare; of the 186 societies in the Standard Cross-Cultural Sample, only three are reported as polyandrous. Almost all polyandrous systems are fraternal: co-husbands are brothers. Of course, the interests of the two brothers differ; and the impact on fitness will also differ by birth order, sex, and opportunity costs (other available options).[65] Polyandry seems to occur under two circumstances, both related to the conservation and concentration of (rare) resources. Among the Lepcha of northern India, for example, brothers marry the same woman.[66] The land is extremely poor, and it apparently takes the work of two men to support one woman and her children. Prince Peter of Greece argued that in Tibet both resources and familial considerations lay at the heart of Tibetan polyandry:

Taking more than one wife for each son of a house would oblige them to partition the property, something which simply could not be done in a difficult environment such as that of Tibet. . . . Polyandry, then, is the ideal solution to the problem, for if the wife were tempted in the absence of her husband to have sexual relations with someone else, at least by having them with his brother, her offspring will always be of the same family blood.[67]

While Tibetan polyandry is common among wealthier families, among the Kandyans of Sri Lanka polyandry apparently occurs among poor families. Yet here, too, polyandry appears to be an arrangement by which two brothers join their land and maintain a common family, minimizing the number of potential heirs and raising living standards.[68]

Polyandry thus can result from brother-brother coalitions in order to combat resource scarcity or from attempts to control the distribution of a resource like land, which is immobile and loses its value when too much divided; it is "a rare but adaptive system for preserving family estates, and hence reliably supporting lineal descendants, across the generations."[69]

5.

Sex, Resources, Appearance, and Mate Choice

I've been called handsome, adventurous, athletic, humble, honest, intelligent, and messy. Seeking woman, 27–34, with most of those traits.
—*Ann Arbor Observer,* personal ad, April 1998

MEN AND WOMEN are not all that different in size or appearance; no naive biologist would take two human specimens, male and female, and think they were different species (as has happened among some bird species). In general, male humans, regardless of current marriage system, are slightly larger than females, consistent with our evolutionary history of mild polygyny. This is because much, perhaps most, male-male competition in humans is not a matter of size, but of other traits: wealth, political savvy, and so on—traits that help in complex social competition more than sheer size.

As we saw in the last chapter, females in many other species simply choose their mates, and while the criteria may vary, the choice process looks relatively straightforward. Depending on the ecology of parental care, females choose different traits. They might seek "good genes" in species with no male parental care; this can involve something as straightforward as expensive displays (see below), or something as subtle as choosing an individual whose genetic makeup is different, so offspring will be heterozygous. Females might, as in some grasshoppers, choose males with the best foraging abilities (and in response, males with poor foraging ability try to force copulations). Or, as in some cockroaches, females may be able to use a male's status and pheromone cues to discriminate against males reared in poor environments. For some species, resource control is so central that females appear simply to choose the resource and get whatever male controls it, as do elephant seal females (choosing a beach on which to give birth), and Red-winged Blackbird females (choosing a rich marsh in which to provision and rear their young).[1]

In contrast, the prospective bride and groom in many human societies may have little to say in choosing their mates. Or the bride's and groom's preferences may count, but in an informal way that is difficult to document. Because marriage and mating in humans involves others besides a male and a female, a whole set of conflicts of interests may exist. Not only might the man and woman seek quite different qualities in a potential mate, so might their families (see below).

What Men and Women Want

Resource control is clearly important to women, or their families, as we saw in the last chapter. Freud asked: what do women want? Resources, as Shakespeare knew,

> Dumb jewels in their silent kind
> More than quick words do move a woman's mind.
>   (*Two Gentlemen of Verona*)

Anita Loos was pithier, if not as poetic: "Kissing your hand may make you feel very, very good, but a diamond and sapphire bracelet lasts forever." In our evolutionary past, women whose mates provided resources for them and their children did better than others.

Even today, women choosing mates are interested in men's resource control. The anthropologist Daniel Pérusse found that among French Canadian men and women, men (but not women) of higher social status had more sexual partners, suggesting that status is important for men in the mating game. He also found that women's (but not men's) number of partners decreased linearly with age, suggesting that women's reproductive potential is important for them. And the psychologist David Buss, asking questions in thirty-seven cultures around the world, found that while some particulars about mate preference vary across these cultures, there are some consistent preferences that are obviously related to evolved sex differences in mate quality. Women rank men's ability to get resources high, and men rank women's youth and health high; both sexes rank social abilities like "sense of humor" high as well.[2]

We don't usually think (after we graduate from high school, anyway) much about physical cues in mate choice, but certain physical cues may matter in the mating game. Anthropologists Doug Jones and Kim Hill find that men prefer neotenic (childlike) faces in women in a number of cultures. Both sexes prefer symmetry in both face and body, reflecting health. I have argued, as have others, that physical sex differences beyond size reflect our polygynous background—that breasts, hips, and buttocks have served as sexual signals when females compete for the attention and investment of powerful, parental, resource-investing males.[3] When males invest parentally, as in humans, males as well as females may profit reproductively from exercising choice in mates.[4] If both sexes can exercise some choice, and if men and women have been reproductively successful through different strategies, they are likely to look for quite different things in their mates.

Of course, ideas about beauty or desirable traits in a mate will surely be influenced by cultural norms. Even in relatively simple species like guppies, females not only choose (costly) male signals and displays that reflect good condition, no parasites, high energy, and so forth, but young females copy the choices of older females.[5] Here, there is an obvious possible selective logic. But what of humans? Certainly we manipulate all sorts of signals: hair, eye, and skin color, body shape; what possible reproductive value could, for example, blue hair and nose rings reflect? We can break preferences down into signals that reflect health (shiny hair, clear skin) or youth (no wrinkles or sags) and current reproductive stage (waist-hip ratio, color of nipples); signals that suggest other reproductively important attributes like wealth; signals that reflect social awareness (stylishness, which may be purely culturally defined); signals of belonging to a certain group. Cross-culturally among traditional societies, the things people describe as attractive in the other sex turns up all of these categories. But just as in David Buss's studies of mate preferences and Doug Jones's and Kim Hill's cross-cultural study of facial attractiveness in contemporary societies, selectively relevant traits consistently rank high.

Signals of women's youth and health, directly related to selective advantage, are universally described as desirable: clear, unwrinkled skin, firm breasts, lustrous hair for women. In men, women favor strength, energy, and vigor. Some traits reflect an interplay of cultural practices and selection; for example, among the Iban people short hair is undesirable: they cut a sick person's hair very short, so short hair in this case reflects recent illness. Some of these "interplay" traits may reflect directly favored traits; in a number of societies, a man's tattoos reflect his status, and, of course, undergoing the tattooing further reflects pain tolerance and fortitude. But we should be cautious; I don't think we have good data to tell us just what all the functions are. Finally, some preferences may simply be "purely cultural": in some societies, relative hairlessness is a sign of beauty in a woman; in others, robust and luxurious hair is desired.[6] If we examine the ethnographies, comments people in traditional societies make are largely related to direct measures of fitness rather than "purely cultural" or conditional culture-interaction traits.[7]

An important widespread physical preference by men—one that would not occur to most of us but which is intriguing in light of this argument—is for a particular relationship between a woman's waist and hip size. Across all sorts of cultures with quite different specific ideas about beauty, both men and women see as most attractive a female waist-hip ratio of about 7/10 or 8/10. This is true whether the preference is for rather generous, Reubenesque proportions, or for slender Julia Roberts builds.[8] Why? Women of reproductive age will, unless they are pregnant, tend to put fat on their hips, breasts, and buttocks. Older women (of lower reproductive value) and pregnant women (not currently fecundable) thicken at the waist, giving a higher waist-hip ratio. The relative size of waist versus hips gives important reproductive cues. A relatively narrow waist means "I'm female, I'm young, and I'm not pregnant." The waist-hip ratio reflects many complex relationships, but they all boil down to: Is she fertile? Is she fecundable?

Hips, breasts, and buttocks are physical signals that communicate age, no prior births, and even, in the case of buttocks, one's ability to metabolize scarce food efficiently. At least some of these physical signals can be deceptive, even without deliberate manipulation. Consider how we put fat on our bodies. Little children tend to have fat on their faces, fingers, and toes, presumably for protection against temperature extremes. All other age and sex categories distribute fat relatively evenly over the body—with one exception. Reproductive-aged women, unlike all other age-sex categories, deposit fat preferentially on the breasts, hips, and buttocks, contributing to that approximately 7/10 waist-hip ratio men prefer. If a woman's hips are broad because she stores fat there, rather than because her pelvic structure is wide, a man gets potentially inaccurate or confusing information (wide pelvic girdles provide a headstart on easy birth of large-headed infants; fat doesn't). It isn't that fat is "bad," but it denotes energy reserves rather than a structurally wide pelvis. Similarly, if a woman's breasts are large not because they comprise mammary tissue for milk production but because fat is stored there, a male gets information not about a woman's lactational capability, but about her stored energy. This seems an irrelevant issue today, when food supplements and medical attention are readily available, but it may well have been an issue in mate choice in our evolutionary history. Even today, insufficient mammary tissue means a woman will have difficulty nursing effectively.[9]

In extreme, food-limited environments, obvious fatty deposits on women's buttocks signal ability to gain sufficient nutrition on a limited diet—a subtle reflection of maternal quality, important in a male-parental species. Thus, it is no surprise that extreme, harsh environments are the context for both steatopygy (the condition in which fat is obviously concentrated on the buttocks; fig. 5.1) and a cultural preference for extremely fat women. Darwin gave a second-hand report of one example of steatopygy as a sexual preference trait. Among the !Kung (then called Hottentot), Darwin's informant reported, a truly sexy woman was one who was unable to rise from level ground because of the weight of fat on her buttocks. Fat on the buttocks is probably a "true" signal of ability to store fat on any particular diet. In contrast, fat deposits on the breasts and hips are likely to be confusing and even deceptive signals; at the least, such fat is likely to be confused with mammary tissue and wide pelvises, traits contributing to two very different aspects of maternal quality.[10]

In Western societies today, a man's mate choice is likely to focus on a woman's health, her reproductive value (which means her youth), and her current reproductive status (fertile, not pregnant).[11] In many societies, the ideal is a healthy young virgin. In an interesting twist, a man's preference might depend on whether he was seeking a short-term mate (in which case, we would expect him to

Figure 5.1. Steatopygy, the preferential deposition of fat on the buttocks, is associated with harsh and unproductive environments. It probably represents an honest signal of nutritional competence: "Even in this harsh environment I can not only maintain myself, but store fat." (Photo courtesy of the Denver Museum of Natural History.)

prefer a woman at peak ability to conceive, perhaps in her twenties) or a long-term mate (in which case we would expect him to prefer a woman with peak reproductive value, about age seventeen).

Certainly cultural and historical factors strongly influence these preferences, but some preferences—healthy, young, not pregnant—are virtually universal. In Thailand, for example, men, regardless of whether they are rich or working class, living in the city or countryside, tend to prefer young virgins (though virginity is not so profound a preference as in some societies); they insist that while they themselves may have extramarital intercourse, their wives may not—their wives are to make a good home, stay there, and be faithful.[12] Women tend to accept that men will be sexually active outside marriage, and their first concern is that any such activity not divert financial resources from the home and children; a good husband, whatever else he does, provides for his family faithfully. An interesting attitude shift is occurring, related to the ecology of HIV transmission. Most women still prefer, when their husband has other women, that he visit commercial sex workers: the cost is modest, and the transaction is complete with the payment. (Traditionally, wives had objected less to commercial sex workers than to minor wives.) But as HIV and knowledge about it have become prevalent, some women are beginning to prefer that their husband have a steady mistress, or even a minor wife: for although these women represent a greater threat to a wife's resources, they represent a smaller disease risk.

Beauty, Resources, and Mate Choice

Even simple physical differences between the sexes reflect that what is valuable in a wife is likely to differ from what is valuable in a husband; differences and preferences are relatively consistent across quite different societies. Put simply, in our evolutionary history, it seems likely that a woman's value was usually her reproductive value, and a man's value was his resource value. Cross-culturally today, while everyone values such traits as a sense of humor in both sexes, women seek signals of resource control in potential mates while men seek signals of youth, health, and "beauty."[13] As we saw, assessments of beauty vary across cultures, but typically they reflect health and youth (and a low waist-hip ratio).

Occasionally, at least for periods of time, very wealthy subgroups may value something that reflects helplessness, but in general "pale and wan" does not become and remain a widespread preference. Exceptions appear when men and women have conflicts of interest and men hold enough power to resolve those differences in their own favor. Consider women with bound feet in Mandarin China; here fathers and families favored a condition that reduced women's fitness under ordinary selection—except that, because supporting an essentially helpless wife who has bound feet reflects a man's wealth, suitors favored it, and fathers helped enforce it. Female circumcision, as practiced in parts of Africa, probably also represents a male-female conflict of interest: the practice clearly does not increase women's general health or fecundity, but so long as men demand it and refuse marriage to uncircumcised women (in a closed society), the practice will continue. Exceptions like these not only reflect important cultural variation but suggest the strength of the general correlation.[14]

A woman's or her family's resources are not irrelevant in marriage choices. In some societies, men with few resources may explicitly trade off reproductive value for resource value (see below)—picking an older, reproductively less valuable woman who controls, in her own right, some resources; perhaps today's pattern of famous actresses "of a certain age" marrying younger men is relevant here. Subtle biases across societies, and fluctuations over time, have typically given an advantage to richer families in marrying. When most women worked in fields in western Europe, the standard of beauty was a pale complexion (which only the daughters of the rich could maintain); when we all began to work indoors, the Caucasian standard of beauty became a winter tan, suggesting that one could afford a trip to warm climates.[15]

SIGNALS OF DESIRABILITY
AND THEIR MANIPULATION

Status signals have a cost and will be maintained only if they benefit the bearer. Some signals make actual confrontation less likely, saving calories and avoiding risk. Other signals serve as sexual attractants and are the source of much physical dimorphism. In nonhumans, the sex competing for mates is the sex that gives such signals. Males in polygynous species are usually big, colorful, and likely to have weapons, because those are the males who win status and are chosen by females (see above). Females in the few polyandrous species are usually a little bigger, and sometimes a bit more brightly colored, than males. Because much of the selection on displays is preference by the choosing sex rather than relative survival enhancement, sexual selection on expensive—and possibly dangerous—displays can "run away."[16]

Thus, males in polygynous species are likely to grow antlers, or large horns, or bright feathers, or long decorative tail feathers—all costly, and sometimes risky, displays that may do nothing more than advertise a male's ability to take these risks: the "Handicap Principle." The message is: I am so fit that I can support this expensive handicap, which would kill a lesser individual.[17] And when females prefer these costly displays, they work. For example, female European swallows prefer to mate with longer-tailed males: these males more often get mates, and get them sooner, than other males. The success of longer-tailed males is high—but these tails carry a cost in terms of survivorship; long-tailed males die sooner.[18]

Most nonhuman examples, including those given at the beginning of this chapter, principally involve male physical (energetic) resources, even when, as in Bower Birds (for whom the criteria are the number and color of decorations on the bower), the display is not simply a physical part of the displayer's body.[19] In contrast, humans invent, augment, and change signals; and females do a great deal of signaling. Bras make our breasts look large and / or young, girdles can imitate an ideal waist-hip ratio, shoulder pads mimic good physical condition, makeup reflects light and hides wrinkles, cheek and lip color make us look healthy and sexually interested. Our manipulations have sometimes been intrusive: in the nineteenth century, for example, some women underwent surgery to remove their floating ribs in order to have a small waist; today we have facelifts and liposuction. These manipulations imitate signals of youth and health.

Cross-culturally, cultural augmentation of sexual signals or ornaments is virtually universal, favored for the same reasons selection favors physical ornaments and displays in other species. Remember the old adage, "If you've got it, flaunt it." Males and females profit by signaling or flaunting different attributes. Humans are actually rather paradoxical with regard to sexual selection and sexual displays: most scholars agree that human evolutionary history, like that of most other primates, is polygynous: 83 percent of societies for which we have information are polygynous.[20] All of these patterns suggest that in humans, males should be the "ornamented" sex, yet most people talk about women's adornments. But ornaments can be what either sex advertises.

Because of our polygynous history, men's and women's cultural augmentation of sexual signals should give information about different characteristics. Men are likely to signal wealth and power status, and members of cultural subgroups with limited real resources seem likely to concentrate those resources in highly visible signals. Sociological studies of wealth and status signals among contemporary poor groups, for example, find the "ghetto Cadillac" phenomenon common.[21] Because humans show male parental investment, a woman's reproductive value becomes important; thus, women should signal reproductive value, things that reflect youth and health. Today, billion-dollar industries exist to do just that: makeup and cosmetic surgery, for example, are designed to signal youth, health, and sexual interest—the products and processes are aimed at making the skin tauter and less wrinkled. Though a few men indulge themselves this way, most clients are women, who get facelifts, and undergo liposuction to obtain an attractive waist-hip ratio. In the nineteenth century, women put belladonna in their eyes, dilating their pupils, ordinarily a strong signal of sexual interest. Women use rouge to make the cheeks rosy, indicating they have either been exercising or are sexually aroused, and lipstick to mimic the dark, blood-engorged state of the lips during sexual excitement. The specifics change across time and societies, but the desired result does not.

Women, like females of other species, can signal interest and availability behaviorally. Patterns of eye contact in flirting appear to be virtually universal and invariant in widely differing societies.[22] The facts that women frequently wear signals of "unavailability" (e.g., wedding rings, styles of clothing or hair worn only by married women), and that in some cultures they undergo treatment that may decrease their general health and vigor (foot binding, clitoridectomy) are suggestive. Men and women's interests often conflict, and women are at least sometimes manipulated by men, (for example by proclaiming unavailability in return for parental investment, or undergoing foot binding to get a mate). Such ornaments of unavailability should be more common in societies in which women are more dependent on men for resources.

There are intriguing cross-cultural trends. Men's wealth or power status is shown by ornaments in 87 of 138 societies studied.[23] Only four societies distinguish men's marital status by ornament, and two of these are "ecologically monogamous,"[24] living in very poor environments where men have trouble becoming successfully polygynous (see chapter 4). In contrast, women's marital status is signaled in all but three societies.

Many anthropologists argue that marriage has the function of building alliances between families; if this is true, women might be expected to signal family wealth, although not necessarily any separate wealth of their own. Indeed, women's wealth is shown by ornaments in 49 of 138 societies. Is it their own wealth, or their family's (father's before marriage, husband's after)? Although men usually have greater resource control than women, there are societies in which women control significant resources or wield considerable influence over resource distribution—but the societies in which women have power and influence are not those in which women wear ornaments of status and power. Put bluntly, advertisement of women's status is less likely to be effective when directed at close kin or at members of the household with whom one interacts daily— *they* can't be fooled. Across most traditional cultures, women's signals of status largely reflect their husband's or male kin's wealth or standing, consistent with the prediction. Such signals also may represent a conflict of reproductive interest between the man and woman, since male resources are used to acquire mates, and signals of "excess" resources, even if worn by a man's wife, can constitute his sexual advertisement, or mating effort.[25] These patterns are consistent with the observation that males seek resources as mating effort, competing against other males to whom they are variously, though often not at all, related, and interacting with individuals they know less well; females, on the other hand, seek resources as a form of parental effort, working at or near home with sisters or co-wives.

Female ornaments of power show one significant relationship cross-culturally: women's ability to hold political posts. However, the relationship is not positive but negative: societies in which women can hold political posts are societies in which women do *not* wear ornaments of power or status.[26] The question then remains: When women do operate independently in the extrafamilial, community sphere, why do they not signal position and power in the same way, and to the same extent, as men?

Cultural specifics obviously can influence the general pattern. For example, ecologically monogamous societies show distinct patterns compared to other societies. It's true, for example, that women signal marital status far more than men. The societies in which men do signal marital status (and thus "unavailability") are ecologically monogamous or large nation-states (fig. 5.2). Ecological constraints mean that men can't profit from polygyny, anyway. Men in 20 percent of "ecologically" monogamous societies signal marital status; men in 1.5 percent of other societies do so. Pubertal and/or age-group status is not discernible for either men or women in ecologically monogamous societies (fig. 5.2), while women in 12 percent and men in 67 percent of other societies signal this. Finally, men signal wealth and power in 67 percent of nonecologically monogamous societies—those in which such signals might be potent advertisements of their ability to take on additional mates; men in only 10 percent of ecologically monogamous societies do so. Women in ecologically monogamous societies do not signal wealth or power; women in 38 percent of other societies do so, though this is not a woman's own wealth or power but rather a reflection of her male relatives' status. When women signal the wealth of men, there is potential for a great conflict of interest.

WHO CAN CHOOSE?

When Juliet was twelve, her father, without consulting her, betrothed her to a man more than twice her age. Because she was in love with Romeo, she complained. Her father's answer was as follows:

> An you will not wed, I'll pardon you!
> Graze where you will, you shall not house with me; . . .
> An you be mine, I'll give you to my friend;
> An you be not, hang, beg, starve, die in the streets,
> For, by my soul, I'll ne'er acknowledge thee,
> Nor what is mine shall never do thee good:
>
> (*Romeo and Juliet*, Act 3, Scene 5)

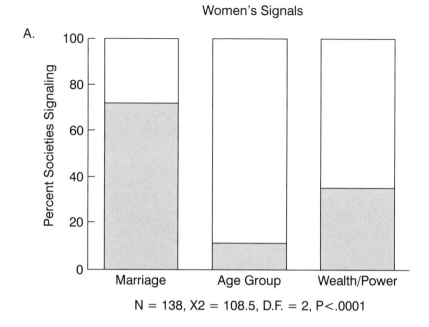

Women's Signals

A.

N = 138, X2 = 108.5, D.F. = 2, P<.0001

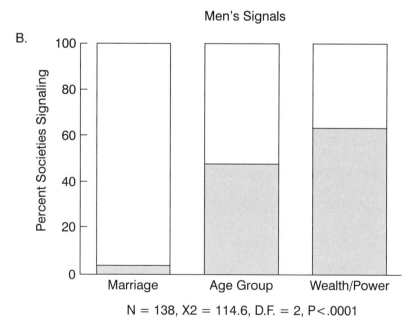

Men's Signals

B.

N = 138, X2 = 114.6, D.F. = 2, P<.0001

Figure 5.2. Across cultures, women (A) and men (B) signal different information by their dress and ornament.

Today, in the United States, Juliet would probably sue her father for child abuse. And she'd be likely to win, though she might not be allowed to marry at twelve. What is common, approved, and thought ethical varies widely across human cultures, temporally and spatially. In other species, mate choice is a relatively straightforward affair: typically, males display and females choose. And in Western industrial nations today, once again individuals are relatively free to choose their mates. But in human history this has not always been—and perhaps seldom was—the case.

Juliet's father was seeking a powerful ally, and Juliet's ability to exercise mate choice was nil. Across human societies, this is far from unusual; in most societies, the reproductive interests of more than the two who mate can matter. In traditional societies, the potential bride had greater say than the would-be groom in marriage negotiations in only 3.7 percent (3 / 81) of societies.[27] Grooms had greater—or sole—say in 39.5 percent (32/81). In most of these societies, the older generation had considerable power in these decisions. Among the Kipsigis of Kenya, for example, two men decide the bride price for a young woman; a younger woman, of higher reproductive value, commands a higher bride price, though a special friendship between the men might lead to a special discount (see chapter 7).[28] In many societies, such as the Arunta of central Australia, a couple is betrothed before at least the female is even born. A boy grows up prohibited from talking to the girl who is his designated mother-in-law and of similar age. Even though older men make the formal decisions, a young woman may have either great or no influence. For example, among the Kipsigis, even though men set the bride price, women do exercise some choice.

Wealth, marriage "market forces," and ecology all influence marriage patterns. Juliet's father was wealthy, otherwise he could not have contemplated marrying her off at age twelve. Recall from chapter 4 that dowry is common in societies that are stratified and monogamous, where wealthy, parentally investing men are at a premium and families of marriageable women compete for them. In such societies, wealthy families are able to marry their daughters off earlier, and brides are chosen not only on their own characteristics (youth, beauty) but on their family's (wealth, father's schooling and occupation). Poorer families, because they not only must pay dowry but are harder pressed to replace the daughter's labor, gain if they can delay a daughter's marriage. Daughters in wealthy families therefore marry earlier.[29] In bridewealth societies like the Kipsigis, wealth makes a difference in a woman's age at marriage—but in these societies, the economic forces mean that wealthy men (or sons in wealthy families) can afford brides who are younger (of higher reproductive value). In both, family wealth contributes to family fertility and growth.

6.

Sex, Resources, and Human Lifetimes

And one man in his time plays many parts,
His acts being seven ages.
—Shakespeare, *As You Like It,* Act 2, Scene 7

A GREAT PHILOSOPHER never married, and, in a possibly apocryphal story I read as a child, on his deathbed he called for all of his works to be set upon his lap. When the works had been brought, he sighed, "All of this is less than the weight of one grandchild." (And, to complete the story tidily, he promptly died.) His insight is an important one: What are resources for, anyway, if not to build our families? As we move through the stages of our lives, our struggle for resources never ceases; in fact, our very lifetimes are shaped by the struggle. And males and females follow different life paths, and struggle differently.

Our individual reproductive costs and benefits depend not only on the ecological conditions outlined in the last chapter, but on our age, our sex, our condition, our conspecific competitors, our resources. Biologists, who are perhaps more sanguine than other scholars about humans as biological creatures, argue that the same rules that apply to other primates apply to our costs and benefits at different ages and stages. But along with the generalities of "being a mammal" and the ecological variation we encounter, we have some traits that are particularly human, and these help set the stage for the things we can do in our lives, and for the patterns of sex differences we see.

Though we are a primate, we are rather an unusual one. In some ways our lifetimes are typical for an ape of our size, yet in other ways our lives are unusual.[1] For example, human babies are quite large—38 percent larger than the expected size for a baby of a primate our

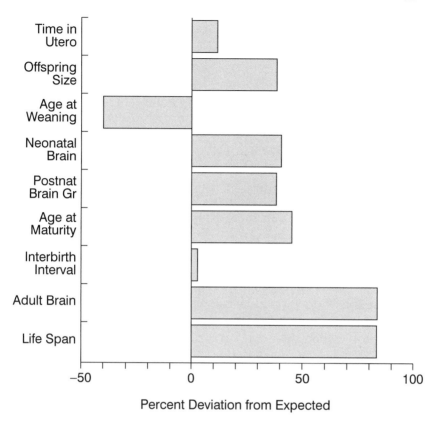

Figure 6.1. Human lifetimes are unusual in a number of regards. Each bar represents the deviation from the value that would be expected for a primate of our size.

size (fig. 6.1);[2] human pregnancies are 11 percent longer than expected. Newborn boys are larger and are carried longer *in utero* than girls, on average. Our babies remain helpless, or altricial, for much longer than other primates while their parents protect and care for them.[3] Human babies roll over for the first time at about the age young chimpanzees are already frolicking in their group, climbing happily up, down, over, and about their mothers, consorts, and playmates.

In all newborn primates, brain and body size are correlated. However, human infants have huge brains for a primate of their size: 83 percent larger than expected from the general primate pattern.[4] After birth, human babies' brains continue to grow rapidly for another year, unlike other primates, in whom brain growth slows significantly soon after birth. This pattern may represent a trade-off between human brain size and our upright posture, for our posture creates great difficulties in giving birth to large-brained offspring.[5]

Because human infants are large at birth and grow rapidly, if they fit the primate pattern they would mature early. But humans reach sexual maturity relatively late. Human females are 45 percent older at sexual maturation than expected—about thirteen years old rather than about nine (fig. 6.1). Further, young women are "subfecund" for several years—they are less fertile and suffer greater infant loss if they conceive than women in their late teens and twenties, suggesting that the resource demands of pregnancy conflict with a young woman's own continued growth. Even in societies in which a girl marries before puberty, and in which there is no evidence of deliberate fertility control, she is unlikely to have her first child before her midteens.[6] This is a real conflict; other things being equal, early reproduction is more profitable than waiting. It's not surprising, then, that human menarche came sooner with better nutrition,[7] or that in societies in which there has been differential access to resources, women paired with wealthy men began reproducing earlier than others. Male reproduction, as in many other mammals, is typically delayed even more, primarily because of the social forces of male-male competition. So both human females and males begin reproduction later than one would expect from size.

We make up for our late reproductive start, however: we are longer lived than, for example, chimpanzees, and our interbirth intervals are short. So we can increase rapidly once we start to reproduce. The average interbirth interval for surviving offspring among hunter-gatherers ranges from thirty to forty-five months, which is similar to interbirth intervals in nineteenth-century Sweden (27–37 months) and nineteenth-century Germany (27–44 months). Chimpanzees, in contrast, even though they are smaller and should therefore have shorter interbirth intervals, average sixty months between births. We thus mature late but produce children rapidly once we begin.[8] Finally, women stop reproducing in most human societies by their fifties, because they no longer ovulate (menopause), though they live for many additional years. Other female mammals spend perhaps 10 percent of life as "postreproductive"; women spend about 30 percent. (This may represent not a shortening of reproductive capacity, but a lengthening of life.)[9] In sum, humans are typical in some simple life-history patterns and unusual in others; we are extreme, but not unique, in a variety of other ways.

The above facts are not unconnected, amusing trivia about humans: they influence what humans can and cannot do. Many of our peculiarities are linked to our complex intelligence, our social competition and cooperation, and our resource gathering and inheritance. We have extensively developed reciprocity, including indirect reciprocity, and often we have complex sets of group-imposed rules, mores, and laws. Here, too, it may be useful to start by treating humans as complex but not qualitatively different mammals until the simple models fail to predict what we see. Our complexity means that the frequencies of different strategies can be influenced both by natural and by cultural selection (chapters 9–15),[10] giving us much about human lifetimes, and sex differences in those lifetimes, to examine.

STARTING OUT: RESOURCE STRIVING IN THE WOMB

Typically, perhaps sentimentally, we view pregnancy as a time of maternal support and care for the growing embryo. But genetic conflicts over resources start here. An infant *in utero* is only half like its mother (this relationship is exact, for the child gets half its genes from its father). A child is only on average half like any other full sibling genetically (half siblings are only a quarter like each other on average). Thus the stage is set for conflict, both with mother and any siblings who share the womb, and later, of course, with the father and other siblings.[11]

And conflict there is. The biologist Robert Trivers first pointed out that any particular offspring gains if it can get more maternal care than is optimum for the mother. Pregnancy, far from a romantic interlude, more closely resembles an arms race, one manipulated by genes in the fetus with interests different from those of the two parents. Fetal genes from Dad that increase Mom's transfer of nutrients to the fetus will be favored. Will such a transfer harm Mom? No matter, so long as she is healthy enough to continue investing, from the fetus's point of view. In fact, if her ability to produce other (competing) siblings is reduced, so much the better. On the other hand, the

mother's genes, both in herself and in her fetus, that limit such transfer when it is detrimental for the mother—for example, when she is
young and still growing physically herself—will be favored.[12]

As early as the implantation of the zygote on the uterine wall, tro-
phoblast cells (fetally derived) invade the lining of the mother's
womb and remodel certain arteries so that they cannot constrict to
shut down blood flow to the fetus. This means several things. A
mother cannot control the blood or nutrient flow to the fetus without affecting herself as well, and the placenta can now release hormones directly into the mother's blood stream. Some of these manipulations are countered by maternal strategies. Fetus and mother
are truly combatants in an arms race. As biologist David Haig has
cogently pointed out, a number of unpleasant accompaniments of
pregnancy (as well as serious medical conditions like preclampsia)
are better explained as maternal-fetal conflict than by any competing theories.[13]

What's a Mother to Do? Optimizing
Maternal Effort among Offspring

Switch viewpoints for a moment, to the mother's perspective: it is clear that maternal investment in one child may come at the
expense of investment in others.[14] Robert Trivers's phrase "parent-
offspring conflict" highlights an important issue: offspring profit by
getting as much, free, from Mom as possible—but Mom, unless she
has only one offspring in her lifetime, must apportion her effort
among all her offspring and herself. Closely spaced pregnancies,
when nutrition or other factors are limiting, may lower a woman's
lifetime reproduction. Among the African Efe, for example, women's
ovarian function and resulting birth schedules show a seasonal pattern that correlates only with food availability—a clear reproductive
response to changing ecological conditions. Even subtle maternal responses (e.g., adjustment of blood flow to the uterus) during pregnancy fit a life-history model of lifetime reproductive optimization.

There may be further conflict for women between maternal investment and the effort required to acquire resources (whether gathering food or finding child care in order to work): What is invested

in work cannot be invested in child care.[15] Ecological factors con-
strain these conflicts. For example, among the !Kung of southern
Africa, women have interbirth intervals of about four years. Predators are prevalent, and !Kung women who depend on bush foods
may carry their children, at least occasionally, for up to six years.

But women must carry the foods they gather, and they can only
carry so much. Women's "backloads" (weight of child plus foraged
material) are good predictors of interbirth intervals and mortality
patterns. !Kung women living in compounds, not dependent on
bush foods, have children close together. For bush-living women,
the number of successful descendants was maximized not by maximizing the rate of births, but by responding to the conflict between
production of a new child and the cost of such production on the survivorship of other children. Women who lived in compounds did
not face these conflicts.[16] The issue of trading off number versus
quality of offspring seems remote in wealthy Western developed nations today, but may in fact still be related to patterns of resource consumption and fertility (chapters 8, 15).[17]

Mothers, more than fathers, face conflicts of getting versus allocating resources. Among the South American Ache, Hiwi, and Ye'k-
wana, nursing women can forage less than others.[18] In some societies, other children, usually siblings, help with child care, and the
availability of such children can have an impact on a mother's lifetime fertility. On the Pacific island of Ifaluk, for example, a woman's
lifetime fertility is correlated with the sex of her first two children:
women whose first two children are girls have greater lifetime fertility than others. Daughters assist in child care on Ifaluk, so mothers whose first children are daughters defray some costs and can
have more children.[19]

In other societies, resources to hire wet nurses could help minimize the conflict. A dramatic example is given by anthropologist
Sarah Hrdy, who found that in eighteenth-century Paris, interbirth
interval, fertility, and infant mortality all varied with the mother's
status.[20] The richest women had very short interbirth intervals, very
high fertility, and low infant mortality; a linear relationship between
the cost of the wet nurse and infant survivorship meant that the richest women, who could afford the best wet nurses, fared best in terms
of fertility. But wet nursing had unintended consequences.[21] Among
the bourgeois, complexities created more variation in pattern. Poor

women had long interbirth intervals, low fertility, and high infant mortality; and the wet nurses fared worst of all, with very long interbirth intervals, very low fertility, and very high infant mortality.

CONFLICTS OF INTEREST: ABORTION, INFANTICIDE, ABANDONMENT, NEGLECT

Parents seldom kill their children. After all, even though a parent's interests are not identical to a child's, an infant's death means the loss of considerable parental investment. Indeed, infanticide in most species is typically committed by reproductive competitors rather than parents; for example, in lions, langurs, and gorillas, when a male takes over a harem he is likely to kill all babies under a certain age. The mother becomes sexually receptive, and the male profits both by eliminating an offspring with a competitor's genes, and by gaining a mating. In humans, also, stepparents (whose reproductive interests do not coincide with the child) are more likely to abuse or neglect children than genetic parents, and, regardless of the old fairy tales, stepfathers are more likely to commit infanticide than stepmothers.[22]

Human parents, like parents in other species, therefore do commit infanticide and abort and abandon their infants. When is infanticide not pathological, but adaptive? Once again, we must remember the trade-offs: if each infant requires great investment, parents must apportion their effort, and parental investment biases, even to the extent of infanticide, can be reproductively profitable: for example, if the mother is alone and without family or resources to help with care, or if the child is unlikely to be successful.[23]

Cross-culturally, deformed or seriously ill newborns are killed most often, and there is evidence that some cultural conceptions of "ill omens" leading to infanticide are real reflections of low newborn quality. Similarly, mothers are more likely to commit infanticide when external circumstances reduce their chances of successful investment; too-close births, twins, lack of an investing mate or stable pairbond—all increase the likelihood of infanticide or neglect.[24]

Historical studies of child abandonment also reflect such considerations: a mother's ability to invest in the child (including her own health, familial resources, economic conditions), and the child's health, legitimacy, and sex. In France, Spain, and Russia, abandonment was related to economic factors and a mother's abilities.[25] Similarly, although he failed to discern any pattern, historian John Boswell's overview of child abandonment reveals that 46 percent (29/63) of cases he examined were, despite great variation in time, country, and other circumstances, related to maternal ability to invest. When resource allocation problems (16/63; 25.5 percent) and offspring quality (4/63; 6.3 percent) were considered, selective reasons were apparent in 49/63 cases, or 77 percent.[26]

Abortion, too, appears more common in circumstances in which the birth of a child is likely to reduce the mother's lifetime reproductive success. As women age and their reproductive value declines (future reproductive opportunities wane), they are less likely to seek abortion. Even attitudes toward abortion in our society are related to the proportion of women in any group who are "at risk" of unwanted pregnancy.[27]

Of course, none of these behaviors, or attitudes about them, are set or in any way "determined"; they can be influenced not only by an individual's own condition, but by the attitudes of those around them. For example, in the United States today, a woman is likely to favor abortion if she is still fertile and thus potentially vulnerable to unwanted pregnancy. Opinions on abortion have also become political as individuals and party leaders influence each other. From 1972 to 1994 in the United States, Democratic and Republican party positions on abortion have changed gradually (as reflected by House and Senate votes); as this has occurred, individual voters have switched party alliances to align with their own attitudes about abortion.[28]

Invest or Desert?

Let us put these empirical data into context. The trade-offs of using resources for oneself (e.g., for growth) versus reproducing are relatively clear in other species. Humans, too, appear to be influenced by resource availability in their reproductive behavior, whether consciously or not, and those trade-offs are very different for men and women.[29] Figure 6.2 shows a combination of reproductive trade-off curves for three people: a man, a woman, and a child. For each parent, there are three families of trade-off curves (A, B, and C) representing different possible trade-offs for the effect of allocating resources to oneself for future mating success. In addition, there are three curves for the detrimental effect of a parent's self-investment on an offspring's mating success (offspring A, B, C in Fig. 6.2); that is, too-selfish parents, in this model, reduce their children's chances of surviving and marrying.

When the "return from child" curve is higher than the "return from new mating" curves for *both* parents at a given resource level, both partners profit from being parental, and both should cooperate in rearing a child (fig. 6.2: shaded area in superimposed curves). Thus, in figure 6.2 (#1), when parental curves are concave (monotonic increasing) and the child curve is convex (monotonic decreasing), both parents profit most by investing in the child, and cooperative rearing is likely. When the payoffs for self-investment are higher for both parents than for investing in the child, the child is in danger (fig. 6.2, #2).

When parental curves are identical, no conflict of interest exists between the parents, whether they are likely to keep or desert the child. But conflicts are likely in some age and resource combinations. Consider, for example, a married couple, both forty-two years old, in which the man is rich and powerful, and the woman becomes pregnant (fig. 6.2, #3). The woman will profit by investing in this child. For the husband in this example, using half of available resources for his own new mating benefits him more than investing in this child.

Because women's reproductive value peaks at the age of first reproduction and declines thereafter, age affects the shape of women's return curves more than men's.[30] Consider the trade-offs for an older versus a younger woman, holding the male and child curves identical: #4a represents the conflicts for a younger woman, #4b for an older woman. The vertical hatching represents the area in which it pays the female to continue to invest. This area is greater for the older woman; she gains more from investing. Because the trade-off curves are shaped so differently, there will always be a greater benefit for the older woman to stay and invest in this child—thus suggesting that women are less likely to abort even an unintended pregnancy when they are older. Another influential factor is resource availability for females: women with considerable resources can enhance their own chances of mating again without significantly altering the child's chances for success; women with fewer resources have a more significant conflict of interests.

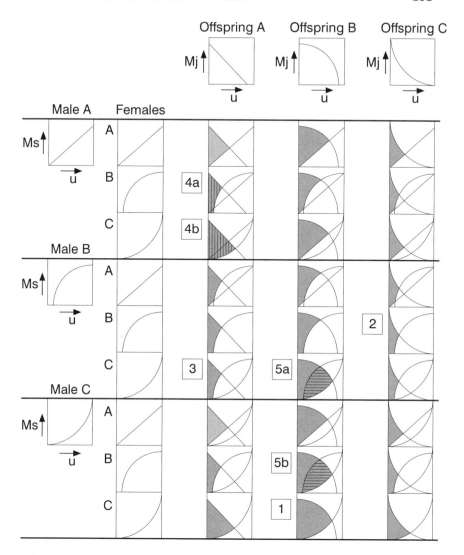

Figure 6.2. Fathers, mothers, and children face different trade-offs, here superimposed. For each parent, there are three families of trade-off curves (A, B, and C) for the effect of allocating resources (u) to oneself on own future mating success (M_s). In addition, there are three curves for the detrimental effect of investment by a parent in self (u) on an offspring's mating success (M_j for offspring A, B, C). (See text for further explanation of shaded areas, numbered cases.)

Sex Differences in Reproductive Lifetimes

When women's and men's lives are compared, some reproductive patterns must, of course, be similar. The average number of children can never be higher for women than for men, for example. Other phenomena, like age at marriage, variance in reproductive success, and the rate and impact of remarriage, can, and do, differ strikingly between men and women.

I have argued that men and women, like other male and female mammals, seek resources toward somewhat different reproductive ends. Much of men's striving, like that of other primate males, centers around the mating aspects of reproductive effort.[31] Male-male coalitions are associated with competition for status or resources,[32] and with resources that can be more effectively obtained and protected by groups of males than lone males: heritable land of lasting value, some large game. These conflicts may become warfare (chapters 13, 14). While coalitions are likely to originate among brothers and be more common and stronger when related men live together, more or less fluid coalitions of this sort arise among men of various relatednesses and among nonrelatives in many societies. Men's coalitions typically are broad reaching and fluid, involve both more risk and higher reproductive stakes than women's coalitions, and exert considerable power and control significant resources (chapter 11).[33]

The principal difference between these expenditure patterns is that mating effort and non-offspring-specific parental effort, unlike true parental investment, may have a high fixed cost: much must be spent before any success is realized, and later successes cost little compared to the cost of achieving the first success. Thus, many males will fail to reproduce, but a successful male may have many times more offspring than the most successful female, just as in the elephant seal and red deer examples in chapter 3.

This fact has profound implications not only for male risk taking and survivorship, but also for *parents* in polygynous societies: parental expenditure may not be optimized by equal investment in the two sexes (see below) but rather by biasing investment toward the sex in which success is more strongly correlated with investment and varies more—for which the maximum possible payoff is higher. Often but not always, in polygynous mammals this investment is biased toward males (for a conspicuous exception in humans, the Mukogodo, see chapter 7).[34] Successful individual responses vary in different ecological and social conditions.

Finally, these individual costs and benefits can lead to different population patterns, depending on the composition of individuals, and environmental richness, evenness, and predictability. That is, population growth rates arise from individual reproductive striving, ecological richness, and predictability—and these vary (chapter 15).

Differential Investment in Sons and Daughters

In a way, we have come full circle, from fetal strategies for resources in the womb, to issues of differential parental investment in children. Some cases (see above) of differential investment or withholding are issues of individual quality of child or parental ability. But some broad, general patterns of differential investment—by sex of offspring—arise from the differences in the return curves shown in chapter 3. The importance of resource value for men versus reproductive value for women means that in many societies it may be harder for a man to get a wife than for a woman to get a husband (even though a really successful man can have many more children than the most successful woman). Age affects men and women differently.[35] If the reproduction of sons is more variable than that of daughters, and especially if wealth or status matters more to men's success than women's, investment is likely to be biased toward sons. Whenever the two sexes succeed by acting differently, interesting consequences follow: parents who invest in and train their sons and daughters differently will prosper reproductively. Investment can involve energy and other physiological costs of carrying a child, differential investment in training and education, and differential inheritance practices.

Mating Market Forces: Fisher's Insights

As long ago as 1928, Sir Ronald Fisher, a British mathematician who identified and clarified several important problems while dabbling in biology (his work centered on statistics), noted that the population sex ratio should influence the value of relative investment by parents in offspring of the two sexes. He suggested:

> The sex ratio shall so adjust itself, under the influence of natural se-
> lection, that the total expenditure incurred in respect of children of
> each sex, shall be equal; for if this were not so and the total expendi-
> ture incurred in producing males, for instance, were less than the to-
> tal expenditure incurred in producing females, then since the total re-
> productive value of the males is equal to that of the females, it would
> follow that those parents, the innate tendencies of which caused them
> to produce males in excess, would, for the same expenditure, produce
> a greater amount of reproductive value; and in consequence would be
> the progenitors of a larger fraction of future generations than would
> parents with a congenital bias toward the production of females.[36]

Fisher's point is not universally true, as we will see in a moment,
but it sets the stage. He focused on the population-wide impact of
the relative average contribution of males and females. Since, in a
sexual species, every child has a mother and a father, if one sex is rare
in the mating market it becomes more valuable. The contribution to
the next generation of *all* the males is exactly equal to that of *all* the
females; so if there are only half as many males as females, each male
is, on average, twice as valuable.

There are several assumptions here that, if not made explicit, will
lead us astray: this model assumes outbreeding and does not consider sexual selection at all except as a function of "numbers available." But it does suggest why, in outbreeding populations, the sex
ratio at birth will tend to equilibrate at about 1:1 (50 percent male, 50
percent female). It also suggests that extreme biases in population
sex ratios might affect social systems. This has rarely been important
for human societies; there have been short disruptions of population
sex ratios (e.g., male deaths in war), and these may have been followed by shifts in the birth sex ratio toward more males, but in most
populations, birth sex ratios do indeed equilibrate at about 1:1. Social and cultural influences, however, might easily interact with natural selection to result in sex ratio curiosities (see chapters 8, 10).

Sex Biases in the Womb: Trivers-Willard Effects

In many polygynous mammals, including humans, because only the
best-invested males have much chance of reproducing, mothers invest more in each successful son than in each daughter. Sons are carried longer *in utero*, they are larger at birth, they nurse longer and

more frequently, and they are weaned later.[37] Robert Trivers (see parent-offspring conflict, above) and Dan Willard argued that in polygynous species under such conditions, females in good nutritional condition should be likelier to bear sons than daughters. A more broadly applicable statement might be: When variance in reproductive success of one sex exceeds that of the other sex (as in elephant seals), or when parental investment can influence the reproductive success of one sex more than the other (as in baboons), there should be a correlation between parental condition and investment in that sex.[38]

Trivers and Willard assumed that a mother's physiological condition (resources available to rear a successful offspring) would decline with age. In nonhuman species, and in many preindustrial societies and developing countries, this may be true. But whenever the nutritional condition of mothers does *not* decline with age, a *male* bias in sex ratio might be found in older mothers.[39] If a female's condition is good, as she nears the end of her reproduction it may pay to invest ever more heavily, with a greater potential reproductive profit if successful; for example, older female gorillas tend to have sons. In some nutritionally rich societies, such as nineteenth-century Sweden, mothers over age thirty-five showed a sex-ratio bias toward sons, and mothers under twenty-five toward daughters. It is interesting, in light of these physiological patterns, that parents can shift their preference for sons versus daughters under different circumstances.[40]

Sex Biases in Inheritance and Survival

Perhaps no other species exhibits the degree of resource transfer that can take place through inheritance in human families; parents invest, often differentially, in their children even after death. Inheritance is typically biased by legitimacy, birth order, and sex.[41] Why? Remember that, in many societies, relatively few males can be highly successful, and resources are central to men's success. Parents can maximize the survivorship and reproductive chances of their sons and daughters—their net reproductive profit—in such societies by allocating resources unevenly.

In societies with heritable goods, cultural factors, the size of a family, and the sex of siblings may influence men's and women's ability to marry somewhat differently. Within traditional polygynous marital systems, inheritance is strikingly male-biased, precisely the pattern you would predict if male reproductive success varies more than female reproductive success, and male success is influenced by resource control.[42] In other situations, for example, the hypergynous Indian societies (see below) and the Mukogodo of Kenya (see next chapter), poor parents may find that daughters are more marriageable (and thus reproductively more valuable) than sons.

Resources are not necessarily allocated evenly, even within sex. In many societies, earlier-born sons tend to inherit the greatest proportion of the resources; they literally have fewer chances to die before putting those resources to reproductive use. Among fifteenth- and sixteenth-century Portuguese nobles, the proportion of ever-married men and women decreased with increasing birth order, as did fertility for married individuals. Among the Kenyan Kipsigis (see also chapter 7), men's reproductive success declines with the number of brothers and increases with the number of sisters, and parents show reduced paternal investment in sons with many brothers and heightened investment in sons with many sisters.[43] Among the Gabbra pastoralists of Kenya, too, investment is biased: men with many older brothers have lowered reproduction.

Sex-biased patterns of investment are common even where more equal distribution is stipulated by law. For example, in nineteenth-century Sweden (chapter 8), first-born sons tended to inherit land, later sons and daughters made do with other goods. Women's lifetime reproduction decreased as their number of siblings increased. For men, only the number of brothers mattered, suggesting that (1) brothers represent resource competitors for men, and (2) as total sibship size increased for women, they were more likely to be drawn into caring for their siblings (regardless of sex), at some cost to their own reproduction.[44]

In nineteenth-century Germany, the overall sex ratio of children born was almost exactly even. A fair proportion of children died in their first year of life, and the pattern of these deaths (many due to parental neglect) were status related: farmers' daughters tended to die, while in other classes sons were more likely to die. For farmers, daughters were likely to be considered less desirable than sons; for other classes, the reverse appeared to be true. Similar sex biases show up in the early (pre-1860) history of the United States. There is thus possible evidence of uneven parental investment tied to the perceived value of each sex for parents in different social classes.[45]

A variety of other Trivers-Willard effects appear in contemporary American society. Interbirth interval, birth weight, and proportion of children nursed are related to income and the presence of an adult male in the household. As income increases, so does interbirth interval and percentage of babies breastfed—for sons, but not for daughters. Daughters receive relatively more from low-investment mothers, and sons get more from high-investment mothers. In Tennessee, sons in higher-status families fare better than others. Among polygynous Mormons, sex ratio and parental status covary as predicted by Trivers and Willard.[46]

Not only does the sex of the recipient matter in terms of how much and what kind of help is received, but so may the sex of the giver. In many societies, not only parents but also grandparents, aunts, and uncles give to younger relatives. Because of the uncertainty of paternity in mammals (giving rise, for example, to the mate-guarding tactics described in chapter 3), we might expect some sex differences in how the two sexes give, and are given, bequests. In the United States, women tend to spread their bequests more widely than men, leaving more but smaller bequests. Men are more likely to leave bequests to their widows, with instructions about dispersal to the couple's children. Perhaps this is because men are more likely to remarry and have more children (creating a reproductive conflict of interest), while women are more likely to leave bequests directly to their children. Son bias is not only associated with polygyny but with larger families and agricultural holdings; this pattern becomes rarer as land becomes less important and as families become smaller. In a study of American college students, aunts invested more in their nieces and nephews than uncles did, and matrilateral aunts (with the most certainty of degree of relatedness) gave more than patrilateral aunts.[47]

In many societies, a sex bias in abortion and infanticide exists; this represents a conundrum if it becomes widespread and persistent, for the rare sex comes to be more valuable in any marriage market.[48] Recent experience under China's one-child policy and the sex preferences in India demonstrate this dilemma (see chapter 10). Many, possibly most, cases of long-term sex-biased infanticide may simply be maladaptive, but there are examples that suggest evolved parental strategies. Among the Inuit Eskimos, for example, the female-biased infanticide pattern suggests that parents may be trying to match their number of sons to local prevailing sex ratios, keeping daughters when the local sex-ratio is male biased, and killing them when females are overabundant.

In hypergynous societies, women may marry "up" and men "down," but the reverse is not allowed. Thus, daughters are valuable to lower-class families, but costly to upper-class families. There appears to be no single across-society sex bias; but infanticide is female-biased in high-status families, and son preference is less strong in low-status families. Male-biased infanticide is very rare. Thus infanticide patterns are consistent with other patterns of fitness-striving in hypergynous societies.[49]

Training Boys versus Girls

By now, it is an obvious prediction that parents are likely to raise children of the two sexes differently, and that these differences should be exaggerated or minimized, depending on ecological conditions and the nature of the society. Cross-cultural research based on children's behavior in natural circumstances suggests that the sex or gender differences we observe in American and English children are not limited to Anglo-Saxon cultures, and that there are patterns to both differences and similarities cross-culturally in how boys are treated compared to girls.[50]

The ecology of mating versus parental returns (fig. 3.2A, chapter 3) makes predictions about raising boys and girls differently across cultures. Because the intensity of sexual selection differs between the sexes in polygynous systems, it seems likely that male and female humans, like males and females of other polygynous species, have maximized their reproductive success through different behaviors throughout their evolutionary history. Cross-culturally, sons are more strongly trained than daughters in behaviors useful in open competition, while daughters are more strongly trained in such values as sexual restraint, obedience, and responsibility—traits widely sought by men in their wives.[51]

The above predicts nothing more than universal sex differences in training.[52] However, further thought suggests that the more polygynous the society, the more should boys be taught to strive. Variance in reproductive success increases for men as degree of polygyny increases, and very few men may be extremely successful while many men fail entirely, just as in other polygynous species. In such situations, the rewards may be great if boys are trained to strive.[53]

But because there is some association between stratification and polygyny (stratified societies tend to be polygynous), and the impacts of polygyny and stratification are opposed, it is important to separate these conditions analytically. In stratified societies, men's potential reproductive rewards for striving are constrained. Whether the stratification is by wealth or is hereditary, the more stratified the society (the less striving and reproductive payoffs covary), the less are boys taught to strive openly.

Cross-culturally, the more women actually control important resources or exercise power, the less daughters are taught to be submissive. For example, the more women are able to inherit property, the less daughters are taught to be obedient. The more formal power women have within the kin group, the more daughters are taught to be aggressive, and the less they are trained to be industrious. In societies in which women can hold political office, daughters are more strongly inculcated in achievement and striving than in societies in which women cannot hold office, although the difference is only marginally significant. The more authority women have over children older than four, the less are daughters taught to be obedient.

In sum, across cultures, sons and daughters are trained differently in ways that relate to the evolutionary history of each sex's reproductive success. Despite the strong inference that childhood inculcation is a parental response to ecological and social pressures affecting reproductive success, I know of no direct data on the relative reproductive success, within any society, of parents training their sons and daughters differently within the society. Here is an ideal candidate for gene-culture coevolutionary modeling.[54] Empirical data do exist for a number of societies showing that male reproductive success is related to resource control and status (see below), and thus that boys who learn to be successful in obtaining resources and/or status grow up to be reproductively successful men.[55] The positive relationship seems to hold generally for a wide range of societies. It is therefore not surprising that across all societies, boys are trained to strive.

In current U.S. school situations, too, there is evidence that boys and girls are treated differently in ways that are consistent with our evolutionary history, but with effects we might neither expect nor desire. For some time, psychologists and educators have noticed that in school, girls are more likely than boys to attribute their failures to low ability and to respond to failure with decreased effort and performance—a "learned helplessness."[56] Yet, on average, girls receive more praise, less criticism, and higher elementary school grades than boys and are viewed by teachers and other adults more favorably. Why doesn't the positive reinforcement result in stronger girls' performances? In one study, the kind of positive feedback teachers gave children differed by sex.[57] Girls were praised for nonintellectual items, like being neat; boys for intellectual cleverness. Eighty-eight percent of negative feedback to girls concerned academic issues (compared to 54 percent for boys). Nonacademic issues were the focus of 46 percent of negative feedback for boys (versus 12 percent for girls). Is it possible that without meaning to, we have been perpetuating a sex difference? (Girls: be reliable, obedient, and neat; boys: be clever and striving.)

SEX DIFFERENCES IN SENESCENCE

Among primates, humans are relatively long-lived, with a lifespan 83 percent longer than one would predict from our size (fig. 6.1).[58] It is still not clear why we are so long lived. Though it is tempting to claim so, it is probably not because we are so smart.[59] Nor are medical advances the answer; most evidence suggests that people in traditional societies also had long lives.[60] The current weight of evidence suggests that while reducing accidental death, for example, has changed human life expectancy slightly, the existing maximum lifespan is not a product of medical advance but of human evolution. One important inference from this is that there is little chance that medical science will be able to change human lifespan significantly.

In humans, as in other primates and most mammals, males die sooner than females; the survivorship curves of the two sexes are quite different. Older explanations argued for "male vulnerability" because males are the "heterogametic sex"—they have an "unprotected" Y chromosome: any deleterious gene on the unprotected portion would cause harm.[61] But such suggestions do not explain the fact that, for example, males in many bird species (in birds, females are heterogametic) still die sooner than females. In fact, what matters is the breeding system and the different payoffs for risk taking by males and females. In polygynous species, especially those with limited or no male parental care, males evolve to be mating specialists, with the attendant high fixed costs and risks of failure. In such systems, males who do not take risks will die without offspring, but risk-taking males will have higher death rates.

Old humans do die eventually, but first, like old soldiers, they begin to fade away. This raises the issue of why organisms senesce, and why they do so at such varying rates. Older explanations like "wearing out" and "toxin accumulation" would not predict rate differences; all species should senesce at similar rates. But that is simply not true. Older group-level "adaptive" explanations (clear the way for new individuals) simply do not make sense; cheaters who stayed alive and well would win. Arguments that senescence is selectively irrelevant do not explain why senescence actually begins at the age of first reproduction, not late in life.[62]

Biologists George Williams and William Hamilton first pointed out that, other things being equal, longer life should be favored by natural selection.[63] Senescence arises in part because of pleiotropic effects (a single gene can have multiple effects). Natural selection favors genes with positive early effects. Because reproductive value declines from the age of first reproduction, at some age selection cannot distinguish between simple early good effects versus early good effects accompanied by later (pleiotropic) deleterious effects, for these later costs affect fewer individuals in any population (many have already died), and they affect a smaller proportion of the reproductive lives of those remaining individuals. In this way, deleterious genetic effects late in life accumulate; and, as we age, if it isn't one complaint, it's another. Senescence, then, is not an evolved phenomenon but a cost, a by-product of selection favoring early positive traits—we senesce because we're stuck with the process.

Human senescence is unusual in another way as well. All physiological systems senesce, and reproductive function is no exception. Yet human female reproductive function decays decades earlier than other systems in either sex, including male reproductive function. Most human physiological systems decay at a steady rate from about age thirty, and function at age sixty-five is about 60 to 70 percent of maximum. Among most mammals, female reproductive function decays at about the same rate as other systems, and even very old females retain some fertility. A female's life expectancy at first reproduction is usually not longer than her expected reproductive life.[64]

In most other mammals, the oldest females might spend 10 percent of their lives after their last birth; in contrast, human females live perhaps a third of their years after menopause, and human females lose reproductive function dramatically after age thirty.[65] Women in natural-fertility (noncontracepting) societies show maximum fertility between ages twenty and thirty; fertility then declines to zero (menopause) between ages forty-five and fifty. In the developed nations today, average age at menopause is about 50.5 years.[66] Such a difference in rate of senescence among systems—heart-lung, for example, versus reproductive—is quite rare. In a very few other species (none of them primate), females live a considerable proportion of their lives either in a condition of very low fecundity, or after something similar to menopause—total reproductive senescence.[67]

Anthropologist Kristin Hawkes and her colleagues note that species with a reduction of fecundity, such as elephants, horses, and humans, are species in which offspring may depend on their mother for some time. Thus, it may pay older females to shift from production of additional offspring to continued high-level care of existing offspring. In traditional societies, grandmothers continue to contribute to the well-being of their families, assisting their daughters and grandchildren. Postmenopausal grandmothers are as efficient as younger women in getting resources. And today, parents support their children across the life course.[68] Perhaps selection has favored not a shortening of human female reproductive function, but a lengthening of active resource garnering.[69] This generates some interesting predictions: for example, that general health and competence are (in contrast to other primates) poor fecundity cues.[70]

The most direct empirical tests of the grandmother hypothesis have been conducted by anthropologists Kim Hill and Magdalena Hurtado, working with the Ache of Paraguay (see chapter 7). They found positive effects of grandmothers' help; men and women with a living mother do experience slightly higher fertility than others, and children with a living grandmother do survive slightly better than others. These effects are small in the Ache, and women who remain fertile longer have higher reproductive success than others. It is, however, quite difficult to be sure we have measured the trade-off between numbers and investment (see also chapters 8 and 15). Here is a truly fine puzzle awaiting further work to obtain a better fit between convincing but largely untested theory and imperfect data sets.[71]

7.

Sex and Resource Ecology in Traditional and Historical Cultures

The fact that a man bears an excellent reputation among men, is no proof that he may not be the worst possible companion for a woman.

He sat for more than an hour, trying to analyze his feelings. When a woman does that, ten to one she is in love. When a man does it, ten to one he is not.
—Ella Wheeler Wilcox, *Men, Women, and Emotions*, 1893

Men and women seek and use resources for reproductive success, but they can differ as much as peacocks and peahens in how they seek resources, what kinds of resources they seek, and how they use those resources. The reproductive ecology of the two sexes in humans, as in other mammals, creates opportunities for quite different uses of resources in reproductive competition by males and females, and different strategies (e.g., coalitions) to get them. In earlier chapters, I explored how our background of mammalian sex differences interacts with ecological conditions to yield different mating and resource systems. Here I want to explore how the two sexes are likely to use resources for reproduction within different systems.

SEXUAL DIVISIONS OF LABOR

It's an odd thing, perhaps, that many of us are not bothered by discussions of male-female differences in other species but find it unsettling to ask about sex differences in ourselves. Although men and women differ little physically, humans are one of the most sexually dimorphic of all the primates in behavior—men and women *do* different things.[1] Except for mating activities, which consume little time, and nursing, male and female chimpanzees or gorillas spend their days in activities far more similar than do men and women. Human behavioral sex differences are not an artifact of Western culture, or recent history: across societies, around the world, men's and women's days are spent differently.

The anthropologist George Peter Murdock, with Catherine Provost, looked at men's and women's resource activities cross-culturally.[2] They found fourteen activities to be exclusively or predominantly male around the world: hunting large aquatic or terrestrial fauna, ore smelting, mineworking and quarrying, metalworking, lumbering, woodworking, fowling, boat building, stoneworking, manufacture of musical instruments, bone setting and other surgery. There was surprisingly little variation around the world: in only seven of 1,215 cases (society × activity) was one of these activities reported as predominantly or exclusively female. These cases are truly scattered, and often something of a misnomer. For example, "mining" is coded as a female activity among the Fur of Sudan; what Fur women actually do is collect dust containing iron ore for sale to smiths.

Around the world, the activities done principally by women included fuel gathering, drink preparation, gathering and preparation of plant foods, dairy production, spinning, laundering, cooking, water fetching. There were no technological activities that widely were exclusively female; in fact, men participated, sometimes equally, in many of these "mostly female" activities. When simpler technology is replaced by more complex machines in daily activities across cultures (e.g., the plow in horticulture), men become involved. And when an occupational specialty begins to involve commoditization, profit, and a larger market—even if women otherwise are the principals—men tend to take over. Examples include male potters among the Aztecs, Babylonians, Romans, Hebrews, and Ganda; male weavers among the Burusho and Punjabi; male mat makers among the Aztecs, Babylonians, and Javanese. As Murdock and Provost have noted, "Even the most feminine tasks . . . cooking and the preparation of vegetal foods, tend to be assumed by specialized male bakers, chefs, and millers in the more complex civilizations of Europe and Asia"—and our own.

I suspect divisions of labor relate to our mammalian heritage: to women's requirements of pregnancy and child care, and to the differential return curves of largely female parental effort, versus the high-risk, high-gain opportunism of typically male mating effort. Women are more likely to do activities that require daily attention, do not require long absences from home, are not life-threatening, do not require total concentration, and can be easily resumed after an interruption.[3] In most societies, throughout much of our evolutionary history, small children were likely to die if their mother died. Under most ecological conditions, the patterns we see were simply efficient and made ecological sense.[4]

Our mammalian background is reflected even in simple daily patterns of sex differences. For example, men and women share food very differently. Consider the Ache. The resources brought in by men tend to be more unpredictable in payoff than those of women, and are eaten more by nonfamily members. Men's foraging appears to be, at the same time, inefficient: men could more than double their caloric rate of return by taking some foods, such as palm starch, which they ignore. Instead, they hunt meat and collect honey, both of which are foods that tend to be shared.

What are the reasons for this behavior? One obvious hypothesis is risk reduction: reciprocity now, when I have meat, might oblige you to reciprocate later when I fail and you are successful. But reciprocity would mean that sharers would discriminate, giving to those who give back, and excluding "free riders" who take much but return little or nothing, and this does not seem to be the case. Two other hypotheses seem more likely. First, when returns (as from hunting) are sometimes too large for one's own family to consume and storage is impossible, "tolerated theft" may be less costly in many ways than its alternatives. Second, "showoff" men who are more successful hunters receive more attention from group members and fare better reproductively; more women are willing to mate with them. I suspect that these patterns will turn out to be common, once we look for them.[5]

Sex and Control of Resources

Not only do resources contribute differently to the reproductive success of men and women, but women's access to resources varies greatly cross-culturally. When do women control significant resources? There is no single meaningful measure of women's power or resource control cross-culturally, though there are some patterns. Across cultures, the more women contribute to subsistence, the more control they have over various resources. In our own society as well, there is evidence that an increase in professional working women has resulted in greater economic independence for them.[6]

Women's ability to get and control resources also differs between monogamous and polygynous societies. Within polygynous societies, as the degree of polygyny (percent of polygynous marriages) increases, women's ability to control the fruits of men's labor decrease and their ability to control the fruits of their own labor increase—women can function as independent economic units, just as in other primates. Not only resource control, but inheritance, is affected: the greater the degree of polygyny, the less likely are women to be able to inherit property, for sons in polygynous societies are better able than daughters to turn resources into grandchildren (see chapter 6). Women are most likely to be able to inherit property in matrilocal societies, when they live among their own kin, and least likely to inherit property in patrilocal societies, living among their husband's kin.

MEN, WOMEN, AND RESOURCES IN TRADITIONAL AND HISTORICAL CULTURES

The arguments so far suggest that any relationships between resources and reproduction in humans will be influenced by ecological considerations. Males with greater resource control, we predict from other species, will show higher fertility than poorer but otherwise comparable individuals. In traditional societies, this typically results from greater polygyny by higher-status or wealthier men.

Men and women are likely to use resources to reproduce somewhat differently. The most fecund woman can have fewer children than the most fertile man, simply because parental investment has different constraints from mating effort. How then do resources affect the reproduction of men compared to that of women, and how are the patterns of families affected? Several studies on cultures that vary in many regards (especially in some crucial resource-control issues) can give us insight both about patterns that are common and patterns that vary.

Yanomamö

The Yanomamö, one of the best-studied of all traditional societies, live in small villages in tropical rain-forest regions of southern Venezuela and adjacent regions in Brazil. Their daily activities revolve around collecting wild foods, getting firewood and water, and cultivating their gardens; about three hours daily are spent in these subsistence activities. The anthropologist Napoleon Chagnon has worked among the Yanomamö for more than twenty years, and most of our information comes from his work.[7] The Yanomamö still wage intervillage warfare, and at least one-fourth of all adult men die violently. Warfare has reproductive causes and impacts, as successful warriors gain status—and thereby reproductive success (see chapter 13). War frequency is ecologically influenced.

Chagnon notes that "social life is organized around the same principles utilized by all tribesmen: kinship relations, descent from ancestors, marriage exchanges between kinship/descent groups, and the transient charisma of distinguished headmen who attempt to keep order within the village."[8] The Yanomamö do not accumulate the sort of heritable resources that would serve as bridewealth. Social dynamics center on the exchange of marriageable girls; marriages are arranged by the male kin of prospective mates and are often highly political.

The Yanomamö, slash-and-burn agriculturalists with little in the way of accumulated resources, are one of the societies held up sometimes as "egalitarian," in which everyone behaves as though the interests of all were identical. As Chagnon has shown eloquently and repeatedly, nothing could be further from the truth for reproductive matters, at least. The Yanomamö are polygynous, and men's abilities to marry and have children vary far more than women's. Wealth is indeed largely absent, but in the lowlands, which are ecologically most desirable, a man's demonstrated skill as a warrior and a large male kin group make a great reproductive difference for him. The variance arises largely because skilled warriors, called *unokai*, are able to marry earlier and more often than other men (see fig. 13.2, chapter 13).

What of women's costs and benefits in this polygynous society? In some societies, polygyny, despite its likely costs, could have real resource benefits for women—if men vary in wealth. But Yanomamö men do not, so a polygynous Yanomamö woman must share the resources of her husband with co-wives, even though he has nothing more than a monogamous man. Furthermore, Yanomamö men have a large say in arranging marriages. Polygynous households tend to have smaller gardens for their family size than monogamous households, but economically, the only difference between polygynous and monogamous households is that polygynous households receive more food from others. A high-status Yanomamö man, therefore, though he does not directly provide wealth, may indirectly create some benefits for his wives.[9]

Ache

The Ache live in eastern Paraguay, in the southwestern part of the eastern Brazilian highlands. This area is generally higher and drier than the rest of the Amazon basin. Most of the rivers in the region flow to the Atlantic or to the Rio de la Plata rather than the Amazon. An excellent study of Ache demography, ecology, and life history is that of anthropologists Kim Hill and Magdalena Hurtado.[10] All Ache groups in recent times have been hunter-gatherers. Apparently, for the last four hundred years they have engaged only in hostile interactions with outsiders and have not traded, visited, or intermarried with the nearby Guarani populations. The Ache live in small bands of fifteen to seventy individuals, moving throughout the forest. Bands comprise closely related kin and some long-term friends. Large sibling groups of both sexes tend to remain together along with additional kin. Only four Ache groups existed in the second half of the twentieth century before they made permanent contact with outsiders.

Daily life centers around hunting and gathering. Men spend almost fifty hours per week getting food. They hunt white-lipped and collared peccaries, tapir, deer, pacas, agoutis, armadillos, capuchin monkeys, capybara, and coatis; they collect honey, which accounts for 87 percent of the calories in the Ache diet. Men often hunt in ways that look inefficient from standard optimal foraging perspectives, but in fact such men seem to be pursuing a high-risk–high-gain showoff strategy that may often fail but can produce big, flashy hunting successes—and, with success, more sexual access to women. Women spend about two hours a day gathering; they collect fruits and insect larvae, as well as extract the fiber from palm trees.[11] Women also carry the family's children, pets, and possessions. Their care of children and possessions constrains their ability to forage. Men may travel with the women's group, but more often they set off in small groups to search for game, spending about seven hours per day hunting.

In the late afternoon, families gather and prepare food (fig. 7.1). Hunters rarely eat from their own kills, and much food is shared, leading early observers to argue that the society was completely egalitarian. While meat is apparently shared evenly under most circumstances, honey and gathered items are not. Further, when a man dies, his young dependent children are far more likely to die than if he had lived. While reciprocal sharing of meat is ordinary, when a man has died reciprocity can no longer be extended, so meat is no longer shared with the widow and children.

The Ache are polygynous, and during young adulthood they may switch spouses frequently. This pattern seems to have changed little after contact with Europeans. After marriage, residence is typically matrilocal. Many children have multiple recognized fathers. Reproductive success is difficult to measure under such circumstances, but despite the fact that the Ache have little in the way of heritable wealth (which in so many societies correlates with reproductive success for men), status matters: the best hunters have the greatest reproductive success.[12]

Kipsigis

The Kipsigis are pastoralists and agriculturalists living in the Rift Valley Province of Kenya. Since the 1930s, the Kipsigis have become more and more involved in commercial agriculture, selling maize.[13] Women do the agricultural and domestic work while everyone, even the children, shares the duties of animal husbandry. The Kipsigis are a polygynous bridewealth society, where resources are given by the groom or groom's family to the bride's family. Half the women are married by age sixteen, half the men by twenty-three. The father of a prospective groom initiates negotiations, making offers of cows, goats, sheep, or, since 1960, cash. The average bridewealth is significant: for a man of average wealth, it comprises a third of his cows, half his goats, and two months' salary. The father of the groom alone

Figure 7.1. Ache men resting. (Photo by Kim Hill.)

is responsible for providing the bridewealth for his son's first marriage. Payment of bridewealth gives a man rights to all of his wife's children (legitimate or not), and to her labor. Once, a woman's labor value was significant; this is less true now.[14]

In contrast to the Ache, Kipsigis marriage is quite stable and divorce is almost unknown.[15] If a mistreated wife deserts her husband, the bridewealth is not returned. Though many things can influence the exact amount of marriage payments, high bridewealth is typically paid for younger women (of higher reproductive value) and plump (healthy, well-nourished) women. Older women, and women who have given birth before marriage, command lower bridewealth. The importance of wealth differences between the groom's and bride's families has varied over time. Once, better-educated and wealthier men had higher bridewealth demands made on them. This has changed, as women's parents seek men with good economic prospects. Another sign of changing times: women with secondary education command high bride prices, and tend to marry better-educated men.

Mate choice operates in both directions: men's wealth (more precisely, land ownership) matters. Men are preferred who are better-educated and who can offer more acres per wife to prospective brides or if (after controlling for land per wife) they have had fewer wives and a successful paternity record. Women's reproductive fate after marriage suggests that the things men and women choose are biologically significant; both fertility and offspring survival matter. Women who are married to poorer men and have co-wives have long (34.2 months) interbirth intervals and relatively higher infant mortality rates than women married to wealthier men.[16] Social and biological factors interact, yielding a rich and complex system, but here, too, traditionally, wealthier men have more wives and more children than others.

Mukogodo

The Mukogodo, in central Kenya, were foragers and beekeepers until early in this century (fig. 7.2). They were, in anthropologist Lee Cronk's words, "neither wealthy, powerful, nor prestigious," and

Figure 7.2. The Mukogodo, neighbors of the Masai, stand at the bottom of a regional hierarchy of wealth and status. Their sons find it difficult to accumulate enough bridewealth to marry non-Mukogodo women, but their daughters, if beautiful, can marry "up" and bring bridewealth. (Photo by Lee Cronk.)

poor compared to several neighboring groups such as the Masai, the Samburu, and the Mumonyot.

Today the Mukogodo sit at the bottom of a regional socioeconomic hierarchy and are somewhat stigmatized by their neighbors. This fact has had curious social and health consequences.[17] As in many pastoral societies, the groups in this region use livestock for bride-wealth. Because the Mukogodo are so poor, Mukogodo men have little chance of raising the bridewealth required to marry a woman from a neighboring group. It also means that as Mukogodo daughters are married to men of neighboring groups, the Mukogodo acquire cattle, sheep, and goats from bridewealth.[18] Daughters are more valuable than sons in important ways.

In chapter 6 we explored the general phenomenon of parental biases in the sex of their children—the Trivers-Willard effect. In many societies, parents prefer sons—they give them better care, more inheritance—because sons under many conditions can turn parental investment into grandchildren more effectively than daughters. Not so among the Mukogodo. Mukogodo parents appear to respond to the different costs and benefits of sons and daughters in several ways—but among the Mukogodo, daughters are preferred. In contrast to the general mammalian pattern, Mukogodo mothers nurse their daughters longer than their sons; Mukogodo caregivers (mothers and others) stay closer to girls than boys, and hold them more. Parents take their daughters more frequently than their sons to the dispensary and clinic for treatment, and enroll their daughters more in the local Catholic mission's traveling baby clinic. Non-Mukogodo parents do not show these biases. And, when Lee Cronk measured the birth ratio of Mukogodo sons and daughters, it was highly unusual: not—as is typical—about even, with a few more boys than girls, but with more than twice as many daughters. Furthermore, probably because of the biased care, girls show better growth than boys: better height for age, weight for age, weight for height.

A Mukogodo son grows up and marries, usually a Mukogodo woman. A daughter may marry a Mukogodo man, but if she can marry a wealthier neighbor, he has to pay bridewealth, and she will bring wealth to her family (which may be used to help any brothers marry). Interestingly, the Mukogodo, living among societies that have a strong male preference, claim, if you ask them, to prefer sons to daughters—but their actions show a clear preference for daughters. Parents among the Mukogodo, as those in a number of societies, thus appear to favor the more profitable sex in their children.

Turkmen

The Turkmen are clearly a resource-based society (fig. 7.3). As anthropologist William Irons has noted, a large part of the daily activities of the Turkmen is devoted to economic production, and they clearly seek to maximize wealth.[19] They have long been immersed in a market economy far larger than their tribal boundaries. Production for trade is important, as are accumulation of agricultural land and livestock and savings in the form of money, jewelry, and the like. Men's work appears more important in wealth building than women's work, and extended families appear to be built around closely related male relatives who remain together. That is, it seems to pay off in inclusive fitness for related men to stay together and build wealth.

Figure 7.3. The Yomut Turkmen are active in a market economy and clearly seek to maximize wealth. Wealthy men have more children than poor men. (Photo by Bill Irons.)

Marriages are arranged by parents. Women are typically married at about age fifteen. Sons of wealthy men also marry at about age fifteen; sons of poorer men marry later, perhaps because the Turkmen are a bridewealth society, and bridewealth is high. A man of median wealth will pay two to four years' income for a virgin bride. In concrete terms, this is ten camels for the bride's father and one camel for her mother, or the bridewealth can be paid in cash or other livestock. One camel equals two horses, one really good race horse, two cows, ten sheep, or ten goats.

The price of a bride appears to be far more fixed and nonnegotiable than among, for example, the Kipsigis. There are no bridewealth "discounts" for hardship, just later marriage. Because the Turkmen are polygynous and most men want to marry, most widows do not remarry, and the sex ratio in the adult population is about 110:100 (male biased), it's a seller's market: a man who wishes to argue down the bridewealth is in a poor position to do so. If a widower seeks a virgin bride, he must pay double bridewealth. If a married man wishes to take a second wife, he is supposed to pay triple the bride price, though in practice it works out to about 2.5 times the bride price. Thus, polygyny is possible for only about the richest 5 percent of men.

As a result of these conditions, age-specific fertility rates are higher for richer men and women than for poorer men and women. Because the Turkmen are polygynous and bridewealth based, richer men have significantly more children than poorer men. Thus, as Irons noted, the Turkmen, in striving for wealth and social success, are striving for the proximate variables that result in greater reproductive success.[20]

Qing China (1644–1911)

Early in the Qing period in China, in 1652, the nobility established the Office of the Imperial Lineage to register births, marriages, and deaths, and to maintain the genealogy of the Qing noble lineage. Membership in the lineage was defined by patrilineal descent, and members were entitled to a wide range of state perks, including a subsistence allowance, titles of nobility, and other state stipends.[21] To ensure complete and accurate registration, there were overlapping sets of household and vital-events registrations, much as in Sweden (see next chapter). These records, commonly called the

"Jade Register," cover thirteen generations over two and a half centuries. Thus the demography of the Qing nobility is remarkably well known.

Marriage in the lineage was relatively early and virtually universal. Mean age of marriage for both men and women was the early twenties; 99 percent of women were married by age thirty, and 99 percent of men by age forty. Yet the Qing's marital age-specific fertility, especially at younger ages, was quite low compared to contemporaneous European societies.[22] The Qing lineage was large, and there is complex variation over time, but in general the pattern was as follows: men who were of high nobility, and whose wealth and rewards were therefore greater, were more likely to be polygynous than the lower nobility—and polygynous men had more children than monogamous men.[23]

Both high nobility and polygyny were strongly associated with higher fertility, but the strength of these associations varied over time. Differences in status-related fertility were relatively narrow between 1750 to 1780 (a period of comparative wealth, when lower nobles could do reasonably well). Differences were greatest during the depressed period 1780 to 1820, when the government cut subsidies to the lower nobility. As imperial lineage rewards and subsidies declined, polygynous men married fewer wives, and polygynous fertility plunged from ten to five.[24] Status, resources, and fertility were clearly linked for noble men in Qing China.

Resources were important to ordinary men in Qing China as well.[25] In the common rural extended-family households, marriage and fertility differed not only with type of household, but with individual status within the household. Among "senior" relatives, household heads married earlier and in higher proportions than anyone except uncles (who, in this system, were even more senior in their own right).[26] Among junior relatives, sons tended to marry earlier and in larger proportions than other men, such as brothers' and cousins' sons. The higher a man's position, the more likely, and earlier, was marriage. Fertility, like marriage, correlated with a man's position; higher-status men had more sons than others (and fewer daughters, since female infanticide was acute among the elite in rural areas).

Another kind of hierarchy, somewhat more related to a man's own effort and capabilities, existed during this period: the banner hierarchy. Here we can separate the effects on a man's marriage and fertility of his household status (i.e., his position rather than his individual capability) from his occupation (e.g., soldier, artisan, official, commoner). Demographic behavior of commoners was more strongly related to banner position (his own efforts) than household position, even controlling for possible effects of overlap between the hierarchies. Men who achieved a banner position were more likely to marry and had higher fertility, than others. In sum, outside the nobility, marriage was not universal for men, and access to marriage and resulting fertility were a function of resource control, partly due to exogenous factors, partly due to their own efforts.[27]

8.

Sex, Resources, and Fertility in Transition

Life history theory deals directly with natural selection, fitness, adaptation, and constraint.

Demography, the key to life history theory, allows us to calculate the strength of selection on life history traits for many conditions.
—Stephen Stearns, *The Evolution of Life History,* (1998)

AN IMPORTANT QUESTION is *exactly what,* in any species or society, contributes to greater or lesser lifetime reproductive success. As we have seen, in most mammals, and in the majority of traditional human societies for which data exist, status, power, or resource control enhance lifetime reproductive success, especially for men. Men's reproductive variation in traditional societies arises mostly through differential polygyny—higher-status men can marry earlier and more often than other men, and they can marry younger women of higher reproductive value.

Does this pattern have any relevance today? Two phenomena make it likely that today we will not see the huge differentials that existed among some traditional societies, or in reproductively despotic societies (e.g., in which a caliph might have a thousand concubines). First, modern nations have for some centuries been (at least nominally) monogamous. Second, in western Europe and North America, total population increased but family size fell dramatically during the nineteenth century—the "demographic transition."[1] Today, societies are large, heterogeneous, and mobile, while family sizes are small; associations between status or resource control and number of children may have become weak, or disappeared altogether.

To discover whether power, sex, and resources still correlate, an obvious next step is to examine the patterns during the demographic transition. We may think that our societies today have little connection with our past, but it is important to explore the history and patterns of resources and reproduction. What did happen in Europe through the demographic transition, the period in the nineteenth century often seen as a beginning of "modern" Western industrialized society?[2]

Nineteenth-century Sweden provides an opportunity to explore the impact of resources on male and female lifetimes—survival, reproduction, and migration (fig. 8.1). By the 1600s, extensive and overlapping civil registers existed in Sweden; although these were sometimes damaged by fires, for example, the completeness of demographic information is remarkable. This is a population for which excellent records exist through the historic demographic transition from large to small family sizes. We are lucky that the Swedish government has funded a demographic database, designed to track individuals and families over time and to make use of the extraordinary historical records as well as modern data. The records are extremely accurate and include longitudinal (lifetime) data for all individuals: sex, date of birth, age at marriage, best occupation, date of record loss, type of record loss (death, emigration), dates of birth of all children, and comparable data for those children.[3] We can thereby follow the fates of families over time.

Let us ask, then, the same questions of Swedish families as we asked of families in the traditional societies discussed in chapter 7. We will also ask some new questions because the ecology and social milieu of nineteenth-century Sweden are different from all the previous societies we examined. Did higher-status men have more children than others? Did children of higher-status parents survive better than others? Did higher- versus lower-status children leave the parish at different rates? As adults, were richer compared to poorer individuals differentially likely to leave the parish? As adults, if they stayed, were they differentially likely to marry? Did the age-specific fertility of richer compared to poorer women differ? We must ask some questions differently from those we posed for traditional societies: in Lutheran Sweden, for example, monogamy was the rule—but we can ask if wealthier men, if their wives died, remarried more

Figure 8.1. In nineteeth-century Sweden, rules were egalitarian; nonetheless, the lives of migrant farmworkers and wealthy families differed considerably. (Photos courtesy of the Skellefteå & Historical Museum, Sweden.

than other men, and whether those second (or third, or fourth) marriages resulted in richer men having more children than other men.

An individual can control only some of the factors influencing his or her reproductive life. By the nineteenth century, life could be complicated, so taking a perspective that begins with individuals rather than populations can also be complicated. Some factors were related to the family into which one was born—occupation(s) of one's father, whether or not he owned land, and one's order of birth; others were related to an external economy.

Here I will follow the lives of some people, their children, and grandchildren from four geographically separated and economically diverse parishes (Tuna, Locknevi, Gullholmen, and Nedertorneå), from 1824 until 1896, when records end for privacy reasons. I'll begin with men married for the first time in these parishes between 1824 and 1840. The four parishes differ in economic and ecological conditions (described below), and people's lives varied greatly, though there are some important commonalities. I will review this variation, but I am also seeking the most general possible answer: did the resource-reproduction connection persist or disappear in nineteenth-century Sweden?[4]

NINETEENTH-CENTURY SWEDEN

Nineteenth-century Sweden was largely agricultural, with emerging proto-industrialization. The beginning of geographically scattered market activity involved transforming raw materials into "made" commodities, but a large part of the labor force worked part time or at home. The family could function as a form of economic enterprise. Proto-industrialization tends to arise in developing regions that have both an underemployed, land-poor population, and urban markets—like Europe in the nineteenth century and much of the developing world today.

Marriage in Sweden occurred fairly late. Women married for the first time in their early to mid-twenties, and men in their late twenties (though this, as we will see, varied among parishes); at marriage the new couple typically set up their own independent household. A relatively high proportion of individuals never married.[5]

Both economic and family patterns differed among the parishes; even in this short summary we will see great variation. Unless we were asking the behavioral ecologist's questions about resources and family patterns, it would be easy to fail to see a "bottom line." Yet I think there is a clear conclusion underneath all the diversity: whenever wealth or resource differentials existed, resources and reproductive success were positively correlated.

Gullholmen

Gullholmen is an island parish, with almost no really wealthy or really poor people in the nineteenth century. Most people earned their livelihood by fishing, and fish catches varied considerably from year to year. The small population of Gullholmen rose steadily during the nineteenth century but was always less than one thousand individuals. Nonetheless, because the island was small, population density was the highest of any parish discussed here. Perhaps related to the unpredictability of fish catches and the costs associated with commercial fishing, people married late. Nonetheless, the lifetime family size of married individuals was the highest of the four parishes. Family size decreased over the study period, from about 5.5 ± 2.7 children in generation 1 to about 4.2 ± 2.4 in generations 2–4. Thus marital fertility was reduced by the 1860s in Gullholmen.[6] However, survivorship improved, and the number of children surviving to age ten did not decline.

Only 12 percent of men living to maturity migrated from the parish. There was little variability in the status men held, and men of different status survived equally well—although men of different occupations were differentially likely to marry, and married men had more children than unmarried men. Men married at about twenty-seven years of age. In most parishes, only the wife's age at marriage mattered to the couple's fertility, but in Gullholmen a man's age at marriage was important, too.

Women married very late in Gullholmen (26+ years), and 58 percent of adult women failed to marry at all while in the parish. Women who married earlier had more children than those who married later. Women almost never remarried, and remarriage did not affect their fertility. A woman's chances of marrying were independent of her father's status, and neither her father's nor her husband's occupation was related to her age at marriage. In contrast to other parishes, a woman's fertility was somewhat related to her husband's occupation.

Great uncertainty in resources (fish catches) seems to have been extraordinarily important in shaping people's lives in Gullholmen, far more important than the slight differences that existed in wealth or class. People married late but had large families; perhaps both of these patterns are related to getting established and becoming able to handle uncertain resource fluctuations. A Swedish historian told me that in the records and contemporary letters from Gullholmen, a frequent strategy was to use some money in excellent fish-catch years to buy "luxuries" like silver candlesticks, which could be sold easily in poor years.[7]

Locknevi

In Locknevi parish, in Småland, only a limited central valley contained fertile fields. A small ironworks in the southwestern part of the parish provided supplemental income for some farmers until the 1880s. At the start of the nineteenth century, there were a few very large landholdings that employed agricultural day workers. The population grew, then stagnated as opportunities waned. Wealthy landowners divided up and sold their large estates and moved out of the parish. As a result, landholdings became progressively smaller, and though more land was brought into agriculture, much of the cultivation was on marginal land. Thus in Locknevi parish during the period of this study, resources shifted from being relatively uneven with some very large holdings, to being more even but limited.

Except for the few richest families, people's economic lives were uncertain: not only did work depend on crops, but people's purchasing power, as reflected by the number of days' work required to purchase a "market basket," varied considerably. Family patterns in Locknevi varied with economic times: marriage and fertility rates fluctuated with crop prices.[8]

In contrast to Gullholmen, where people married late, Locknevi folk married early but delayed having children. A man's best occupation influenced his chances of marrying: 74 percent of agricultural workers and servants living their entire life in the parish failed to marry, compared to 20 percent of lower-middle-class men. The age at which a man first married was not related to his lifetime fertility. Remarriage was common in Locknevi, and it influenced men's fertility: men who were married more than once had more children and more children surviving to age ten than men married only once. Wealthy upper-middle-class men were most likely to remarry (25 percent).

In generation 1, when landholdings were uneven, a man's occupational status predicted the number of his children who would survive to age ten; wealthy men did best. But when the wealthiest landowners sold off their manors and left the parish, remaining men showed few fertility or class differences.[9]

Sixty-one percent of women failed to marry while in the parish, and those who married delayed their fertility remarkably, perhaps because of growing resource constrictions.[10] Women who married earlier were somewhat more likely to have more children than those who married later. A woman's age at marriage was not correlated to either her father's or husband's occupation. Further, women's lifetime production of children and number of children alive at age ten were not related to their father's or husband's occupation. The resources controlled by the fathers and husbands of women in Locknevi did not directly influence women's reproductive patterns. Men's patterns were connected to wealth, but women's patterns (perhaps because of the high remarriage rates) were not.

Tuna

The population of Tuna parish, in Medelpad, rose from approximately 1,200 in the early nineteenth century to about 3,300 in the late part of the century. Tuna was largely a farming parish, though forest and mining industries were also present in the early 1800s. Many men worked in the local iron foundry as well as in farming. Industrialization increased rapidly beginning in 1850. The iron foundry closed in 1879 and reopened in the mid-1880s; in response, people moved out of the parish, then back in. The economy (forestry, ironwork, mixed-crop agriculture) of Tuna was more diverse than Locknevi's, and, perhaps as a result (and the availability of nonmarket alternatives such as hunting and fishing), population measures did not vary with economic fluctuations in Tuna as they did in Locknevi.

In Tuna, occupation provided no clue about family formation and reproduction. Land ownership, however, did. Landowners were almost certain to marry (95 percent), in stark contrast to other men (35 percent); they married women about 2.5 years younger than other men, and had about one to 1.5 more children. In sum, landowners had larger families no matter what the times, and their families were not only larger, but less variable in size than those of nonlandowners. Even within families, the reproductive lives of brothers—landowning and nonlandowning—differ in this way. Landownership appears to have provided a buffer against hard times, over and above the nonmarket alternatives.[11]

Nedertorneå

Nedertorneå, in the far north, was a farming parish, although land was generally of poor quality. The population of Nedertorneå rose steadily during the nineteenth century.[12]

In Nedertorneå, a man's occupation mattered to his lifetime reproduction: 36 percent of migrant worker and servant men married, compared to 57 percent of those in the upper middle class, 44 percent in the lower middle class, 57 percent of *bönder* ("farmers," often small landowners), and 44 percent of cottars (roughly, tenant farmers). The interaction of resources and reproductive patterns was influenced by historical particulars. First, there were very rich and poor people (in contrast, e.g., to Gullholmen); this was because the central Swedish government early in the nineteenth century moved upper-level civil servants to the outpost town of Haparanda, to shore up the local economy. Second, infant survivorship was low, because parents fed their infants often-contaminated cow's milk, rather than breast feeding them. In the 1840s, a doctor began a campaign to reinstitute breast feeding. Because he worked mainly with upper-middle-class families and in the town of Haparanda rather than in the surrounding countryside, there is great variation in interbirth intervals, survivorship, and thus fertility and family size, that is tied to location, class, and time. Fertility and survivorship differed both with class and residence (town versus countryside).

Sixty-four percent of women failed to marry while in the parish—but Nedertorneå had the highest peak age-specific fertility of the four parishes. Women's likelihood of marriage was related to their father's occupational status; daughters of upper-middle-class men and farmers were most likely to marry.

The four parishes differed in size, population levels and growth patterns, economic bases and stability. All these are reflected in the complexity of demographic variations. Considering family fertility and net family size (number of children surviving to age ten) for married individuals, and without considering possible occupational differences, both fertility and family size fell significantly as resources constricted in Locknevi between generation 1 and generations 2 to 4.[13] In Gullholmen, fertility fell, but family size did not, reflecting changes in infant and child survivorship; in Nedertorneå, fertility fell slightly while survivorship increased; and in Tuna, no change was apparent in either fertility or family size.[14]

Occupational status was related in all parishes to a man's likelihood of marrying, although the relationship was only marginally significant in Tuna, where land ownership was crucial. The relationship between occupation and a man's lifetime fertility and family size varied considerably among parishes.

People in the four parishes responded somewhat differently to ecological and economic fluctuations. In many areas in southern Sweden, single crops dominated the economy. Rye and corn were major crops; corn was particularly labor intensive. Bad harvests created real hardships, reflected in prices and purchasing power. In the north, barley was an important grain crop, but agriculture was more mixed and was consistently supplemented by fishing and forestry. Thus, in the north, failure of any particular crop was likely to have less impact on people's lives. Harvest and price information alone are insufficient reflections of conditions in such areas. In areas with single crops, famines and high food prices predict, for example, theft rates, but in counties in which agriculture was more varied (like the one that included Locknevi), or in the northern "forest" areas (like Nedertorneå) where less market force was in effect, this was not true.[15] All of these differences are important, particularly in the examination of historical and parish-specific patterns. But they cloud answers to the very basic question we started with: Did the resource-reproduction correlation persist or disappear? We need more than statistical analyses of the parishes to reach a general answer.

SEX, RESOURCES, AND LIFE HISTORIES

How can we subsume temporal and spatial differences without ignoring them? Although the particulars of reproduction varied in time and among parishes, broad comparisons are possible without ignoring this variation among parishes and through time. Let us classify each individual's wealth as "richer" (owned land and/or had an occupational status of upper middle class, lower middle class, or *bönder* [= farmer]) versus "poorer" (occupational status of cottar or proletariat and no land ownership record). We can then compare each adult individual's lifetime reproduction to the median

---

Figure 8.2. Some paths (heavy lines) were more likely than others for individuals born to richer or poorer fathers. Reproductive comparisons are relative to all adults who reached age twenty-three in the same decade in the same parish for nonmarried individuals, and relative to all individuals marrying in the same parish during the same decade for married individuals. Strong within-sex differences are highlighted by asterisks. Extrinsic factors (indicated by valve symbols) could matter, for example, in the probability of outmigrating or marrying. These greatest-likelihood pathways simply track, for all individuals born in the sample, the percentage of individuals at each comparison point who follow one or another fate. This is a visual representation, and the numbers diminish at each juncture, so the percentages will not always suggest the results of the statistical analysis (e.g., a statistical difference may be great, while the percentage is small, or vice versa, because numbers are large or small). (A) A daughter born to a poorer father was more likely than her richer cohorts to leave the parish before age fifteen; if she stayed, she was about equally likely to marry. She was overwhelmingly likely to marry a poorer husband. Though her fertility could be great (33 percent had the median number of children for their decade of marriage and parish), her sons were likely to leave the parish, and, if they stayed, to do poorly reproductively. If she did not marry, there was an 86 percent chance she had fewer than the median number of children for all adults. A daughter born to a richer father had a higher chance of remaining in the parish, and an equal chance of marrying. If she married, there was a 77 percent chance her husband was richer. If she did not marry, there was a 48 percent chance she would have greater than or equal to the median number of children for all adults. (B) Sons born to poorer fathers were likelier to leave the parish before age fifteen, compared to sons of richer fathers (15 percent versus 8 percent); for sons of poorer men who stayed, there was an 89 percent chance they would be poor, and a 40 percent chance they would migrate out as adults. Men who stayed were likely never to marry (57%), and 97 percent had fewer than the median number of children compared to all adults. Sons born to richer fathers, once they reached age fifteen, had an excellent chance of becoming richer themselves (91%). These men were more likely to stay as adults (69%). Their chance of marrying was about 48%. Those who married tended to have the median or greater number of children compared to other married individuals (59%); those who did not marry were likely to have fewer than the median number of children, compared to all adults (55%).

A. Females born to poorer fathers

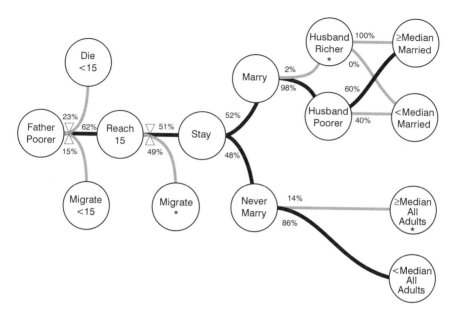

Females born to richer fathers

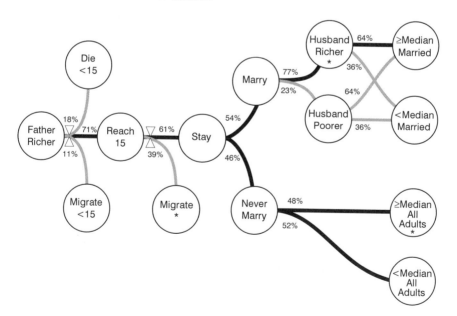

Figure 8.2.

B. Males born to poorer fathers

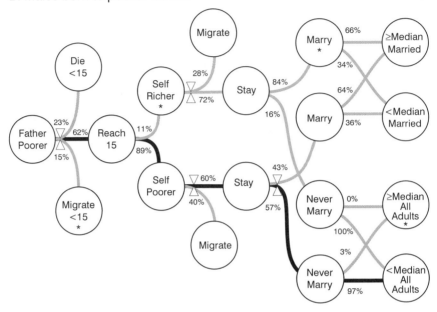

Males born to richer fathers

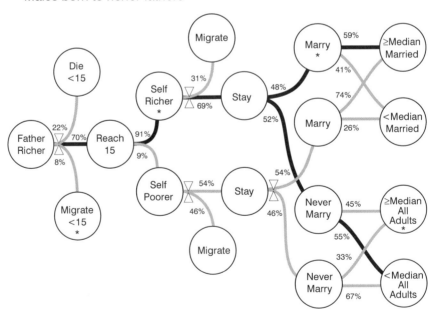

Figure 8.2. (Continued)

SEX AND FERTILITY IN TRANSITION

for (1) all individuals reaching "maturity" (twenty-three years) in any decade in any parish, or (2) all individuals marrying in each decade in each parish.

Thus we compare any individual's lifetime reproduction to the median for his or her parish and decade of maturation or marriage, as appropriate. These data form a picture of life prospects for people born in different conditions, and relate to the general problem of resources, family decisions, and demographic transitions, taking regional differences and historical particulars into account but still yielding a general picture.[16]

Figure 8.2A highlights in bold lines the likeliest lifepaths followed by sons and daughters of rich and poor men to greater (above-to-median marital fertility) or lesser (below-median fertility for all adults) reproduction.[17]

FEMALE LIFE PATHS

Despite the rich variation discussed above, there were dominant patterns. Daughters of poor fathers were about 10 percent more likely to die or migrate from their parish of birth before the age of fifteen than were daughters of rich men.[18] Women were also more likely to leave as adults if their fathers were poor. Overall, daughters of richer men were more likely to remain in the parish and marry than others; however, of the women who stayed in their parish of birth to reproductive age, daughters of both rich and poor men were equally likely to marry (fig. 8.2A,B).

The likely remaining life paths for married women and unmarried women differed, depending on their father's wealth. Of women who married, virtually all women (98 percent) born to poor fathers married men who were poor, while 77 percent of women born to rich fathers married rich men. Married women's fertility was compared to that of other women in the same parish of birth who married in the same decade. Fertility of unmarried women was compared to all adult females in the same parish of birth who reached age twenty-three in the same decade. Women who were born to poor fathers and never married were far more likely (86 percent) to have fewer than the median number of children born to all adult women. Never-married daughters of rich men nonetheless had a 48 percent chance of having greater than the median number of children, suggesting

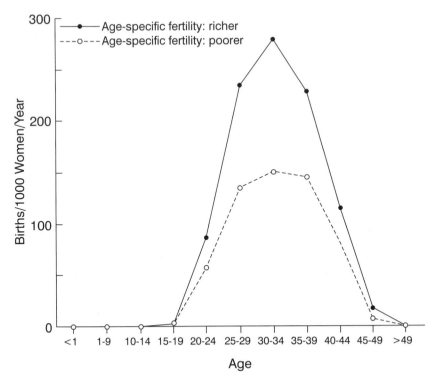

Figure 8.3. In nineteenth-century Sweden, wealthier women had higher age-specific fertility at all ages than poorer women.

that illegitimate births occurred at higher rates among the rich. The result of these pathways meant that, at all ages, "richer" women had higher age-specific fertility than "poorer" women (fig. 8.3).

MALE LIFE PATHS

Sons of poor fathers were 7 percent more likely to migrate before age fifteen than were sons of the rich (fig. 8.2C,D). As adults, sons of rich men were very likely (91 percent) to become wealthy themselves, while sons of poor men were likely to stay poor (89 percent). As adults, poor sons of rich men were 6 percent more likely to migrate than poor sons of poor men.[19]

Of men who stayed in their birth parish, poor sons of poor fathers were most likely to remain unmarried (57 percent); 97 percent of such poor, unmarried men had fewer children than other men. Rich sons of rich fathers had an approximately equal chance of marrying or not (48 percent versus 52 percent), and once married had a 59 percent chance of having relatively large families. Rich sons of rich fathers who did not marry were, like poor sons of poor fathers, likely to have fewer than the median number of children. Although 97 percent of poor sons had this fate, only 55 percent of rich sons did.

Two rare but interesting paths occurred. Most dramatically, the few men (11 percent of the whole sample) who were able to attain wealth even though born to a poor father had an 84 percent chance of marrying, and once married, a 66 percent chance of having at least the median number of children. Although we have only birth, marriage, tax, and death records (no memoirs), it looks like personal effort and ability mattered for these men. On the other hand, poor sons of rich fathers did not fare as badly as the poor sons of poor fathers. Poor sons of the rich still had a slightly greater probability of marrying (54 percent versus 46 percent), and once married had a 74 percent chance of having at least the median number of children. Perhaps these men benefited from their relationship within a wealthy family in ways not reflected by their own occupational status.

Despite all ecological and large-scale economic variations among the parishes, a man's lifetime reproduction still varied with his, and his family's, wealth. Sons born to poor fathers were most likely to remain poor themselves, never to marry, and to have fewer than the median number of children. Sons of rich men were most likely to become rich themselves. Rich or poor, if a man married, he was likely to have at least the median number of children.

Wealth mattered throughout life, and "richer" men, like "richer" women (and wealthy and high-status men in a variety of other societies), had higher age-specific fertility at all ages. The most dramatic pattern of children's lives with father's wealth is that sons of poorer men, even in egalitarian nineteenth-century Sweden, had almost no chance of becoming wealthy and having more than the median number of children. As figure 8.2 shows, father's wealth is certainly not all that matters: some percentage of both sons and daughters of wealthier men fail to marry and have children. But statistically, wealth and status still mattered, even in late-marrying, monogamous, demographic-transition Sweden.[20]

SEX, RESOURCES, AND FERTILITY

The "bottom line"? Even in monogamous egalitarian Sweden, and even during the demographic transition, wealth augmented fertility. Indeed, wealth, viewed simply as "richer" (upper middle class, lower middle class, land owners) and "poorer" (no land owned, and composed of cottars, migrant workers, or household servants), influenced the life patterns of both men and women (figs. 8.2, 8.3). From decade to decade, as fertility shifted with external conditions, richer families showed less variation in their reproductive success—they not only reproduced more, they did so consistently, and they were less at the mercy of external ecological and social influences.[21] I find it striking that, in a Western, late-marrying, monogamous, relatively egalitarian society, wealth differentials still promoted fertility differentials. When resources became constricted, family reproductive differentials shrank; but whenever possible, individuals (especially men) converted resources into children.[22] Family wealth and status made some life paths far more likely than others. In Sweden, most of the locally obvious variation in wealth-fertility relationships occurred among men. For women, the conflict between investment capability and fertility seems to be sharper than for men, both in this transition society and in traditional societies (e.g., women discussed in chapter 7).

When monetary resources become central to children's success and women shift from traditional maternal patterns to market employment, fertility typically declines, and this may well be exacerbated today, when a woman's resource-earning abilities are finally approaching those of men. Are men's and women's patterns converging? Perhaps. I'll discuss this more in chapter 15.

Once, for humans as for other species, the sheer number of children was the best predictor of a human lineage's success; this may have begun to change in the demographic transition. External conditions that affect the competitiveness of the environment influence the relative potency of investment versus production. *Any* force enhancing the effectiveness (for net reproduction) of increased investment in individual children will have a cost, but may be more effective than large numbers of insufficiently invested children.[23]

Thus, fertility trends with local, reversible patterns (as in Sweden) seem more likely than a singular demographic transition with a singular cause. The primary components of population change (fertility, mortality, and migration) respond to ecological conditions at the individual family level in ways that are predictable and familiar to students of nonhuman populations.

Even in the demographic transition in Sweden, certain predictable ecological rules underlie patterns of fertility, mortality, and migration, although these may be constrained by a variety of cultural complexities and interactions. If these patterns are so clear here, why do they not show up in the standard aggregate data for nineteenth-century Sweden? First, it is a question seldom asked. Second, if we fail to measure the appropriate resources, or if we look simply at aggregate measures or only for conscious decisions as mediators, we may fail to uncover the pattern.

Suppose that poor people have fewer children. Even with the intensive record keeping in nineteenth-century Sweden, more poor people (who had lower fertility) were "lost" to the records than wealthy land and business owners. Now remember that statistically, it is those who fail to reproduce who affect the overall pattern most; these are the individuals lost, and they are nonrandom in the population—they are poor. Then, too, census data are not designed to elucidate information about family lineages, but about households. This has two impacts. First, it increases the proportion of homeless poor who are not counted. Second, within households in most censuses, it is impossible to tell "own" children from stepchildren, from other relatives living in the household, and from nonrelatives living in the household.[24]

In nineteenth-century Sweden, perhaps women exchanged what information they could about avoiding unwanted pregnancies; various forms of fertility limitation, including sex taboos, abortion, and infanticide are widespread in "natural fertility" populations. More than ever before, today the existence of effective, cheap, contraception is bound to complicate any relationship between wealth and fertility. Today wealth still correlates with copulation frequency for men; this used to correlate with children produced, but now it may not.[25]

These are complicated and interacting phenomena; no wonder our typical analyses fail to capture what is happening. I argue in chapter 15 that now, more than ever, we need to be concerned about exactly what to measure and how to measure it. We need to understand the functional patterns if we are to imagine how current population-environment issues—continued population growth combined with increased consumption—will affect standards of living around the world.

FERTILITY TRANSITIONS: WHAT, IF ANYTHING, DO THEY MEAN?

When resources are relatively ineffective in lowering children's mortality or enhancing their competitive success, fertility will be high. Such conditions obtain in many traditional societies, and probably in a good many pre- and proto-industrial societies as well. In several of the traditional societies discussed in chapter 7 (the Ache, the Yanomamö), few physical resources are owned; for these people, status is most important. In none of these societies was there a correlation between children's survival and father's resources; the correlation existed between wealth/status and number of wives, and wives and their children were frequently relatively independent economic units. In Sweden, too, for quite different reasons—and in one of the earliest and most advanced social and medical welfare systems known—a man's wealth also had no effect on children's survival, although it did affect his children's likely wealth and fertility. What if wealth can make a difference to children's survival and success? Then, higher per capita investment directly increases children's chances of surviving and becoming successful.[26] Wealth and fertility remained related in the Swedish demographic transition, even while fertility fell. Fertility shifts were local and reversible and responded to local conditions (to local crop failures, for example, not to a global effect of, say, industrialization).[27]

A very general pattern exists in other species that is relevant here. When offspring must compete for limited resources, parental shifts from *production* of offspring to *investment* in offspring will be favored.[28] Unless parental resources increase, true parental investment, specific to particular offspring, must reduce the number of offspring; one cannot simultaneously maximize both numbers and per capita investment. Typically, then, as investment becomes more crucial to success, fertility (and the range of variation in fertility) will fall. In Sweden, it did not reduce the correlation between fertility and wealth, but under other circumstances it might.

Following this logic, across human societies *complexities in either the ecological or social environment that result in increased effectiveness of parental investment should result in more investment, even at the expense of fertility itself.*[29] Thus, it is not surprising that wealth differentials promote fertility differentials, even in Western societies that are monogamous and attempt to be egalitarian.[30]

The extent to which we can predict fertility shifts as a result of economic conditions or purchasing power will depend on a number of factors, but especially on how much parental investment assists individual children. Technological advances may require more education or better training to enter the labor market, and thus more investment to produce each competitive child. And if even a few families increase investment to enhance competitiveness, the stakes are raised for all families in the population.[31]

Surely the relative costs and benefits of children themselves are influential: better-educated children have a greater chance of marrying a high-status partner, earning a higher income, and migrating than other children.[32] Even the richest family's wealth could be dissipated through continued even investment in large numbers of children. Unequal investment, even in the face of legal mandates for equal inheritance, is unsurprisingly common.[33] When increased investment in individual children enhances their ability to survive, marry, and reproduce, net lineage success can be enhanced by shifting more resources into investment in children (education, savings, health insurance, resource gifts, etc.). Unless there is a net increase in total resources, the allocation of available resources must be into fewer children.[34]

9.

Nice Guys Can Win—In Social Species, Anyway

The real great man is the man who makes every man feel great.
—G. K. Chesterton

Simply stated, an individual who maximizes his friendships and minimizes his antagonisms will have an evolutionary advantage, and selection should favor those characters that promote the optimization of personal relationships. I imagine that this evolutionary factor has increased man's capacity for altruism and compassion and has tempered his ethically less acceptable heritage of sexual and predatory aggressiveness.
—George Williams, *Adaptation and Natural Selection* (1966)

WE STARTED with very simple and general hypotheses about how resources affect reproductive success, and why men and women typically have quite different resource strategies. But complexity has crept in: By the demographic transition society discussed in the last chapter, market forces, governmental rules, and societal mores clearly influenced men's and women's costs and benefits in resource, mating, and marriage decisions. Simple hypotheses about selfish genes, favoring themselves and their copies in kin, can't explain what we see. The world may be nasty, but it is much nicer than selfish genes in solitary animals would make it. As biologist Richard Dawkins put it, "nice guys finish first" under some conditions.

How do simple systems of kinship and reciprocity generate the complex sociocultural patterns we see? Why do we persistently help others, often others who cannot repay us? Do we do so for selfish or selfless reasons? Do we need theories of "group selection" that are more complicated than the striving of individuals and coalitions in their social conditions to explain apparently selfless behavior? In this and the next chapter, I explore these questions before returning to issues of sex differences in coalitions, competition, even warfare over resources.

ARE WE LEMMINGS? A CAUTIONARY TALE

If only we were as selfless as the lemmings! Lemmings, small mousy rodents, breed like crazy and frequently face population crises even more severe than the one some people think we humans are facing now. However, some lemmings, so the story goes, sacrifice themselves willingly in the face of these traumas: they leave their home and soon fling themselves over cliffs, dying so that others may live, thus averting disaster. There's a problem here: the facts are quite different. Ecologist Dennis Chitty, in his book *Why Lemmings Commit Suicide*, referred to the myth of the selfless lemming as "beautiful hypotheses, ugly facts" and raised an interesting question: Why do we get so upset at suggestions that self-interest might be rearing its ugly head? In fact, why do we think self-interest is "ugly"?[1]

Humans are like lemmings, and other living organisms, in some basic ways. They must solve a set of ecological problems—in which individual and group "interests" are likely to diverge.[2] Group living introduces subtlety into the evolutionary rules. We cooperate daily, in many ways. The rule "genetically profitable behavior is likely to persist and spread" must be fleshed out considerably before it can explain much that is interesting, including much of our cooperation and sex differences in that cooperation.

WHEN AND WHY DO WE COOPERATE?

One step in fleshing out our understanding involves honing our definitions. Some interactions have a true genetic cost to the "doer" while others that appear costly are actually profitable to our genes (chapter 2)—much of our helping behavior arises precisely from this fact. Yet our very language makes this distinction difficult to remember. A recent New York Times crossword had the clue "was altruistic"; the answer was simply the word "shared." This common usage creates problems both in theory and in everyday life. Do we mean that costly and cheap sharing will be equally common with friends and strangers? What do we really mean by "altruism"? Any and all sharing? Helping our relatives? Trading favors with our friends? Sharing equally with everyone, strangers and kin alike, without reciprocation?

The distinction between *apparently* costly versus *genetically* costly behaviors lies at the root of much misunderstanding. If we do not distinguish between actions that benefit our genes through reciprocity and kin selection, on the one hand, and true (genetic) altruism on the other, we will not be able to make sense of many human behaviors. Because the functionally important distinction is the effect of behavior on genes, *I will restrict the meaning of "altruism" here*,[3] and will call behavior that incurs a net genetic cost "genetically altruistic" (table 2.1): things like giving your life for strangers under conditions in which no help comes back to your relatives, becoming celibate in a way that doesn't help your family,[4] giving costly help to nonrelatives repeatedly without reciprocation, and so on.

We expect these latter behaviors, termed "genetically altruistic, phenotypically altruistic" in table 2.1, to be rare, and they are. Most of the examples we can think of sound a little silly. Because they benefit reproductive competitors *at a reproductive cost to the doer*, they cannot evolve through ordinary natural selection and are always vulnerable to competition from genetically selfish behaviors.[5] This fact, of course, is more easily seen in simpler situations. In chapters 4 to 8 the male and female strategies of other primates, men and women in traditional and transition societies, appear to make ecological sense.

Hamiltonian kin selection, and reciprocity, mutualism, and their "relatives," should, in contrast, be common because, while they *look* costly, they help their causative genes; in table 2.1 they were termed "genetically selfish, phenotypically altruistic." Thus, many "nice" behaviors—behaviors that look costly—actually help ourselves or our relatives enough to be genetically profitable: nursing children costs calories but is important in successful maternity; helping my neighbor of twenty years when a storm damages his house means he will help me if and when I need him. We are such a social species that to ignore such interactions, to define "self interest" as narrow economic interest, is misleading.

Individuals in many species do phenotypically costly things, and sometimes we do not see reciprocation. Lion females nurse one another's young; since they are relatives, we are watching inclusive fitness maximization. But Mother Theresa was real; she had no children, devoted her life to helping nonrelatives, and did not accumulate benefits to give any of her own relatives. Are these comparable observations? In short, no. Analytically, the problem is to distinguish true genetic altruism (predicted to be rare) both from genetically profitable (though apparently costly) behaviors like helping friends and relatives, and from simple poor observation or inference (as in the lemming case). Often, we *assume* a "helping" behavior to be costly when we have, in fact, no idea whether it is. It is important that both our definitions and our observations are correct.[6]

Matters get more complicated as groups—societies—get larger, more fluid, and more heterogeneous. The complexity of group living in highly social, intelligent primates like humans raises new possibilities. How can we generate societies of Nice Guys? If selection worked to favor the good of the group rather than genes, true genetic altruism would become common, as common as nepotism and reciprocity. As I explain below, there is, in fact, no evidence that any organism has evolved to assist unrelated individuals without reciprocation. But that doesn't mean there must be a dearth of Nice Guys; we do a whole array of nice things for one another. At the most basic level, the help of Nice Guys sorts into three main classes: (1) helping kin; (2) reciprocity, and (3) a set of behaviors such as "by-product mutualism" called pseudoreciprocity. In each, the doer gains some advantage.[7]

In a nepotistic or familial interaction, one individual pays a cost and benefits an individual who also shares genes; there need be no apparent (phenotypic) return because the returns are genetic, as my son so often reminds me. The most general statement of Hamilton's rule (chapter 2) is that there is a cost-benefit trade-off to all behaviors, depending on the extent to which they help or hurt others, and how closely those others are related to us.

Cooperating with nonrelatives can also affect one's genetic success and generate "nice" behaviors. In the simplest sorts of mutualism, the presence or interaction of two individuals is profitable to each—the mutualism is just a by-product of individual profit seeking. Mutualism occurs between cleaner fish and their "client" fish, for example; cleaner fish, usually boldly marked, set up "stations" in obvious places (an isolated clump of coral reef). Client fish come and assume a striking head-down "ready" position; the cleaner approaches and picks parasites off the client's body. The client is rid of parasites, and the cleaner gets a meal. It is even possible for mutualistic interactions to evolve from originally parasitic ones—for win-win to arise from win-lose.[8]

SIMPLE STRATEGIES IN WINNING GAMES

The cleaner fish–client mutualism is a fairly simple kind of cooperation, by-product mutualism. But cooperation is sometimes complex and indirect. One way to gain insight into when to cooperate with others is to simplify the problem into strategic games, in which we can examine individual costs and benefits. Robert Axelrod and W. D. Hamilton explored a classic example, the "Prisoner's Dilemma."[9] They stripped the problem "when does cooperation pay?" to its essentials. In this game each of two players chooses between two options: "cooperating" with the other player, or "defecting." Their payoffs depend on the combination of strategies the two players choose. If both cooperate, they get, say, three points apiece ("Reward," R). If both defect, each receives only one point ("Punishment," P).

The game gets interesting when the two players do not choose to do the same thing. If one player defects while the other cooperates, the defector gets five points ("Temptation," T) and the cooperator gets nothing ("Sucker's Payoff," S).[10] Thus the dilemma: when is cooperation worth trying? If player 1 cooperates, player 2 gets three points for cooperating and five for defecting. If player 1 defects, player 2 gets zero for cooperating and one for defecting. The obvious nasty solution is that whatever the other guy does, it pays each player to defect. The perverse result is that rational players defect, getting one point each, when, had they only cooperated, they could have gotten 3—but the large payoff is possible only if each player were to cooperate. So it pays any player to defect. And this is the key. Cooperate-cooperate is highly unlikely in a single interaction. In fact, Prisoner's Dilemma is a game defined by the fact that $T > R > P > S$.[11]

Of course, many interactions are not one-shot deals. If you are likely to meet someone repeatedly and you can communicate, you both may be able to cooperate and reap the "Reward" of three points. When the players are likely to see each other repeatedly, defection on either side is less likely ("Repeated Prisoner's Dilemma"). In game theory terms, the repetition of interactions changes the value of Reward (benefit, b, minus cost, c); it becomes $(b-c)/(1-w)$, when w is the probability of meeting again. So when w is small (little likelihood of meeting again), R is still not very enticing; but when w is large, the Reward grows.

It is important in Repeated Prisoner's Dilemma that players cannot predict the length of the game.[12] If the game is known to be ten rounds long, the last round is a noniterated Prisoner's Dilemma, and each player will be tempted to cooperate for nine rounds and defect in the tenth—but once that is apparent, it is really only a nine-round game, and the temptation arises to defect in the ninth . . . and so on. The only solution appears to be the uncertain length of the game: if one doesn't know when it's the final round, and if one is likely to meet again, then the best thing to do is cooperate. The "Shadow of the Future" helps keep us cooperative. Discrimination is a key element, for continually cooperating with a defector is a losing strategy. Discrimination can become, even in nonhumans, a "social contract."[13]

In a repeated Prisoner's Dilemma game of uncertain length, what's best depends on the opponent's strategy. Based on an extensive computer tournament followed by analytical proof, Robert Axelrod has argued that a class of very powerful strategies has two rules: (1) cooperate (be "nice") on the first move of the game, and then (2) with nearly perfect regularity, mirror your opponent's moves. The simplest such strategy is "tit-for-tat," in which you cooperate, then do unto your opponent just what your opponent did to you in the last round: cooperate if he or she cooperated, defect if not.[14]

In single-interaction games, the order of the payoffs (is $T > R$?) determines the likelihood that players will cooperate or defect. In Prisoner's Dilemma, the Sucker's Payoff (cooperate with a defector) was the worst that could happen to you, and Temptation (cheat on a cooperator) was the best. Other games have different relative payoffs, and there is a rich array of strategies.[15] We can, in pure game theory, set the payoffs to whatever values we wish. In the real world, additional considerations affect trust and cooperation: whether interactions are likely to be repeated, whether risks are high or low, and whether one is among kin or strangers. Cooperation can be a highly effective competitive strategy, if help goes to kin and / or is reciprocated.

Prisoner's Dilemma makes it easy to see why reciprocity is discussed entirely in the context of long-lived, repeatedly interacting, social animals, in which there are many unknown-length series of repeated interactions.[16] All the examples I can think of also involve organisms that learn, so that a continued "defection" by an individual seems likely to cost the defector: it will be detected, and responded to. If A helps B repeatedly, but B fails to return that help, A can redirect his or her efforts, other individuals in the group can see that helping B is a bad deal, in some species A can tell others—and the stage is set for tracking reputation, for discrimination, and for ostracism.

This sort of discriminatory reciprocity increases the costs of defection in a way not reflected by most games. It is not a particularly human affair; chimpanzees and dolphins, for example, do this regularly. While undergraduates may find it onerous to learn the definition of a Nash equilibrium (if either player changes strategy, he or she is worse off), people, as well as many other social-living species, clearly achieve Nash outcomes without knowing definitions.[17]

Reciprocity may be indirect (A helps B at some cost, B helps C, C helps A, and so on). Indirect systems, however, are hard to evolve, for at any moment, the balance of costs and benefits is likely to be uneven (A has helped B, at some cost; B has not yet helped anyone). If individuals interact only rarely or occasionally, such indirect reciprocity is extremely vulnerable to cheating: perhaps you help me, but we may never again interact and I never reciprocate. If we are close relatives, your defection may still help your genes, and some of mine. However, any individual who consistently helps nonreciprocating nonrelatives will have little effort left for kin.

Elegant examples of reciprocity can develop when risks are high and interactions are repeated. Vampire bats (long-lived, repeated interactors), for example, face uncertainty as they seek blood from livestock, and if a bat fails to eat, it can be in danger of starving after as few as sixty hours. These bats tend to drink more than they need and share the surplus with unlucky neighbors. Perhaps this system is facilitated by the fact that the same physical amount of blood makes a smaller "hours-to-starvation" difference to a well-fed bat compared to a hungry bat: the cost of helping is low, and the benefit to being helped is great.[18]

Simple reciprocity, without additional considerations, is limited in its ability to explain much of what we see in large societies. When risks are high and individuals move about so that the next interaction is uncertain, expensive helping behaviors may occur only or primarily among kin, or individuals may mirror the behavior of others, for example, in a tit-for-tat manner: I'll start by cooperating, but if you default, I will, too. As biologists Anne Pusey and Craig Packer point out, simple, clear-cut, direct behavioral reciprocity is probably far rarer in the real world than we realized.[19] In many systems, cheating is further discouraged by the rule: punish not only cheaters but also anyone who fails to punish cheaters.[20] This is one of several cultural additions enhancing the spread of cooperative behavior.

The idea of Tit-for-Tat (TFT) is familiar, but it does not seem to be really common, at least in its pure form.[21] It can be beaten, not only by defection but by surprising variations. TFT is very vulnerable to mistakes or imperfect information, both rife in the real world. If I think you have defected, I respond by defecting, and a downward spiral has begun; but if I am wrong, I have cost myself. Mathematical modelers Martin Nowak and Karl Sigmund developed a probability-driven tournament in which there was no certainty, only a likelihood, that your opponent acted in a certain way. The winning strategy? "Generous" Tit-for-Tat (GTFT), which "overlooks" single defections about a third of the time. GTFT wins over TFT (which can be too harsh) and does well against a teachable opponent, and in a noisy environment in which mistakes occur. It is also less vulnerable to exploitation than Tit-for-Two-Tats (which always forgives single defections and is resolute in punishing two defections). Now, a complication: GTFT holds its own against nastier versions, but is easily invaded by really sweet variations like "Always Cooperate" (AC)—but of course, as AC increases, it will lose to any defecting strategies.

So among these strategies, there is no evolutionarily stable solution. Is there no chance for stability? Add one slight improvement in mimicking reality: memory. The rule is fairly simple: remember what your opponent and you did; if the last thing you did let you win, repeat it; if you lost, change strategy. This sounds remarkably like what people do, and describe doing, in personal relationships.[22] It is evolutionarily stable in these simple two-actor games.[23]

In all the games and models so far, an individual's payoffs (even if constrained, as in coercion) are what predict cooperation or defection. Kin, short-term mutualists, long-term repeated interactors are likely to cooperate. Natural selection favors genetically profitable behaviors. Even our powerful human intelligence may have evolved in the context of cooperative competition for resources and mates.[24] These examples are usually very simple, two-person games, and we face the same difficulty as we do in genetic models: we must, to understand individuals in societies, build more elaborate models (though still as simple as we can get away with, and match what we observe).

FROM FAMILY TO DYADS TO GROUPS TO CULTURES

In the real world, some relatively simple additions to the two-person games may make a world of difference to promoting cooperation in large groups. First, do nothing unusual; remember what others did, and copy both your winning behavior and the winning behavior of others. Thus, copy common behaviors (they are likely to be profitable if they have become common), and copy behavior that has clear profitability (this could be judged indirectly as well: copy high-status winners even if the behavior you are copying is not obviously the source of the status). Second, watch others; do not offer cooperation to defectors or to those who give to defectors. Third, punish cheaters. This means that cheaters, in stable-membership societies, lose more than the game at hand (this strategy, elaborated, involves reputation issues). Finally, advertise your future behavior ("commitment"), both negative (if you cheat me, I will make you very, very sorry) and positive (I am willing to make great personal sacrifices to belong to your group)—and exhort others to make expensive commitments as proof of future cooperation.

These relatively simple additions can lead to important influences: (1) strong and widespread norms, (2) strong socially imposed costs and benefits, (3) punishment of cheaters (including those who do not enforce the norms), and (4) low cost of imposing the norms. In fact, any behavior can become stable if these conditions are met; thus we can see great variation in social norms.[25]

It is clear that organisms in long-lived social species, including humans, are likely to do things that benefit others, either because the actor also benefits or because there is some likelihood that there will be future interactions. "Watchers" may assess us by watching how we treat others, so that our future costs and benefits with additional individuals may be affected. Cultural norms become strong forces influencing how people can profitably act. We still know little about how particular rules come to exist in particular situations, though we can say, I think, that in relatively small and very stable groups, the costs and benefits will be more predictable and less avoidable than in fluid groups. I will return to these issues in the next chapter.

THE GROUP SELECTION MUDDLE

There are four sets of arguments that have been invoked to explain how we could have evolved to be truly (genetically) altruistic (table 9.1). They work quite differently—their functional pieces and predictions are quite distinct—but all are confusingly referred to by the term "group selection." They include (1) interdemic selection, (2) cultural transmission and culture-gene interplay, (3) coercion, the imposition of costs on some individuals by within-group coalitions, and (4) Wynne-Edwards's benefit-of-the-group formulation.[26]

Classic Interdemic Selection and Its Descendants

Simply stated, this theory predicts that if certain conditions are met, the structure and composition of groups can influence the frequency of various genes (table 9.1).[27] The conditions under which deleterious alleles (such as one for genetic altruism) could theoretically spread are very restricted: small, viscous (i.e., largely inbred) groups, close to each other (so you can assume that the ordinary selective pressures are similar); occasional "mixing" followed by restricted interchange. Within each group, the more advantageous (selfish) alleles increase at the expense of altruists; thus, the timing and type of mixing between groups are critical. Because differential success occurs at birth and death (and the group equivalents of these), and because the turnover of individuals within groups is typically more rapid than the turnovers of groups, a bias exists for selection to be more powerful at the level of individuals.

This model is logical but unlikely to be a strong force generating

TABLE 9.1.
Summary of selection theories.

	Natural Selection			Social Selection			Group Selection	
						Cultural Transmission Boyd & Richardson Cavalli-Sforza & Feldman Lumsden & Wilson Durham	Interdemic Sewall-Wright Sloan-Wilson Hamilton Wilson	Wynne-Edwards
					Coalitional Alexander Irons			
	Individual Selection	Sexual Selection	Kin Selection	Reciprocity				
OBSERVATIONS								
	Organisms seem well suited to their environments.	Individuals sometimes have extremely risky, dangerous behaviors.	Individuals sometimes behave in ways that seem to be a cost to themselves but benefit other individuals.					
HYPOTHESES								
	Heritable variation occurs. Environments are not infinite; thus there is competition among organisms. Some variants will be more successful than others, and will leave more offspring. Thus, over time, the proportions of the population comprised by different variants will change.	In sexual species, the evolution of anisogamy (unlike gametes) favors specialization by individuals into either mating or parental effort. The reproductive return curves for mating effort involves a high fixed cost (much expenditure before any return), resulting in high variance—which favors risky competitive behaviors.	Relatives carry genes identical by descent. Thus, acts that appear to cost the performer may, in fact, profit the performer genetically if $rb - c > 0$, where b = benefits to recipient, c = cost to donor, and r = degree of relatedness.	Individuals may profit genetically by assisting non-relatives when the assisted individual in turn reciprocates. Conditions that must be met are (a) individual recognition, and (b) repeated interaction. Such reciprocity is most likely to evolve in species with long-lasting social units.	Within-group coalitions can impose their interest, constraining the actions of others.	Mechanisms of cultural and genetic intergenerational transmission differ. Cultural transmission and selection may favor traits that natural selection alone, acting on genes, might not favor.	The organization of individuals into groups can affect gene frequency. Conditions that must be met are: (a) small viscous (restricted gene flow) populations; (b) close together (same selective pressure); can replace each other; (c) population turnover must be rapid, approaching the rate of turnover of individuals.	Groups have characteristics that are not simply the sum of individual characteristics. Group selection operates on the groups and is thus very different from individual natural selections. If there is any conflict, group selection should prevail. Altruistic behavior should be common.

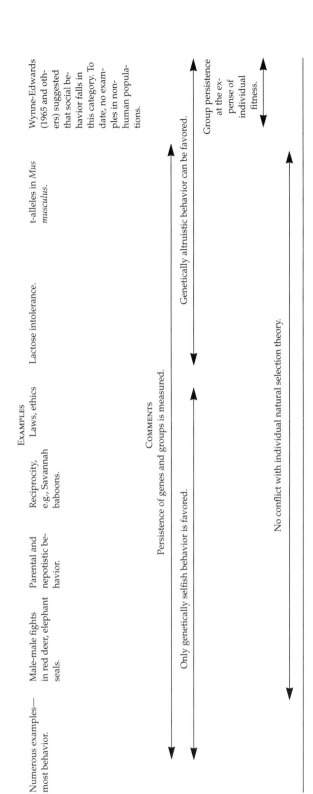

EXAMPLES

Numerous examples—most behavior.

Male-male fights in red deer, elephant seals.

Parental and nepotistic behavior.

Reciprocity, e.g., Savannah baboons.

Laws, ethics

Lactose intolerance.

t-alleles in *Mus musculus*.

Wynne-Edwards (1965 and others) suggested that social behavior falls in this category. To date, no examples in nonhuman populations.

COMMENTS

Persistence of genes and groups is measured.

Only genetically selfish behavior is favored.

Genetically altruistic behavior can be favored.

Group persistence at the expense of individual fitness.

No conflict with individual natural selection theory.

indiscriminate altruism.[28] In game theory terms, it is quite possible for "always cooperate" to lose within a group but for groups of "always cooperate" players to grow and spread more rapidly than other sorts of groups. The catch is that this is true *only if no defectors sneak in;* this is a fragile system. As anthropologists Richard Boyd and Peter Richerson summarized, versions of this model of group selection intended to show how altruistic genes can spread "clearly [are] not verified by the data concerning ethnic cooperation among humans."[29]

Culture-Gene Interactions: Coalitions and Coercion

These interactions[30] include the derivation of the "additional rules" suggested above: conform, watch and punish cheaters, advertise your commitment to action, and exhort others to commit. In any group-living animal, others in the groups are part of the selective environment, with some interesting outcomes.

Individuals within a group may form coalitions to gain advantages. We find it useful to gather in small groups working together to advance our interests. Think of laws, for example, which are inflictions of constraint on individual behavior by coalitions of others in the larger group.[31] Less formally, ethical and moral systems do the same thing; for example, religious groups can make demonstration of conformity, and sometimes costly commitment, essential to gaining entry. In many situations, advertising commitment can be important, from painful cicatrice scars (reflecting initiation into manhood in some societies), to military recruits' haircuts, to the "cheap talk" of bumper stickers (this last is an inexpensive, easily faked advertisement, and we give it little currency).

The fairly simply rule "copy those around you who are successful" (above) can further advertise commitment and belonging and promote group cohesiveness. Even in simple situations, copying is unlikely to be random.[32] Consider the Mukogodo (chapter 7); by any objective measure, they care for their daughters better than their sons. Yet they claim to prefer sons. Why? The Masai, who hold higher status among the local peoples, prefer sons, and the Mukogodo claim to be Masai (the Masai often reject this claim). So the Mukogodo are "copying" (at least in their statements) high-status people. But their behavior reflects the selective reality that among the Mukogodo, daughters can rise in the world better than sons.

When groups consist of close relatives who share genetic interests, or when group members are threatened by external forces such as competing groups, individual and group interests converge. In a small group of hunter-gatherers, for example, sharing one's catch looks costly—but failing to share one's catch can literally be deadly. Coalitions of individuals within any group can change the costs and benefits of actions like sharing, especially when memory and punishment exist.[33]

Cultural rules and those of natural selection interact; we have a "dual inheritance" of both genetic and cultural information. Because the rules for cultural and genetic transmission may differ, even under the same selective pressures, we find that different genetic equilibria could result from culturally transmitted versus genetically transmitted traits. That is, cultural and genetic selection can interact, and culturally transmitted preferences can affect gene frequencies— behaviors can be favored in social systems that we wouldn't expect in the absence of social pressures.[34] So, indeed, genetically altruistic behavior by some individuals can, at least for a time, be favored or forced.

The Global Group: Wynne-Edwards
and Population Regulation

The biologist V. C. Wynne-Edwards made a truly distinctive argument about how selection acts on individuals in groups that is relevant to other, more likely cultural evolution arguments.[35] He noted that groups have traits such as dominance hierarchies that are not a simple statistical sum of individual traits; he argued that in any conflict between what was good for the group and what was good for the individual, the good of the group would prevail.[36] Thus, he expected individuals routinely to pay reproductive costs for the good of the group. He did not count genetic costs, and in his view, group selection would swamp ordinary natural selection.

To date, concerted work has uncovered neither logical nor empirical support for Wynne-Edwards's argument: the empirical evidence suggests that any system depending on individuals to perform genetic sacrifices for the benefit of strangers collapses quickly as the altruists are out-competed and out-reproduced by reproductively selfish individuals. So why even consider such an arcane and proven-incorrect argument? There are two reasons. First, even though Wynne-Edwards's formulation is biologically naive and easily dismissed after examination, it is widely accepted in parts of the social and management sciences today.[37] Second, the most recent disinterments of "group selection" have, I think, insufficiently clear definitions to distinguish among the above arguments.[38] There is a distinct and disturbing possibility that policy suggestions will be based on unrealistic assumptions that we all regulate our personal behavior for the group's good.

ALTRUISTS OR GOOD NEIGHBORS?

The bottom line? The more genetically costly a behavior is, the rarer it is likely to be. Culturally transmitted traits (like celibacy among Shakers) that are clearly costly tend to diminish over time.[39] But culture interacts with natural selection in myriad ways, and while the Shakers will go extinct, many other cultural practices influence both our behaviors and our gene frequencies. We are probably using heuristic rules of thumb like "copy my successful neighbor," "hope to have a child if I'm doing better than my parents or peers," or "be especially demonstrative of my commitment." These will be local and variously successful. Even after we have checked to be certain we are correctly assessing costs and benefits, behavior we think is selectively stupid may take a long time to decrease and disappear.

If group good at the expense of individual fitness were relatively powerful, we should expect genetic sacrifice to be common. In fact, it is so rare as to be undetectable in populations of any organism. Yet the term "group selection" has been muddled to become almost meaningless; all the arguments above have been called group selection although they have almost nothing in common.[40]

The general consensus is that individual, kin, reciprocity, and social selection will be relatively much stronger than inter- and intrademic selection (in which metapopulation structure creates differences in genetic representation) under all but the most restrictive conditions—and especially so in fluid modern conditions.[41] Since George Williams's classic work, most biologists have come to agree that genes, carried about by individuals and shared by relatives, are what matter.[42] As biologist Graham Bell put it succinctly: "Group selection is a fallacious concept when it is held to cause the evolution of characteristics that benefit populations, or species, or communities, as a whole, without distinction of ancestry."[43]

COOPERATION AND FREE-RIDERS

We expect, then, that genetically costly behavior will be uncommon. Even if we can create groups of altruists who may last longer in dire straits than selfish, or simply reciprocal, groups, this advantage will persist only so long as altruists alone comprise the group and outsiders cannot also profit. If a cheater—a "free-rider"— invades and takes benefits but does not do the costly help, the altruists face extinction.

Real-world examples of this difficulty abound, from traditional societies to modern global warming and acid rain. In many traditional societies, for example, the "fruits of the forest"—berries, mushrooms, wood, and so on—are common-pool resources: any villager can gather them.[44] If the village is small and close-knit, if villagers have lived there a long time, if exploitative cheaters are easily detected and punished, and if there are no external markets, the villagers' use is liable to be shared for subsistence, and to be sustainable over long periods. But if the village has many transients, or if it is hard to catch and punish cheaters, or if anyone can convert forest products into lots of hard cash and move away, "take the money and run" defection is likely.[45] In these problems as in most, reciprocity and local optimization are the rule.

Today, we see new and larger-scale but similar problems. Ozone hole depletion, acid rain, global warming, open-ocean whaling declines—all are common-pool resource problems in which free-riders do best. The difficulties of solving large-scale free-rider problems among nation-states are truly daunting. We can, for example, promote CO_2 emissions standards to reduce the degree of future global warming around the world; the difficulty is that such action requires some very local costs now for the sake of some very dispersed future benefits—so it's hard to get cooperation (see chapter 15).

In today's large heterogeneous societies, individuals are not all subject to the same ecological or economic constraints. Cultural rules and the actions of others are important constraints in our environment. How we act daily in different situations to accomplish genetic selfishness varies greatly and can be strongly influenced by others, directly and indirectly, in the cultures in which we live, as we will see in the next chapter.

THE FAR SIDE © 1980 FARWORKS, INC.

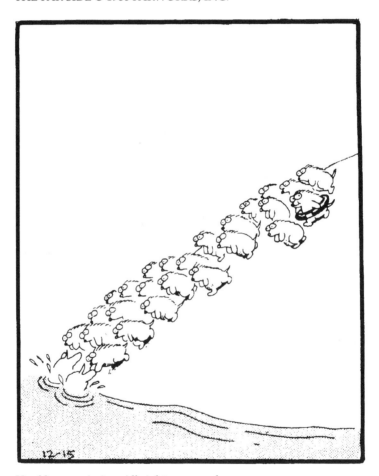

10.

Conflicts, Culture, and Natural Selection

Tragedy is when I cut my finger. Comedy is when you walk into an open sewer and die.
—Mel Brooks

KATHERINE HARRISON was a witch. After the death of her rich and powerful husband, she was accused of witchcraft, based on events occurring years before. When accused, she was a rich widow who chose not to remarry, and she had no sons to protect her property. She continued to be accused for years, and was even forced to leave her community, until, in 1670, her older daughter became engaged to a man from a powerful family. The future father-in-law, previously an attacker, became her protector, and the accusations ceased.

The demographic and economic particulars of witchcraft trials show a pattern that is a logical, if curious, example of conflict over resources and reproduction in a particular culture. The communities in which accusations of witchcraft flourished were communities long torn by internal strife. Witchcraft accusations often originated in property disputes. Women owned almost no property, but they were three times as likely to be accused of witchcraft, seven times as likely to be tried, and five times more likely to be convicted of witchcraft than men.[1]

Property disputes between men tended to go through the ordinary court system, so why did property disputes involving women segue into witchcraft accusations? Land was becoming a scarce resource in colonial New England, concentrated in few hands, and the circumstances under which women could hold property were limited. Suppose a woman's husband died and she had only young sons or no sons at all. Men's routes to gaining the property she held were limited: marry her or remove her. Two things appear to have mattered: (1) Did the woman have powerful male protectors? (2) Was she of marriageable age—or did she have low reproductive value?

Unless they were single or widowed, accused women from wealthy families could be fairly confident that the accusations would be ignored; women alone, with no powerful male protector, were likely to be accused, tried, and executed.[2] Particularly if she had no sons and her daughters were the inheritors, a woman without a powerful male protector became very vulnerable. If she were young enough (of sufficient reproductive value), she would be courted; if she were older, she was likely to be accused of being a witch. Even a rich woman without a husband could be vulnerable: because of the way inheritance laws worked, widowhood did not bring women independent economic power.[3]

Most "witches," except in the heat of the Salem outbreak, were women who owned some property but had no strong male allies. The tales of witchcraft are rich and varied, but a significant pattern holds true: women who held resources alone, and were not likely marriage candidates because their reproductive value was low, were significantly more often accused, tried, and executed as witches than others.[4] Men's and women's interests were in conflict over resource allocation. The history of witchcraft is a fascinating set of interactions among genetic self-interest, coalitions over resources in that context, and cultural and historical particulars such as social norms. Moral systems are obviously far from immune to perversion by the strong to meet their own ends. I suspect, however, that with some thought we can predict much about the direction of such twists, even in quite particular situations.

COOPERATION, COMPETITION, AND GROUPS

Conflicts of interest, whether between parents and children, in children's games, or at legal trials, are rife at many levels. Why then is rampant self-centered behavior not apparent at many levels? As we began to see in the last chapter, there are several answers.

First, in many cases, such behavior exists. Mice are a well-known example of simultaneous conflicts within and between individuals. Male house mice may have a gene, the t-allele, that causes sperm with competing alleles to clot up: the sperm are at war. Males who are heterozygous for *t* (and should make 50 percent normal sperm, called +, and 50 percent *t* sperm) make 85 to 99 percent *t*-bearing sperm—that is, *t* wins.[5] But since males who are *t-t* either die or are sterile, *t*'s victory within the individual in heterozygous males results in disaster for homozygous males.

The *t*-allele is a rare exception, for obvious reasons. Within an individual, there are usually limits to the profit an utterly self-interested gene can garner. Think of a mutation resulting in cancer: if it kills its bearer, it too dies (although even here, cytoplasmic warfare and genes like the *t*-alleles remain lethal exceptions). The phrase "the parliament of the genes" and the prevalence, for example, of suppressor genes reflect the fact that there are common interests shared by the total group of genes (keeping their bearer alive, healthy, and reproductive) that cause coalitions of genes to prevail over "outlaw" genes.[6]

Some of the same difficulties face groups of genes sharing a body and groups of humans sharing resources. In this chapter I want to explore how cultural rule making and transmission might play out, and how group living has shaped the social nature of our intelligence.

Working Out Our Conflicts: Moral Systems and Group Life

The parliament of genes works because there really is a unity of interests once genes are stuck in the same body: what destroys other genes is likely to be self-destructive. Among groups of individuals, as we saw in the last chapter, things are more complicated. Several forces appear to limit the degree of self-interest that could harm others. Kin selection, reciprocity, mutualism, and coalitional restraints such as laws and social mores can limit rampant individual self-interest.[7]

For an individual in a group, other individuals constitute a strong selective force; in many situations *other individuals* are the strongest predictor in the environment of the costs and benefits to any action.

Coalitions of powerful individuals can make it really costly to display unbridled short-term self-interest. Laws, mores, taboos, shame—all are mechanisms to help subdue self-interest in the interests of tolerable group living (a cynic would say, in the interests of keeping your self-interest from impinging too much on my self-interest).[8] The social contract can be powerful.

The more we learn, the more likely it seems that the biological underpinnings of human group living strongly influence moral systems.[9] Years ago, biologist Richard Alexander first identified the pressures of group living—and the resulting between- and within-group competition and cooperation—that underlie many of our cultural models. If our behavior toward our fellow humans is at all strategic, as behavior in general seems to be, there should be some common patterns (e.g., rules restricting cheating within the group) and considerable variation in other patterns (rules about who can choose marriage partners).

Now here I am stirring up a mare's nest, no doubt, and I urgently refer you to the growing literature on this topic. I simply want to note that (1) the "naturalistic fallacy" (imagining that what *is* reflects what *ought to be*; chapter 2) is a particularly dangerous trap in this inquiry; (2) some patterns are already predictable as being universal except in pathologies (parental and kin caring, not murdering innocents within the group); (3) men and women are likely to hold somewhat different beliefs in societies in which their roles, costs, and benefits are different; and (4) within-society differences seem likely when people's costs and benefits diverge; thus small kin groups will teach different and more uniform mores from large highly mobile societies. (Perhaps because of the importance of small kin groups in our evolutionary history, there seems likely to be a bias toward retaining "small-group" sorts of rules.)

Lawrence Kohlberg structured a progressive list of moral "stages" he suggested people went through (from stage 1, "preconventional," in which "right" was whatever didn't get you punished, to stage 6, in which one upholds universal ethical principles as a matter of commitment and principle).[10] His thought was that moral sentiment developed toward stage 6 as a matter of human maturation, though some of us never made it. Not surprisingly, not only children scored relatively "early" on Kohlberg's scale, but so did adults who live in small, traditional, kin-based groups with competition from outsiders. I am far from certain that I would assume from this a certain moral "progression" as Kohlberg did. The ecology of social behavior is far different when you live among your relatives and the same friends all your life, then when you spend your life in a large and fluid society, and the rules should also be different.

Not only does social complexity matter, but, once again, what people say may not reflect what they do. It is easy to say that we believe in A, or B, or in environmental protection, or civil rights in China, but the real test is what we will sacrifice to achieve those ends. And here we are likely to discriminate; while all of us are in favor (perhaps) of healthy children and a clean environment, what we are willing to pay personally to get some defined version of these goals is less clear.[11]

I'm Committed, Are You? The Deception and Manipulation Problems

What you and I are willing to pay for our own benefits compared to others' reflects an important kind of social selection: the "commitment problem." Because we live in groups, and because we, individually, do need help from our neighbors to thrive, we profit from discerning who is reliable, who will give us what we need (if it's cheap, if it's expensive; if we reciprocate, if we do not, etc.). Exhorting others to expensive commitment, requiring costly displays of commitment by outsiders, and forcefully proclaiming our own commitment to the group are common strategies.

In many subgroups in most societies, from military training to fraternity initiations, from financial dues to physical sacrifice, we require proof that new members will do what's required when the time comes and will help others in the group. As in sexual selection (in which females are likely to require expensive, unbluffable displays of quality and willingness to commit), in social selection we frequently require evidence of a willingness to be a good group cooperator, even—or especially—when that is likely to be expensive.[12]

Much manipulation and deception centers on commitment. The concept of "Machiavellian intelligence"—that we evolved to manipulate our cooperators and competitors—is highly relevant, and it represents a fascinating area for new work. When we leave a 15 percent tip in a restaurant, do we do so because we are altruists? Do we do it because we can use such "cheap talk," in game theoretic terms, to advertise consistently our goodness?[13] Or is it because we have internalized such cheap strategies, since they are more convincing than having to think about when and how expensively to advertise our goodness? Testing among such alternatives will be difficult but exciting.

Intertwining Cultural and Natural Selection

Cultural transmission is a great complication, far more than simply one of several possible mechanisms to shape coalitions and conflicts and to foster socially acceptable behavior. Consider: genetic inheritance can only be vertical and in one direction—down from parent to offspring. But cultural transmission has several possible routes: not only do we teach our own children, but we can learn from our children, our children can learn from nonparental adults (we formalize this in the school system), and we can learn from others in our own generation.[14]

Like other human complexities, cultural transmission has antecedents in other species. It is true that many of these have very simple social lives, and I doubt that any of us would imagine that thinking in a biological framework alone could tell us what we need to know about human cultural diversity; our framework would be woefully incomplete. Yet other species certainly have intergenerational transmission that interacts with environmental conditions to change gene frequencies; though the cases are surely simpler, perhaps they can show us something about how genes and environment can interact.

For example, optimal foraging theory postulates that foraging efficiency increases relative reproductive fitness. In ground squirrels, optimal foragers survive better and have more offspring than nonoptimal foragers.[15] Foraging optimality is heritable: babies are more like their parents than like others in the population. However, learning is important. Heritability is about 60 percent genetic; the other 40 percent of the parent-offspring correlation in foraging optimality comes from babies foraging near their mothers and learning what to eat. The functionally important facts are that heritable variation exists; that one can predict, in a specific environment, which strategies (learned as well as genetically transmitted) ought to result in an increased reproductive fitness for their possessors, and an increased proportion of the possessors in the population; and that one can test and falsify these predictions. Humans, of course, are extreme in their reliance on culture as a means of responding adaptively to environmental pressures.[16] Although many cultural practices appear superficially to work in ways that enhance the success of the practitioners, this is an area of inquiry that isn't very well organized or complete; we could use a really critical assessment of cultural practices.

Several groups of scholars have quantitatively studied the interactions of inheritance through cultural and natural selection. They have different foci, but some generalizations emerge.[17] When the two modes of transmission interact, the result can be an equilibrium in gene frequencies that differs from those expected from ordinary natural selection alone. Humans, like other learning animals, may have been influenced by natural selection to accept socially transmitted ideas that "increase pleasure, reduce pain, reduce anger, reduce fear or increase cognitive consistency," and these pressures, arising from our group members through social as well as physical mechanisms, have shaped us.[18]

The course of cultural evolution is altered by the decisions people make as they learn what behaviors to adopt and decide whom to imitate.[19] Anthropologists Robert Boyd and Peter Richerson concentrate on the formal analysis of dual—cultural and genetic—inheritance.[20] They suggest two main decision-making forces: "guided variation" (because individuals acquire information socially, there can be rapid cumulative change that may be adaptive), and "biased transmission" (people do not imitate others at random). Because these changes can be rapid and cumulative, cultural transmission has a particular advantage in environments that change moderately rapidly.

Cultural transmission is like genetic transmission in the sense that information about how to behave is transmitted from individual to individual, but because cultural transmission can be both vertical and oblique or horizontal, a practice can change almost instantly. Nonetheless, vertical (parent-offspring) transmission remains important in cultural models; it is enhanced by assortative (non-random) mating in models of the transmission of sign language. Among African Pygmy hunter-gatherers, 80 percent of specifically-identified skills were learned from parents. This is likely to be a typical pattern in many cultures, though empirical testing needs to be done. It has been found, for example, that among Stanford undergraduates there is strong parent-offspring transmission of political and religious affiliations.[21]

But problems remain. If culture were purely ideational (the piece of culture of most interest to many workers), we might have fewer problems. But people live in groups that experience conflicts, and culture contains material aspects. One general class of model incorporating this aspect suggests that, in general, individuals strive for rewards that have typically correlated with reproductive success. What if we view "culture" as not only the context of human action, but also as a tool for social manipulation? Perhaps if we recognize individual conflicts of interest and think about cultural manipulation, we can gain further insight.[22]

Social and Lactic Acid Cultures

Cultural and natural selection interact in a potent way and can influence things we seldom think about. Consider the physiology of milk digestion. Around the world, there is systematic variation in how well adults can digest milk products. Many adults lack the ability to digest fresh milk; they have no lactase (which allows digestion of lactose), and so are intolerant of lactose. Anthropologist William Durham, comparing alternative hypotheses, found a strong interaction between cultural practices and genetic inheritance.[23]

The genes responsible for adult lactose absorption have evolved to high frequencies in populations that have a tradition of dairying and fresh milk consumption and live in environments of low ultraviolet radiation (where vitamin D and metabolic calcium are chronically deficient). Populations without dairying traditions, but with low UV exposure, have low lactose absorption (e.g., Eskimo and Saami groups, who live at high latitudes with low UV but have diets rich in vitamin D). Similarly, populations with dairying traditions who live at low latitudes (high UV), as in the Mediterranean, have low lactose absorption. Milk-dependent pastoralists, in contrast, no matter at what latitude they lived, had high lactose absorption as adults. Thus, cultural practice has mediated selection on ability to absorb and digest lactose.

Religion, Government, and Sex Ratios

Worldwide, the average sex ratio at birth is about 105 male births to every 100 female births. But sex ratios at birth among orthodox Jews and the recent "one-child" policy of China (where widespread cultural preference for sons exists) provide examples of cultural-natural selection interaction.

Among orthodox Jews, marital intercourse is prohibited during menstruation and for seven days thereafter, and the husband is not to masturbate or seek other sexual outlets. At the end of the seven days, the wife takes a ritual bath, and the couple is directed to have intercourse at that time, and twice a week during the rest of the month, with the exception of men in unusual occupations. There is additional advice if the couple wishes to conceive a son: intercourse should take place twice in succession.

It is difficult to obtain birth sex ratios for orthodox Jews independent of nonorthodox Jews, and conception biases are certainly difficult to measure, for the obvious reason that important parameters are difficult to control. Nonetheless several things are true. Y-bearing sperm, which combine with the egg to make an XY (male) fetus, are slightly pointier-headed (hydrodynamically better) than X-bearing sperm; they are also smaller, with fewer resources to stay alive if the egg is not immediately ready. As a result, in humans as in most mammals, conceptions close to time of ovulation tend to be male-biased. The orthodox cultural practice of abstinence for about twelve days per month, combined with frequent intercourse near the time of ovulation, appears to interact with biological biases in conception probabilities: sex ratios for Jews in a number of traditionally orthodox locations historically average 137 males / 100 females, while for nonorthodox Jewish populations, and nearby secular populations, they average 105, the worldwide average.[24]

Historically, throughout much of Asia, a widespread preference for sons has existed, reflected by the old Chinese proverb: "It is better to raise geese than girls." But as population numbers have increased worldwide, so has concern about population growth. In the early 1980s, the government of China instituted its "one child" policy in an attempt to slow China's population growth rate. Couples were restricted to one child per family, with some exemptions.[25]

The cultural history of son preference has interacted with the limits on family size, and possibly with marriage preferences.[26] The proportion of families with only one child did increase, and the birth sex ratio became more male-biased. An obvious first thought would be that perhaps couples about to have their first (and possibly only) child might engage in sex-specific abortion and infanticide, but this was not generally the case. First, the real concern over an infant's sex was in rural areas, where parents preferred boys to work in agriculture and therefore might have interest in sex determination, but in these areas the technology for sex determination was largely absent. In the cities where the technology existed, cultural son preference was not so pronounced. Further, the government decrees were seen by most people as perhaps "good" for the whole nation in terms of population limitation, but as running counter to one's own family's interests. Thus many people continued to have more than one child, even when that carried financial, and sometimes social, costs.

Occasionally, an effect could be seen in first births: in two provinces in 1985, the sex ratio of children in single-child families soared to over 129. But it was primarily in *later* births that the sex biases became most pronounced (fig. 10.1). In a nationwide study in 1989–1990, the sex ratio of first births was 105.6, right at the worldwide average, but the sex ratio of later-born children depended on how many older brothers and sisters already existed. For example, the sex ratio of third-borns when there were two older sisters was 224.9 males per 100 females, and the sex ratio for third-borns with two older brothers was 74.1. Some daughter preference did exist when several older brothers were already born, and the "empty" spaces in figure 10.1 suggest that families do avoid some patterns of birth numbers and sex ratio.[27]

Religion, Fertility, and Celibacy

The interactions of natural selection and culture can be complex.[28] Ideas, attitudes—all can be transmitted between individuals as "semiautonomous units" and may, but need not, be adaptive.[29] Individuals in any society can choose to do things that are detrimental to their survivorship and reproduction—but whether such patterns

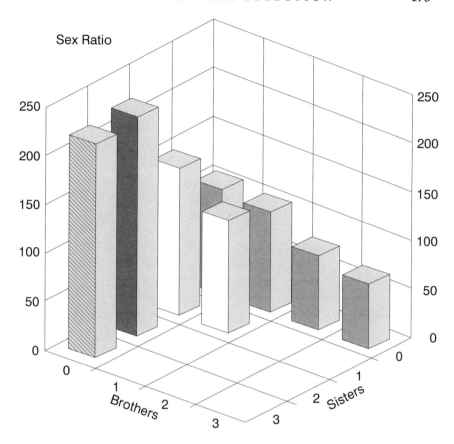

Figure 10.1. In the face of China's one-child policy during the 1980s, the sex ratio of births changed, though the impacts were seen not in first births but in the sex ratio of later births, in response to the sex and number of existing children. Later births in families with more than one existing daughter were extremely male-biased; later births in families with several sons were somewhat female-biased.

will persist and spread, becoming and remaining the mode for longer than a generation or two, depends on how detrimental the behavior is. For this reason, in the coevolution of genes and culture, we do expect that, by and large, natural and cultural selection will interact to favor behaviors that benefit those who do them.

It doesn't surprise us, then, that most stable cultural institutions with a long history do not promote genetically costly behaviors among their members. Consider the positions of religious institutions on fertility: most religions, unlike the Shakers, have rules that urge the reproductive success of their members.[30] Mormonism strongly promotes both financial striving and reproduction (God created all human souls long ago, and they must wait to be born, so having children is an altruistic act for these souls); polygyny was a long-time Mormon social norm, and it may still be practiced in some locations of the United States today, despite federal laws against it. The Hutterites, who are used as the standard among demographers for maximum "natural" fertility, prescribe that all adults in a group must assist in the raising of children. This expanded and shared adult care, in small and close-knit groups, is quite possibly what allows the very high Hutterite fertility of twelve to fourteen children per woman.[31]

But what of the celibacy required of Catholic priests?[32] While fertility is promoted for parishoners (even if indirectly, through abortion and contraception bans), men and women bound for religious life are required to be celibate, something that is an issue of conflict within the church today. Here is surely a case of imposition of rules by more powerful group members on the less powerful.[33]

The rules of celibacy in the Catholic Church appear to have had their origins in property disputes. By 1139, under Pope Innocent II, priestly marriage was forbidden, but not until the dictates of the Council of Trent in 1545–1563 went into effect were married men no longer able to become priests (widowers, however, still could). If families had been small, it might have been more difficult to require celibacy of priests, but giving later-born sons to the church often helped both the church and the family in times of very large families when too many sons complicated inheritance.[34]

When celibacy rules were not absolute, they were typically easier on the younger and lower-status clergy, who had the least property open to dispute between the church and possible heirs. Yet high-status men in the clergy could contravene the rules. Rank, as ever, tended to have its privileges: for example, Rodrigo Borgia, who became the Renaissance pope Alexander VI, not only was married, but had several children by his mistress Vanozza de' Cataneis, whom he brought openly to church celebrations. His son Cesare was his oldest son by Vanozza, and in the traditional manner was destined from youth for a career in the church.[35]

Today, with very small family sizes, the church is having trouble recruiting priests in the United States. A major reason appears to

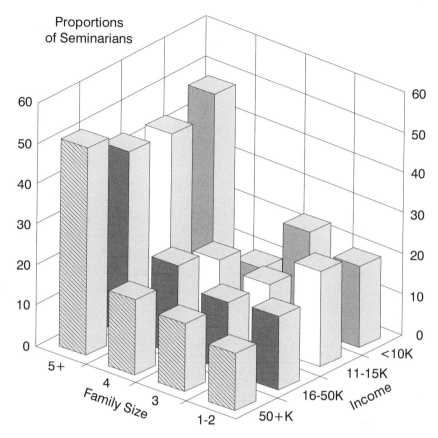

Figure 10.2. In the United States today, Catholic seminarians tend to be later-born and from large families. Some bias toward lower-income families is also apparent.

be the stricture against marriage. A complex set of culturally derived behaviors—giving younger sons of large families to the priesthood—once worked for large Catholic families. The trend today is still for sons of larger, poorer families to become priests (fig. 10.2), but the interactions of devotion and church requirements cause some proportion of only sons of devout Catholic families to become celibate.[36]

The complex and repeated interaction of culture and biology means that today the behaviors we see may or may not be adaptive. When we find apparent maladaptive behaviors, we have interesting further work to do: what will happen to the frequencies of these strategies? We cannot assume, for humans today, that any particular behavior is, or is not, adaptive.

LOGICALLY INEPT, SOCIALLY ADEPT:
THE SOCIAL CONTEXTS OF INTELLIGENCE

The inherent conflicts of group living—among kin, cooperators, and competitors—may well have shaped how we think as well as what we do. Most of the time, we think of our intelligence as something that liberates us from the exigencies—the selective constraints—of the environment. We can turn that consideration on its head and ask about the forces that may have shaped our intelligence. What if our social environment, as well as our physical environment, has been central in shaping our intelligence, our cultures, and the ways we go about seeking resources?

The evidence suggests that we're not qualitatively different from other organisms in getting and using resources to survive and reproduce—just more subtle and complex about it. We must beware of falling into thinking that somehow if it's conscious, it's different. As Robert Smuts has noted:

> That choices made unconsciously in the midbrain normally have a biological function is taken for granted, but choices made consciously in the neocortex are often thought to lie in the realm of social values and culture, and to be free of biological constraints. The neocortex, however, is as much a product of natural selection as the hypothalamus, which means that it evolved because it served the most basic of biological functions: it helped human ancestors to perpetuate their genes.[37]

A strong argument can be made that our extraordinary intelligence, and much of our culture, derive from selection for ability to compete against *each other* (rather than against the physical environment), often as coordinated, cooperative groups.[38] Among the primates, big brains (which are costly and unlikely to be favored unless they confer an advantage) seem to be associated with both environmental and social complexity. Primates with large home ranges tend to have larger brains; primates that live in larger social groups also tend to have large brains. For example, arboreal travel may require some cognitive map making, and tool making for foraging requires some intentionality.[39] Machiavellian intelligence, the smarts to manipulate one's social cooperators and competitors, clearly requires considerable complex understanding: to manipulate, to deceive, even to keep track of others' social lives.[40]

How to sort out the relative impact of these different forces? Neocortical enlargement seems to reflect the extent to which primate brains are specialized for intelligence, and the relative size of the neocortex can be compared. When anthropologist Robin Dunbar did this comparison, he found that measures of environmental complexity (range area, day journey length, amount of arboreal fruit in the diet) showed no relationship to neocortical enlargement. In contrast, the size of the social group is correlated with all measures of neocortical enlargement. What's more, the ability to deceive conspecifics ("tactical deception," surely a measure of Machiavellian intelligence) is strongly related to neocortical enlargement. So, whatever the role of environmental complexity in shaping primate intelligence, it pales in comparison to the role of social complexities. In fact, scholars are leaning more today toward the view that communication may be best understood if we view it as evolving to manipulate and deceive, rather than to transmit, information.[41]

Over our evolutionary history, then, other humans have been one of the strongest selective pressures molding our behavior. Conflicts of genetic interests, played out in social situations, may well have shaped our problem-solving abilities in special ways. We pride ourselves on our intelligence, but as any decision theorist or operations specialist will tell you, we are in fact lousy logicians. Daily we face political, procedural problems which have—and we know it—logical, intelligent solutions. Nonetheless, even when we know a great deal about the technology of solutions, we frequently find ourselves unable to agree on how to proceed, caught in "social traps"—that is, situations in which our decision rules might make social (but not logical) sense.[42] Perhaps we could profitably explore them as "evolutionary traps"; we can ask whether patterns appear systematic in ways predictable from our ecology and our genetic conflicts, in the context of our social history.

The sorts of illogical things we do are legion, and they are consistent with an evolutionary history of (1) life in social groups, and (2) environmental uncertainty that we remember and seek to understand (and impose a pattern on). We consistently interpret single or isolated events as predictive of a whole category, even when the event is portrayed as atypical or explicitly not representative. We do this most emphatically if it represents something dangerous, noticeably different, or (especially) a social defection. We attribute someone else's behavior to his or her "character," but we explain our own behavior by circumstances (that way, we ourselves never behave badly, it was the circumstances, but those other rats . . .). We are rotten at reasoning from the general to the particular, and we overdo reasoning from the particular to the general. In fact, as Robyn Dawes, an operations research specialist, noted "a great deal of our thinking is associational, and it is difficult indeed to ignore experience that is associationally relevant but logically irrelevant"—the vampire lore discussed in early chapters is an example. We also are likely to adopt wrong (sometimes blatantly wrong and illogical) explanations if they are proposed by someone "of status" in a group situation.[43]

We estimate probable frequency of events on the basis of many things other than actual frequency data; yet logically, frequency is what matters. A famous example is the "compound probability fallacy." Consider some rare medical condition. Now, is it more likely if it is associated with something common but not correlated, like freckles? Well, no. But if asked about the likelihood of a rare condition, medical internists estimate it as more likely if another, more common symptom is listed with it. That is illogical: the combination of two events cannot be more common than either one alone. Nonetheless, illogical as it is, when we see the less common problem tied to a common problem, we tend to estimate the combination as more likely. We are apparently especially prone to this sort of thinking in the context of (perceived) dangerous or costly events and identifiable ethnic or socioeconomic groups.[44]

This is illogical associative thinking, and it looks like *perceptions of individual interest in the context of a group have been an important force in shaping it.* We are logically inept, but socially adept. One experience at being cheated, and we are likely to generalize to future interactions with individuals of that category. One dangerous event witnessed, and we fear it ever afterward. We remember and overestimate the occurrence of rare (especially dangerous or socially harmful) events and conditions. When we lived in small groups and interacted with the same people repeatedly, this may have been a reasonable predictor. Today, it means that if we are asked to estimate the relative frequency of murder versus suicide, most of us imagine murder as more common; it is not, but it's dangerous and dramatic and gets more media attention. We have more (usually vicarious) experience with it.

Although we can certainly learn logic, we nonetheless typically solve problems, at least initially, in the context of our social history. And herein lies an interesting clue: some of our illogical thought makes sense if our history has typically involved competitors. One common solution seems to be not "what violates the logic?" but to "look for cheating against me." This could be simply because we are more stupid than we want to admit, but I doubt it. I suspect rather that, in our social history, we have been concerned with protecting our *own* rights under the social rules rather than protecting "rights" in general.[45] People solve concrete examples of logical dilemmas, perhaps especially if it is in the context of detecting cheaters, much better than the formally identical abstract versions of the problem.

Sometimes people solve concrete examples in utterly wrong ways logically, but ways that make sense if the heuristic were simply to "look for cheating." First, consider the problem: "If Harry has a ticket, he should have paid for it"; people are superb at discerning this problem. But what if the problem is: "If Harry has paid, he must take his ticket." One knows Harry has "paid"; thus we should ask "does he have a ticket?" to see that he has not been cheated—but many people ask instead if Harry indeed did pay (did he cheat us?), even though it was specified that he did. The two problems are logically, but not socially, parallel.

Not surprisingly, training people to solve logical problems works better if abstract theory is linked to concrete examples. Taking this perspective helps us approach reasoning in new ways. Twentieth-century psychologists are, on the whole, pessimistic about teaching reasoning. And if one imagines that people solve problems using abstract algorithms and tries to teach in the abstract, the results will reinforce that perception. But people do seem to use rough cost-benefit rules in daily life, and if one teaches reasoning by tying abstract examples to problems people see as applicable to daily life, they learn much better.[46] Are there further patterns? Here is an area of inquiry ripe for cross-disciplinary sharing of concepts.

Undoubtedly, our evolutionary history has shaped our male and female brains in at least slightly different ways. Spatially, women rely on "landmarks" rather than compass directions more than men (chapter 3); girls solve disputes more often by avoiding the context of the dispute while boys are more likely to argue about the rules (chapter 11). The very structure of our brains differs by sex.[47] Few behavioral sex differences are consistently strong and clear, but in many traits there are strong trends (with overlap) for the sexes to differ. In the United States, girls tend to score higher on verbal tests, boys on mathematical tests; these differences increase as the children grow and learn (despite our best parental efforts) that boys are going to be the engineers and physicists.[48] Once again, interaction of natural selection and culture yields a rich diversity.

Our evolutionary history has shaped our intelligence in ways we would never suspect if we thought of ourselves as socially neutral logicians, as living computers. In fact, we solve quickly some problems that are extremely difficult and costly (in algorithms) for computers; but in very special, largely social ways, our logic is liable to be skewed. Though this is a fascinating proposition, our thinking in this area is in its infancy. So far, the main value of changing our perspective is that we have uncovered new ideas to explore.

11.

Sex and Complex Coalitions

A Prince is likewise esteemed who is a stanch friend and a thorough foe,
that is to say, who without reserves openly declares for one against the
other, this being always a more advantageous course than to stand neu-
tral.
—Niccolò Machiavelli

COALITIONS, like so many other phenomena, can be a reproductive strategy; and if this is true, male and female coalitions will tend to be different. Among bottle-nosed dolphins, females often swim with each other. Dolphin society is (like chimpanzee groups and many traditional societies) a "fission-fusion" society, in which individuals travel and feed together, splitting up and reforming. But the sexes differ. A female might swim with her mother today, her mother and sister tomorrow, and her friends next week. Males, however, are a very different story. They travel in small and stable groups: you will always see the same two or three males together. However, sometimes you see mega-groups of two "usual" alliances swimming together; these mega-groups are predictable in their membership. What's going on?

When a female dolphin comes into season, she is a reproductive resource—but a male has a problem (as is true in deer, caribou, most primates). Perhaps he can lure her away by dramatic aquatic display, or even kidnap her,[1] but how can he control her and prevent other males from fertilizing her? He might run into another group at any time. A single male simply cannot keep a group of males from taking away the female. What dolphin males do in the face of this predicament is form coalitions: the same two or three males will swim together, day in and day out, displaying together, and cooperating in getting females.

But a two-male coalition can be displaced by a three-male coalition, for example. Dolphin males have a flexible system of stable coalitions and temporary supercoalitions (fig. 11.1). When group H (three males) has a female, group B (three males) may not be able to capture her from H; so they might leave, recruit group A (two males) and together steal the female. There are stable alliances at both the coalition and supercoalition level: males within each group cooperate; coalitions A and B cooperate; H cooperates with G and D, and so forth.

In primates, the cooperate-to-compete strategy is often starkly evident. Baboon males form coalitions specifically to steal females from other groups. For the chimpanzees of Arnhem Zoo, studied by primatologist Frans de Waal, male coalitions were clearly reproductive strategies. The three big males (Nikkie, Luit, and Yeroen) played power politics for reproductive ends.[2]

Sometimes struggles were straightforward. Yeroen dominated, not only allying himself with males, but seeking the help of Mama, the dominant female. Luit (with some help from Nikkie) eventually dominated Yeroen, displaced him, and took over. Before he took over from Yeroen, Luit cultivated other males who had just won fights; but as soon as he displaced Yeroen, he switched to supporting less-powerful males who typically lost, dissipating any centralization of power. Yeroen cultivated Nikkie (who was the least powerful of the three males on his own) and together they displaced Luit.

The rewards were clear: copulations with females. In the beginning, while Yeroen was the single (and most powerful) dominant male, he got about three-fourths of the matings, and almost all of the matings with mature, prime females. So far, this is simply another example comparable to those in chapter 4, suggesting that dominance tends to be related to reproductive success for male primates. However, Luit and Nikkie, neither able to displace Yeroen alone, formed a coalition and revolted against him. During the power takeover by Luit with Nikkie's help, Luit's share of matings jumped from 25 percent to well over 50 percent (while Nikkie went from perhaps 5 percent of the matings to over 30 percent). During Luit's dominance he (~37 percent) and Nikkie (~30 percent) together gained two-thirds of the matings. When Yeroen and Nikkie displaced Luit, both Yeroen and Nikkie gained while Luit's share declined to under 25 percent of the matings. In the next power takeover, Nikkie displaced Luit, and in his second year's tenure, gained over half the

Figure 11.1. Dolphin males travel in stable coalitions; occasionally, as here, two coalitions will form a temporary supercoalition to gain access to females. (Photo by Andrew Richards.)

matings himself. Not only is dominance related to reproductive success in most species, but in social species cooperative coalitions are useful in gaining dominance and reproductive resources.

COALITIONS, RESOURCES, AND REPRODUCTION

Group living is a powerful force in human lives; our neighbors influence what we can do successfully. Much of our behavior is coalitional. Anthropologists Lionel Tiger and Robin Fox have commented that "the political system is a breeding system. When we apply the word 'lust' to both power and sex, we are nearer the truth than we imagine."[3] Not surprisingly, human sex differences are similar to those of other primates: if coalitions had reproductive ends in our evolutionary history, and if men and women succeed reproductively through different strategies, then men's and women's coalitions should differ.

While it is clear that men strive for power, the reproductive consequences of power have seldom been investigated, and it may not be obvious that men often use power and resources as mating effort. But as we have seen in earlier chapters, across cultures men form coalitions to gain resources, and use resources to gain reproductively, through formal reproductive rewards associated with status, or simply through polygyny. It is probably true that *coalitions evolved as reproductive strategies,* and may remain so. Because reproductive return curves differ for mating parental efforts (chapter 3), the potential reproductive gains, as well as the variation in success, are likely to be greater for men than for women. In our evolutionary history, men had more to gain from risky coalitions and high-stakes politics than women.

It should be no surprise that the payoffs differ for mating-effort versus parental-effort coalitions, and that defection (e.g., leaving a coalition) and open aggression are more frequent in (higher-risk, higher-gain) mating-effort coalitions than in parental-effort coalitions. Parental-effort coalitions are more likely to be among relatives, to be longer term and relatively more stable than mating-effort coalitions. Coalitions among males supporting one another in dominance fights often involve nonrelatives and are likely to be much more fluid. When resources are sought to raise offspring, the struggle is likely to be individual or involve close relatives; when higher-risk, higher-gain mating effort is involved, coalitions among nonrelatives are common. Men and women, normally expending as they do greater proportionate mating and parental effort, respectively, are predicted to form different sorts of coalitions, at different risk levels, and to be involved differentially in community and in wider levels of politics.[4]

Male-Female Coalitions

In all sexual species, there is at least a temporary male-female cooperation to produce offspring; these may be simple and temporary. In environments requiring the parental effort of both parents in feeding offspring, males and females form lasting pair bonds and cooperate in getting and distributing resources. Some male behavior will be directed toward mate guarding or other behaviors that increase the certainty of paternity. In polygynous species with coercive males or mate guarding, females may seek out powerful males as protection against other harassing males.[5]

The game "Battle of the Sexes" (BoS) reflects the tension in male-female coalitions nicely, because the two players have some convergence of desired outcome and some differences. The classic example (and the reason it is called BoS) begins with John and Mary wanting to spend the evening together. However, John wishes Mary would come with him to the baseball game, while Mary wishes John would accompany her to the Stravinsky String Quartet recital. (Today, for political correctness, the game is sometimes called Baseball-Stravinsky, but the rules are the same.) The players "want" to coordinate their behavior, for they have somewhat convergent interests, but they also want to spend their time doing different things.[6] In game-theory terms, the real "battle of the sexes"—over mate desertion—is a form of "Chicken," not "Battle of Sexes."[7] In this game, in other species, the payoff for defection can be asymmetrical between the sexes, as in mating-parental conflicts of interest—the female often gets stuck with sole parental care, because it would cost her more if she had to start over.

In species in which a single parent is markedly less successful than two parents in raising offspring, the male-female coalition may persist, sometimes for the lifetimes of the individuals. When it does so, obligately monogamous breeding systems result. In contrast, in any species in which one sex can raise offspring alone, the payoff for defection is high, and the cost of cooperating when defected upon is not devastating. When either parent can raise offspring alone, desertion is likely; the two parents will monitor the appropriateness of desertion (can the other raise the offspring now? Will the other stay?).[8]

Same-Sex Coalitions

In many mammalian species, same-sex coalitions exist relatively briefly. In primates, however, long-term cooperation, in order to compete, is common. Individuals form coalitions of varying stability for reproductive reasons, and some coalitions, as in the Arnhem chimpanzees above, are sufficiently elaborate that they become indistinguishable from politics.

The basic differences between the sexes in the kinds of coalitions formed are further influenced by ecological factors. In many primates, males disperse from their natal group, joining a new coalition in a group of strangers.[9] In "male-bonded" species, in which males remain in their natal group while females disperse, male politics can become highly developed; groups of related males form coalitions; they have a power structure similar to patrilocal societies among hunter-gatherers. In "female-bonded" species, in which females remain in the natal group, matrilineal coalitions likewise can be extremely powerful. Coalitions of females may assist males, resist male sexual coercion, forage together, and even harass subordinate females and female coalitions, so that low-ranking females experience more reproductive failure than high-ranking females.[10]

MALE COALITIONS

Struggles among male primates for status, territory, or harem control often involve coalitions. The context is typically mating effort. Males may or may not be related, and coalitions may be temporary and fluid, particularly if they involve nonrelatives. Many species of primates have "one-male" groups; however, even in those species, male coalitions in the mating context occur occasionally. In red-tail and patas monkeys, several males may cooperatively invade a single male's group and mate with the females. In gorillas and gelada baboons, a subordinate male may share breeding rights with the dominant for long periods.[11]

In most primates with multi-male groups, males tend to be simply competitors—outside of coalitions in sexual selection, tolerant or cooperative relationships among males are rare (see chapter 4).[12] This is particularly true for species in which males leave their natal groups as they reach adolescence, so that adult males in a troop are seldom related. In male-bonded groups, in which males remain in the natal group, co-operation is more common; in red colubus monkeys, for example, only natal males are accepted into adult coalitions.[13]

In chimpanzees, it is clear that reproductive resources are at stake in male coalitions and political interactions, as in the Arnhem Zoo study.[14] Frans de Waal argued that "the unreliable, Machiavellian nature of the male power games implies that every friend is a potential foe, and vice versa. Males have good reason to restore disturbed relations; no male ever knows when he may need his strongest rival." Indeed, in the Arnhem study, unresolved male-male power tensions eventually erupted with lethal consequences, and one of the males was killed by a coalition of other males. Anthropologist Richard Wrangham noted that chimpanzees and humans are quite alike—and highly unusual—in having not only male-bonded groups but also intense male intergroup aggression. In humans we call the end result warfare (chapters 13, 14).[15]

FEMALE COALITIONS

Female-female coalitions tend to center around parental effort. Bonobos, or pygmy chimpanzees (*Pan paniscus*) are not only one of our closest relatives, but show an apparently unique and strong female coalition. Even though bonobos are a species in which females disperse and males stay in their natal groups (and thus adult females in a group are not related), female bonobos are skilled at establishing and maintaining strong bonds with one another; they perform a behavior called G-G (genito-genital) rubbing, which appears to help reassure and cement friendships in a variety of situations. Females control and share with one another (but not with males) highly desirable food, and sometimes cooperatively attack and injure males. Food control appears to allow female bonobos to mature relatively early and, in the end, translates into enhanced lifetime reproductive success. Their coalitional power is remarkable, and reproductively profitable. Females are receptive for longer periods than in common chimpanzees; copulations are common across groups of bonobos, and mating interference is rare. Most scholars working on bonobos attribute the relatively low level of male aggressiveness in bonobos to the strength of female alliances.[16]

Female coalitions in some species can have serious and bloody results nonetheless. In a number of primates, groups of related females may cooperate to get resources useful to their offspring (and prevent others' access), or to harass subordinate females in reproductive condition.[17] In female-harassment species, less-dominant females have difficulty raising their offspring successfully. In fact, early descriptions of "allomothering" (a non-mother associating with an infant) often included incidents of female harassment ending in the injury or death of an infant. So there are reproductive impacts of female-female cooperation and competition. But remember that female mammals are limited in the maximum number of offspring they can raise. The variation in reproductive success of females is less than that of males, so the impact of dominance on reproductive success for females is less marked than for males.[18]

Most examples of female reproductive coalitions involve preventing the success of others. In yellow baboons, mid- to high-ranking females about to ovulate or give birth form coalitions and attack low- to mid-ranking females, suppressing their ovulation or causing spontaneous abortion.[19] After a lower-ranking mother gives birth, her infant is likely to be roughly handled by more dominant females during its most vulnerable first three months—again, by those in reproductive states most likely to gain from the death of a competitor's infant. Dominant bonnet macaque females, often working in coalitions, also handle the infants of subordinate females roughly. Further, they differentially direct such handling toward female offspring of subordinate females, forming a sort of "corporate matrilineal hierarchy." Female mate choice tends to increase the coefficient of relatedness within the matriline, consolidating the coalition. Harassment of infants of other lineages may be a method of decreasing the size and strength of competing coalitions. In at least some baboons, however, we are discovering that female dominance struggles may have reproductive costs for the aggressors as well (chapter 4).[20]

In common chimpanzees and macaques, female coalitions can influence male competition, because female coalitions come to the aid of preferred males in power struggles. In chimpanzees, female coalitions are quite stable, and overlap with the pattern of social bonds.[21] Female chimpanzees interact much more with other females than with males, and the frequency of aggression among females is about one-twentieth of that among males. Females also show far fewer submissive and other reconciliation behaviors. Although females do interact with males and sometimes enter into aggressive encounters in support of a particular male, there are relatively separate spheres of male and female life. Overt strife is more characteristic of males. This generalization appears to hold true in both wild and captive populations.[22] In the wild, furthermore, while the male patterns appear to be similar to the Arnhem Zoo population, the females appear to be even less sociable and less involved in bonding than the Arnhem females.[23] Once again, the ecological conditions favoring group formation have a ripple effect, influencing social patterns as well as simple dispersion.

Male chimpanzees are both more aggressive and more social than females because they are more political, particularly in the wild. But males have at stake access to matings with estrous females, and the possibility of fathering the majority of offspring in any mating period. For females, reproductive success is not a matter of matings, but of access to resources such as food for raising offspring. For females, a strategy of avoiding competition by dispersing, becoming solitary females with their offspring, works—but it does not widely lead to coalitions, negotiations, and politics.[24]

In gorillas, female kin are friendlier to and more tolerant of one another than they are with nonkin; coalitions involve kin against nonkin. In a number of species, there is evidence that females form coalitions to intervene in aggressive encounters (against either males or other females) on behalf of maternal kin far more often than on behalf of other individuals, and to incur greater risk for maternal kin.[25] Reciprocity does occur among nonrelated females, although it does not appear to be intense. Coalitions among nonrelated females are rarer overall than coalitions involving nonrelated males, and the context of female cooperation is typically parental effort. Nonkin coalitions are often directed against lower-ranking females within the group and involve little risk for coalition members.[26]

In nonhuman primates, then, there are important sex differences in the formation and function of same-sex coalitions. Male coalitions largely serve mating effort competition; male rank and male coalitions tend to be fluid, and nonrelated males may form coalitions. Female coalitions involve parental effort such as nursing, foraging, and the support of particular individuals—kin and friends (reciprocators)—as well as the harassment of competitors in some species. Female coalitions, though they may involve both kin and nonkin, appear to be more kin biased than male coalitions.

Both male and female coalitions in nonhuman primates appear to form in the context of reproductive gain, either through heightened mating or parental success, or through hurting competitors' success. The major sex difference is that the potential reproductive impact of coalitions is greater for males: in any season, a male's success can range from zero to many, while a female's success can range from zero to one or two. For males, cooperation with nonrelatives can be highly profitable, and male coalitions, more often than female coalitions, involve nonkin and high risk.

Informal Human Coalitions

Coalitions are central to human life and society, are potent forces both within societies and between groups—and they clearly can be more complex and flexible than other primate coalitions. I will not look at formalized coalitions such as organizations, but only at the costs and benefits for individual alliances. Games like Prisoner's Dilemma that help us explore the costs and benefits of cooperation also highlight the fact that one cannot simply assume that cooperation is reproductively profitable; in fact, the conditions under which cooperation should evolve are relatively restricted.[27] So it makes sense to ask how the costs and benefits play out in the real world.

Remember that in Prisoner's Dilemma (chapter 9) and in related games like Hawk-Dove, a tension exists: if both players cooperate, they do best, but a player who cooperates while her opponent defects loses big, and a defector whose partner cooperated wins big. Thus, a common (and "rational") solution is for both players to defect, getting a small payoff. The outcome from defect-defect is worse than the outcome from cooperate-cooperate—but better than the payoff to the sucker whose partner defects.

If one could count on cooperation, all would be rosy. As economist Ken Binmore has pointed out, this has led to a common fallacy: because the payoffs from defect-defect are arguably the worst possible, isn't it not only immoral, but even foolish, to be "rational" and defect in games like this?[28] This fallacy of ignoring that we can't assume that individuals with nonconfluent interests will cooperate in case others also will, is sometimes called "Fallacy of the Twins." When you hear it invoked, you have a clue that the author is falling prey to the fallacy "What if everyone were to behave like that?"— without recognizing that strangers have no basis for cooperation. This, as Binmore notes, misdirects attention to the outcomes only if everyone does the same thing. Since hawk-hawk (defect-defect) has a worse payoff than dove-dove (cooperate-cooperate), one could be led to the mistaken conclusion that surely everyone should—*will*— act identically and cooperate.[29] But neither player can count on the other, so the players act independently. The payoffs for defecting if your opponent cooperates are the highest of all. If the players were identical twins, we would predict cooperation, and indeed (chapter 2) twins and other siblings typically act more cooperatively than nonkin. Similarly, individuals who interact repeatedly in contexts in which the number of interactions is uncertain cooperate more than strangers.[30]

COALITIONS AND COMMUNICATION

Important clues to the dilemmas of cooperation versus defection in group decisions come from a recent series of experiments by the political scientist Elinor Ostrom and her colleagues, who manipulated the payoffs for defecting, how much people were allowed to talk about decisions, and how much they could discuss penalties for cheaters.[31] The games had the same structure as "commons"—resources to be allocated among group members.[32]

"High-endowment" games, in which it paid handsomely to defect (take more than your share), had more defection and lowered everyone's "yield" (profit).[33] But even in high-endowment games, communication mattered: the more that people could communicate, and the more they could punish cheaters (though extremely harsh strategies, called "grim triggers," were not tried), the more they cooperated and the better their payoff.[34] Communication—even once—among participants in these experiments could significantly increase cooperative behaviors.

In low-stakes commons, average net yield was 35 percent if no communication was allowed; when repeated communication was allowed, the yield rose to 99 percent. When the payoffs for cheating were higher, people behaved differently: in high-endowment commons, average net yield increased from 21 percent (no communication), to 55 percent (one-time communication), to 73 percent when communication was allowed repeatedly.

Just being able to talk should, in contemporary noncooperative game theory, make no difference—but it does. Further, when the participants had the right to choose a sanctioning mechanism, even with only a single opportunity to communicate, groups achieved a net yield of 93 percent (compared to 21 percent); when the costs of fees and fines were subtracted the yield was 90 percent. Thus, while there are no easy answers to solving large-scale ecological "commons" problems, there are some clues and some useful strategies to try: monitoring for cheating, identifying cheaters, and punishing them—all are examples of coalitional selection (see chapter 10).[35]

In an intriguing extension of these experiments, Ostrom and her colleagues eliminated face-to-face communication; people knew each other only through a computer network, and only as numbers.[36] If a strategy had to be unanimously accepted, it didn't take long for someone to propose, and others to accept, an evenhanded allocation to all seven people. However, if it took only a majority (four of the seven players) to implement a strategy, a common outcome was that someone would pick three others in the group using the numbers of their computer terminals. Proposals tended to be for computers 1, 2, 3, and 4, or 4, 5, 6, and 7, or 1, 3, 5, 7. So they used an abstract logic to make these proposals and propose a one-fourth-each allocation. This formed a coalition that effectively shut out three of the seven people in the group.

Coalitions and the Ecology of Trust

As you would guess from the game outcomes discussed above, cooperation is likely when cooperators can expect that others will continue to cooperate. In nonhumans, evidence is accumulating that only rewarded cooperation is likely to continue.[37] Conditions that make it easy to exclude individuals within the group, as in the computer-commons game above, promote very specific coalitional cooperation, which may constitute "defection" against the rest of the group.[38]

The issues of cooperation, discrimination, and payoff show up in places we would not expect from traditional paradigms. Just as in other species, cooperation in humans begins among kin, for defection is less costly genetically. How far it is extended beyond kin depends on the costs and benefits. For example, among the Ache (who share much food and are often called egalitarian), when a man dies his small children are likely to die: since he can no longer share, other men do not share with his widow and children (chapter 7).[39]

Does trust teaching vary cross-culturally? If, for example, cooperative strategies can function to reduce risk of failure (e.g., food sharing), then we expect such strategies in environments in which food return is uncertain and frequently low. Though we still do not expect indiscriminate cooperation, external extremeness and uncertainties seem likely to promote some kinds of cooperation.[40]

Cross-culturally, as population density increases, the degree to which children are trained to trust others decreases somewhat. In societies with communities of more than five thousand people, children tend not to be taught to trust others.[41] This is hardly surprising; as the proportion of strangers, neither friends nor relatives, increases, the less fruitful trust is likely to be a first approach.[42] As the degree of risk of infection from serious, life-threatening pathogens increases, children are less strongly taught to trust others (see also chapter 4). Here, the issue is not the social implications of "stranger," but the risk from any individual.[43] Patterns of marital residence—and the relative strength of male-male versus male-female coalitions—are strongly related to the intensity with which children are taught to trust. In patrilocal societies, in which related groups of men co-reside, children are less strongly taught to trust others, compared to other living arrangements.

When internal warfare (among villages within the same society) occurs more frequently than every ten years, children are taught to trust others much less than in societies in which such warfare is rare;[44] no significant pattern emerges with the frequency of external warfare. These patterns carry over into other social behaviors: when trust inculcation is low (as in polygynous societies, or societies with frequent internal warfare), boys are trained to be more aggressive and competitive.

To move beyond simple kin coalitions and small-society reciprocity requires either trust or a rather Machiavellian approach to fail-safe devices.[45] International trade specialists have hoped that trade might create coalitions and develop trust.[46] Societies trade when they have complementary needs and perhaps the need for repeated interactions. In the societies of the Standard Cross-Cultural Sample, trade is more common when temperature and rainfall conditions vary greatly. Hunter-gatherers trade less than others; fishing and agriculture are slightly positively associated with trade; and trade increases strongly with the importance of animal husbandry. Despite these patterns, the inculcation of trust is not at all related to the importance of trade, nor to any of the subsistence types.

Sex and Human Coalitions

Once again, the fact that males and females evolved to get and use resources somewhat differently in reproduction influences the patterns we see in coalitions. Two very different conditions seem likely to lead to strong male-female bonds.[47] In harsh environments in which a man and a woman can form effective resource coalitions, they are likely to do so, and the coalitions will follow confluence of reproductive interests (e.g., husband-wife).[48] Such male-female coalitions center around garnering resources for offspring and are likely to be strongest in situations in which male and female interests are identical, that is, in monogamy. Brother-sister coalitions may exist in conditions in which certainty of paternity is low. Male-female coalitions are likely to be relatively weaker in polygynous situations, in which the male's reproductive interests are identical to those of all his wives and children and overlap with the interests of any particular wife inversely with the number of wives.

Male-male coalitions are associated with status competition and with resources that can be more effectively obtained and protected by groups of males.[49] Thus, heritable land of lasting value and some large game (especially if used for brideprice) are often the focus of male-male coalitions. At their extreme, these can become warfare (chapter 13). Such coalitions are likely to be among brothers and are thus associated with patrilocal residence,[50] but more-or-less fluid coalitions of this sort arise among men of various relatedness and among nonrelatives in many societies. These male-male coalitions may exert considerable power and control significant resources. Male-male coalitions exist in spite of conflict of reproductive interests of the men involved—but only if each member can gain sufficient resources or influence to better his position compared to operating alone. The fluidity of such coalitions may be related to this factor.

Female-female coalitions, like male-female coalitions and unlike male-male coalitions, tend to operate in the familial sphere and are seldom powerful outside the household. Cowives, even if sisters, have less convergent interests than monogamous husband-wife coalitions, and in fact a major cause of divorce cross-culturally is conflicts among co-wives (the level of conflict is lower when wives are sisters). Female-female coalitions may arise among female relatives or cowives. Most appear to function for the exchange of information (e.g., location of good foraging spots), child care, and subsistence-related work. Resources women gather are used primarily for offspring and family, and significant female resource control is unlikely. These coalitions are seldom significant beyond the household boundaries; even female solidarity groups tend to be among relatives.[51]

For a woman, the potential conflicts of interest over resource distribution in the household seem likely to be distribution to her relatives, distribution to her female reciprocators (related or unrelated), distribution to her husband's relatives, distribution to her husband's reciprocators, and distribution to his other wives in a polygynous situation. The use of resources for the man, the woman, and their joint offspring are unlikely to be a source of conflict.

All these sex differences—in the potential reproductive risks and returns, in the usual degree of relatedness of cooperators, and in the sphere of activity—seem likely to be reflected by differences in men's and women's behavior. Consider the options open, if they are dissatisfied, to the following individuals: a monogamously married woman, a polygynously married woman, a man in a coalition of other related and nonrelated men. A man's options depend on his resources and his connections in other coalitions; he can fight for power openly within the coalition, try to manipulate the rules so that the coalition operates to his own advantage, or leave (change coalitions). Even though a woman's natal family may have some importance as support, neither of the women is likely to gain by changing coalitions. Within families, open conflict is seldom successful for a variety of reasons, and avoidance of conflict is common.[52] A monogamously married woman loses the only ally whose reproductive interests are virtually identical to hers, and if she remarries she must begin reconstructing the same confluence of interests.[53] A polygynously married woman also has few options and must operate in a coalition with a lesser confluence of reproductive interests with her husband. Such a woman may control few resources, cannot easily find a new coalition, and is unlikely to gain by open confrontation within the family; she cannot easily gain status and resources in a second, more desirable marriage.[54]

When do women compete directly for resources? When do women's coalitions function more like men's coalitions? There seem to be two such conditions, both more prevalent in Africa than elsewhere. In complex matrilineal or dual-inheritance societies, a powerful woman, while not increasing her own number of children, may increase her grandchildren by passing wealth and power to her son, as in the Ashanti (see chapter 12). Even when such systems are changed by contact with industrialized societies, traditions of women's independence and power, even though they no longer yield reproductive gains, may persist and thrive, as in West Africa today.

In sum, the sexual dimorphism in human coalitions follows from the arguments of chapter 2 (mating and parental return curves) and parallels that in other species to a surprising degree, particularly in light of cultural diversity.

Coalitions in the Play of Boys and Girls

The two sexes begin to diverge in their coalitional behavior early in life (chapter 6). Girls, but not boys, are more likely to rely on adults to settle disputes, both in the United States and cross-culturally.[55] Similarly, boys, when challenged, tend to respond directly and aggressively, while girls give signals of "disengagement" to possible conflict.[56] Even in nursery school, girls compete by cooperating rather than confronting.[57]

Today, in Western nations, sex roles are far more similar than in most societies, but important sex differences remain in competitiveness and striving. The psychologist Piaget found girls more "tolerant" in their attitude toward rules, and more willing to make exceptions.[58] In general, boys play more often than girls in large age-heterogeneous groups; they play more competitive games; and their games last longer than girls' games. Boys' games last longer apparently not because they are more complex or less boring, but because continual disputes over rules arise, and boys engage in conflict resolution. In a wide variety of settings, boys tend to cover more space in their play, to play more roughly, and to be "more chaotic, more disorganized, and less neat."[59]

In one study, boys were "seen quarreling all the time, but not once was a game terminated because of a quarrel, and no game was interrupted for more than seven minutes."[60] It almost appeared that the negotiation of rules (politics?) was as important as the game itself; certainly it was more constant—whatever the game, boys argued about the rules. For girls, the occurrence of a dispute tended to end the game, in a sense sacrificing continuation of the game for continuation of the relationship, as if no options for negotiation within the coalition or for changing coalitions existed. Put bluntly, boys get more practice than girls in negotiating their best interests, starting early in life.

These differences are understandable if, in evolutionary history, women have enhanced their reproductive success by cooperating in the familial sphere, with female relatives and cowives—that is, in situations in which they could not gain through open conflict, or in attempting to change coalitions. Men, on the other hand, have enhanced their reproductive success by cooperating to get greater resources and power with both related and unrelated men—situations in which open assertion of dominance (with greater risk) may frequently gain. The patterns we see suggest that different coalitions may indeed have led to reproductive success for each sex.

12.

Politics and Reproductive Competition

Politics—strife of interests masquerading as a contest of principles.
—Ambrose Bierce, *The Devil's Dictionary*

HUMANS, like other primates, move almost imperceptibly from "coalitions" to "politics." We have come far from our starting consideration of the ecology of sex differences in reproduction. But a variety of current issues have precisely the characteristics that make this logical stretch worthwhile: simple rules, interacting with environmental and historical particulars, create outcomes of increasing scale and complexity. If we can understand how men and women profited reproductively from differing resource strategies in the past, perhaps we can follow the emergence of complexity and begin to make sense of likely differences—and how to separate those from issues of equity—today.

Primatologist Frans de Waal's definition of politics as "social manipulation to secure and maintain influential positions" is a biologically useful one. The art of politics, defined this inclusively, is older than humanity itself. De Waal, describing his work on chimpanzees at the Arnhem Zoo, commented that "whole passages of Machiavelli seem to be directly applicable to chimpanzee behavior"[1] (and surely earlier chapters of this book bear out this sense). All chimpanzees appear to strive for power and influence, but there are major differences in the ways male and female chimps strive and in the ways they use their influence and power. There are also important interpopulation differences. In chimpanzee society, the final reproductive goal of political strife is clearer, perhaps, than in human societies, but it may be useful to examine human politics from this broad behavioral ecological perspective.

Here, I want to follow the kinds of coalitions male and female mammals are likely to form, up to that most rarefied of coalitions—formal politics. In most countries, women are rare in national and international politics (Norway, India, the United States, and England are recent exceptions), even when there are societal and legal mechanisms to mandate that women as well as men have access to power and control of resources. Women may spend most of the money and outnumber men in professional jobs, but they hold relatively few top-level positions.[2]

In some societies, politics and reproduction are overtly interwoven. Among the Tiwi of northern Australia, for example, there is an extreme form of gerontocratic polygyny.[3] Older men monopolize women; the difference in age between spouses in primary marriages is 23.6 years, quite different from the two- to seven-year age difference we think of as common in our society.[4] Men get wives through demonstrating various skills and negotiating with other men, trading favors and female relatives. Highly polygynous men are old and have demonstrated skill and power.

Even if coalitions have been central in the evolution of social behavior, the connection to complex modern phenomena like international politics may be far from obvious. Yet I am suggesting that both politics and war (next chapters), if defined in a biologically functional way, are outcomes of highly developed coalitions and have their roots in prehuman history.

Not only primatologists like de Waal have viewed politics as the seeking of rewards. Disraeli claimed that politics is nothing more nor less than "the possession and distribution of power": power over others, clearly an issue in behavioral ecology. That this is not a new idea is reflected by the wealth of informal definitions of politics, including:

- Nothing but corruptions; the madness of many for the gain of a few. (Jonathan Swift)
- The conduct of public affairs for private advantage. (Ambrose Bierce)
- The possession and distribution of power. (Benjamin Disraeli)
- Economics in action. (Robert M. La Follette)
- The science of how who gets what, when, and why. (Sidney Hillman)
- Nothing more than the means of rising in the world. (Samuel Johnson)

To be sure, these are informal and ironic definitions; any connection to power is hardly obvious from most formal definitions of the word (and few nonbiologists, except perhaps Henry Kissinger and François Mitterand, openly relate power and sex).[5] Because most definitions concern the formal practice of politics in nation-states with powerful official positions gained by election, or in societies with political inheritance, they obscure the functional significance of political behavior. They lack a consideration of *why* we have evolved to strive or to gain influence.

Under the formal structure, however, two phenomena in all of politics are particularly interesting to a behavioral ecologist: coalitions and contests about resource control and status.[6] Sometimes these are relatively straightforward matters of local "horsetrading" of influence and support; sometimes the situations are more complex.

MEN, WOMEN, AND POLITICS CROSS-CULTURALLY

With notable exceptions—Margaret Thatcher, Indira Gandhi, and a few others—national politics in the developed world is largely a man's game. However, comments abound that in tribal and band societies, women had significant influence and power. Whether this is true depends largely on one's measures of power and influence. Cross-culturally, though women's rights vary in many domains, men hold power almost uniformly in politics. About 70 percent of societies cross-culturally have only male political leaders; just over 7 percent have both sexes, though men are more numerous and more powerful.[7]

In very few societies are women coded as equally powerful as men (and men are always more numerous).[8] Women can hold kin leadership posts and have more say and influence than men in six of ninety-three societies. Both sexes have roughly equal influence in four other societies.[9] Women can attend and participate equally in community gatherings, though they may be segregated from men, in over half the studied societies. While women may widely have an informal "voice" in community affairs, formal political influence is rare for women, both within the kin group and in the community at large.

We think of men's and women's interests as converging most strongly in monogamous societies, but women are not more likely to hold political office in monogamous societies, in societies in which men are absent, in those with biased sex ratios, in particular subsistence bases, or much of any other correlate I can find—I think women as leaders are too rare to see any pattern.[10]

The rarity of women as major political figures is consistent with the hypothesis that men have evolved to make reproductive gains from striving in coalitions, but it is important to predict the exceptions as well as the general case. The very rarity of women's political power makes thorough statistical analysis difficult. The societies in which women are coded as holding political power in the odd-numbered societies of the Standard Cross-Cultural Sample occur on several continents, are settled or migratory, patrilocal or matrilocal, can be highly polygynous, and include subsistence forms that are generally not associated with women's ability to control resources (e.g., animal husbandry and agriculture). Four of the nine societies are African, suggesting a possible geographic bias, but it is important to remember that African societies comprise over a fourth of the half-sample. In fact, the geographic association appears weak.[11]

There is one strong pattern with women's large-scale political activity: descent rules. Five of the nine societies in which women are said to have such political power are societies with matrilineal or double descent, precisely those societies in which women's power is predicted to enhance their son's (not their own) reproductive success. Over half the cases of women having political power occur in one-fifth of the sample, the matrilineal and duolineal societies. Of the eighteen matrilineal and double-descent societies, five have women in political office (we would expect two); of the fifty-six patrilineal, ambilineal, and bilateral-descent societies, four have women in political office (seven would be expected). Thus, despite the difficulties of recording descent systems in great detail, there does appear to be an association between descent systems and women's ability to hold public office.

There may be virtue in examining in more detail those societies in which women are coded as having significant power, to see if there are patterns in the *type* of offices and power women have and what the obvious independent variables are. As you will see, there are problems with the coding of women's power in some of the nine societies; in an effort to recognize women's participation fully, three societies were coded as "women hold power" when the reference was a mention in conversation or a rumor.[12] When these societies are excluded, no statistical result is changed. No factor other than descent has even a marginally significant relationship to women's ability to hold political office, and descent is significant.

Nama

The sociologist Martin King Whyte, in a cross-cultural study of women's status, coded the Nama of the central and northern Kalahari as having women in political positions, but women were neither as numerous nor as powerful as men.[13] If anything, this is an overstatement. The Nama were patrilocal and patrilineal;[14] thus we would not predict they would have women in political office. There were few goods to inherit. Men's weapons were usually buried with them, and women sometimes gave their ornaments to their daughters.[15] In their small family-based groups, there was no organized system of government; older men simply had the greatest influence over daily affairs.[16] In the northwest, where bands were larger, each band had a recognized chief, usually hereditary in the male line. The heir succeeded only after he passed through the puberty ceremonies and killed his first big game.

If the office fell to a man while he was still a minor, there was no regency except among the more northern Naman group of the Nama. The one supposed example of female political power is recorded as follows: "The successor to the office, under normal circumstances, is the eldest son of the last chief, and as such he is usually accepted without question by the tribe. If he is a minor, his father's brother or some other near relative in the male line acts as regent, although one instance is recorded, in the case of the Gei // Kuan, of a woman reigning on behalf of her young brother until he came of age."[17] Thus, while the code is technically accurate, most Nama groups had no political offices. Only one example of a woman holding power is recorded, and the real force of women in politics was nothing like that found in other societies coded as having female politicians.

Mbundu

The Mbundu, in south-central Africa, were polygynous and patrilocal but had a double system of inheritance. Land descended patrilineally, from father to son, and movable property was inherited matrilineally. In one sample village, there were twenty-seven monogamous marriages, and nine men had two wives, and the headman had four wives and a concubine. Other households were occupied by formerly monogamous widowers now living alone.[18]

The village headman functioned as both priest and legal authority. Succession to headman was patrilineal, from father to son, or from elder to younger brother. Villages were grouped into twenty-two chiefdoms, but almost half of these chiefs were tributaries of one of the more powerful chiefs. Chiefs had religious, legal, and external-affairs duties. The report that women could hold political office apparently arises from the following:

> The Sambu kingdom has long specialized in the training of medical practitioners and witch-doctors. The tradition is that it was founded by a woman who seems to have come either from Bailundu or from the stock from whence came the Bailundu royal family at an even earlier date. From the family of Wambu-Kalunga, in addition to Ciyaka, came those who set up the kingdoms of Elende, which had a woman ruler early in its history.[19]

Thus the Mbundu, like the Nama, have a historical tradition of only one woman having held political power at some time, and there is no evidence that women routinely held public office. In fact, the ethnographies are quite explicit that public office is patrilineal. Women's public activities and power in these two societies are quite clearly different from, for example, the Ashanti.

Ashanti

The Ashanti were polygynous, matrilineal, and avunculocal or virilocal, with wives occupying separate quarters.[20] Both women and men could own land, but there were complexities in property inheritance: a woman could only inherit from a woman and a man from a man. Political power was associated with the "stool." In each political unit there were two stools: the chief's and that of the queen mother. The classic study on the Ashanti states that "the recognized seniority of the woman's stool is no empty courtesy title. In fact, but for two causes [physical inferiority of women, menstruation and ritual avoidance] the stool occupied by the male would possibly not be in existence at all."[21] Menstruation involved ritual avoidance; this meant that premenopausal women would be barred from war. Several queens did historically accompany armies to war, but all were postmenopausal.

The queen mother had considerable say in the choice of the chief, and her veto could not be overruled. After the chief took office, her place was at his left hand wherever he went, except to war (and there, too, if she was postmenopausal). She alone could rebuke the chief, his spokesman, or his councilors in open court. She could address the court and question litigants. She did not, however, attend court when menstruating. Each queen mother had the right to choose the chief's senior wife, and to replace her if she died. The senior wife, or her daughter if the senior wife was old, became regent when the chief went to war.

The queen mother thus had direct political power, influence over coalition formation, and indirect power through her choice of senior wife. Further, since the senior wife was likely to have little or no political training or experience, and no knowledge of customary law or court procedure, the queen mother often assumed these duties, deciding cases with the full powers of chief. The queen mother was also entitled to a share of the court fees.[22]

Matrilineality influenced women's importance. As the anthropologist R. Rattray commented: "A king's son can never be the king, but the poorest woman of royal blood is the potential mother of a king."[23] A queen mother's resources, accumulated through her reign, could enhance her sons' abilities to get wives in this polygynous society.

Rattray's comments about women's power and the clan system of the Ashanti are poignant in the light of historical development. Europeans had never recognized the women's stool or the queen mother's power. Noting the comments of other authors on the "troublesome" activities of the queen mother (while not recognizing her position), he asked the Ashanti about her and found that they recognized that women held no such positions among the European visitors, so they assumed that Europeans held women to be of little importance. Rattray argued that the queen mother could be an ally, but Europeans' treatment of her was guaranteed not only to alienate her but to be destructive as well. Later reports seem to have proved him correct.[24]

Bemba

The Bemba, occupying the high plateau of northeastern Rhodesia (Zambia), considered themselves warriors. They practiced shifting cultivation and traded relatively little.[25] Elephant hunts were a source of wealth to chiefs. Descent was matrilineal, and marital residence was at least initially matrilocal. The Bemba were polygynous, but the degree of polygyny declined; by 1951 it was rare to find more than one man per village with two wives, in part because of women's resistance to polygyny.

A striking Bemba political feature was a centralized form of government under a hereditary "paramount chief."[26] The paramount chief ruled over his own country, and also acted as overlord to a number of territorial chiefs who were drawn from the immediate family of the paramount chief. Some sisters and nieces were chieftainesses with authority over villages. The chief's mother had a territory of her own and was important in tribal councils. Succession to office was matrilineal. When a male chief died, the office passed first to his brothers, then to his matrilineal nephews, then to the children of his sisters' daughters. A female chief was succeeded by her sisters, maternal nieces, and granddaughters.[27]

Rome

The Roman Empire represents a very different level of societal organization and technology than most other societies in the sample. Much more information is available on certain aspects of women's and men's political lives. Since early in the empire, wives, who legally were in wardship, were in fact emancipated. Also quite early, marriage shifted from a purely parental arrangement to a requirement that a woman's consent be obtained. Married women could, by Hadrian's time, write their own will. Upper-class women were well educated.[28]

Women were active in society, and their titles, privileges, and distinctions were apparently as closely graded as those of men.[29] A woman generally shared her husband's station, but emperors occasionally gave women themselves consular rank, if they were relatives and not married to consuls, and, rarely, might let them retain their status if they then married lower-status men. Such women had great privileges, even though they may simply have been individual privileges. Women in formal political bodies were unknown, except for the *convenius matronarum*, an ancient guild of religious origin that became the empress's maids of honor. This body determined allowable fashion and etiquette: what costume was appropriate for women of various rank, who had precedence and the "right of the kiss," who ought to have sedan chairs and whether they should be adorned with silver or ivory, who could have gold or jeweled shoes, and so on.[30]

Women had considerable personal latitude under the Roman Empire; they were active in public social life and in a sort of women's senate. But they apparently did not hold senatorial appointments of the same sort as men, and the women's senate made decisions solely on matters of etiquette. This description arises from the restricted sources allowed in the Standard Sample analysis; if one looks further, the impression that women's political power was largely illusory is strengthened. Though wealthy Roman women may have had considerable personal autonomy, their formal activities were largely honorific, and their influence on public policy came primarily through informal influence and association with powerful men.[31]

Marquesans

The Marquesans, of Polynesian origin, lived on a series of islands in the central Pacific. The tribe was strictly a localized group, without a hierarchical political structure. Many anthropologists treat data on the Marquesans with caution, arguing that they are a society too changed by contact to trust reports. Marquesans are well known for being polyandrous, but the circumstances are important.[32] The adult sex ratio was about 2.5:1 (male-biased), although the Marquesans said that infanticide was not practiced. A household's status depended primarily on the number of men available to work for it, building houses and manufacturing items (men produced 85 percent of manufactured work). The eldest child was considered the heir, and all sons who were not first-born children were expected to find alliances as secondary husbands. Eldest-born sons tried to marry attractive women; secondary husbands were acquired by offering an attractive wife to share. "Average" households comprised a household head, a group of other men, and a single wife. While officially all husbands had equal access, one ethnographer commented that "the first husband ran things and distributed favors, although it was to his advantage to see that his underlings were sexually satisfied so that they would work for his house and not wander off with other women."[33]

Polyandry appears to have been an outgrowth of male-male alliance, in this way: status was achieved by the work of men. Men were common, women were scarce. Powerful or rich men sometimes offered the hospitality of their wives to other men (although they apparently controlled access). The chief's household had not only several husbands, but also several wives. When a family became powerful, it could challenge the current chief; to avoid this, the chief might arrange a marriage between his eldest son and the first-born daughter of the rising family, or he might adopt their eldest-born son. Such adoptions and marriage alliances required wealth, and wealth required manpower. It is unclear how skeptical we should be about this argument, and there is no written analysis of the conflicts of interest arising from such arrangements.

Women had some choice in sexual arrangements, but their accession to public power was rare—and it is unclear just what "rare" meant, or whether women chiefs had the same duties and perquisites as men chiefs. The relative impact of chieftainship on men's versus women's reproductive success is also a matter of speculation; however, women could not gain in direct reproduction, and if male chiefs had more wives than others and could control others' access to their wives, men at least had the potential for direct reproductive gain through the chieftainship.

Saramacca

The Saramacca of Guyana were matrilineal. Polygyny was general and residence was not patrilocal. Kin clans had ties and claim to particular areas, with fishing, hunting, and forestry rights. Lacking any centralized political system, villages and clans were relatively autonomous. Law-making rested with the chiefs and the councils (krutus) over which they presided; law enforcement was the task of other officials, the *bassias*.[34]

Most villages had at least one woman *bassia*; she was responsible to the village head for the behavior of the women. In many tasks, she organized the women and was responsible for seeing that work got done. If two women were involved in a dispute, she informed the village head and transmitted his decision if the affair was not of "sufficient gravity" to warrant calling the old men together. When anthropologists asked if women talked in the big council, they were told, "No, not in krutu. But they talk plenty at home," suggesting informal influence.[35] Nonetheless, since *bassia* had duties in council, and women were *bassias*, presumably the women *bassias* were active in the councils.

In theory, women *bassias* had "equal power" with male bassias; but the ethnographies make it clear that women's and men's spheres of influence were separate. Since the Saramacca are coded as women having "equal power" as men, it is important to note that their power was restricted to women's matters.

Montagnais

The Montagnais (Naskapi), in and around the Labrador Peninsula, were a sub-Arctic hunting people, living in low density and hunting either in family groups (in the southern part of their range) or communally (in the less populated northern areas).[36] Marital residence was sometimes matrilocal, sometimes patrilocal.[37] They gathered in the summer, but were dispersed for the rest of the year. One southern informant suggested a chief was irrelevant—it would be absurd to depend on the authority of a chief hundreds of miles away.[38] The authority of the chief was more developed but still weak in the northern areas, where the Montagnais hunted communally, than in forest bands.

Political organization was loose. There was usually a chief, and if the band was large, a council of older men assisted him. Political units were loosely associated with territories, and not strictly kin based; there was no political organization beyond the band. In the early days, chieftainships, when there were any, were patrilineally inherited. In 1927 a government Indian Act required the election of chiefs, though many bands did not bother. One ethnographer reports that chiefs were chosen on the basis of personal characteristics: a man must be an excellent hunter, with a well-developed sense of responsibility toward the others in the band and (because of the interactions with the Hudson Bay Company) a shrewd negotiator. No chiefs were bachelors; in part, informants felt this was because a large kin group made for a strong support base in attempting to attain the chieftainship.[39]

One informant, Joseph Kok'wa, remembered that Tommy Moar's mother had been a chief in the Nichikun band for several years, but Tommy Moar responded that, while she had been held in high regard by her tribespeople, she had not actually been a chief.[40] The issue of women and public office is thus cloudy in the Montagnais. An informal system of advice seeking was common. Tommy Moar's wife, Maggy, for example, was highly regarded, and both women and men submitted their problems to her judgment and followed whatever suggestions she offered. Thus women could have considerable informal influence in the "community" sphere, though it is not clear that they were ever "chiefs."

Creek

The Creek, once ranging throughout the southeastern and southwestern United States, were warriors, traders, and agriculturalists, growing corn, beans, squashes, and other crops. Descent was matrilineal.[41] By about 1850, inheritance rules usually depended on parents' wills, but intestate cases were relatively equal with regard to both sex and birth order.

The Creeks had a nobility, but the chieftainship did not attain the level of power among them than it did, for example, in the Natchez. In the eastern part of the Creeks' range, particularly near the eastern Siouan area, women were frequently chiefs. Though this was an uncommon occurrence toward the west, some women in the western tribes did become war leaders. After the end of the Civil War, chiefs were elected.[42] But consider this description of the annual assemblies of the Creek nation: "Only the chiefs of the warriors are admitted there; the subordinate chiefs who are present are intended to serve the others, but they have no voice in the deliberations. The women are charged with the duty of preparing the necessary food and drink for the assembly; the subordinate chiefs go to fetch the provisions and place them in their turn in the grand cabin for the members of the assembly."[43]

Women at least sometimes were awarded the rank and title of warrior, but in at least one case a woman's bravery was rewarded by conferring a war title on her son. It is unclear how frequently women chiefs were war chiefs, though it is clear that they could hold this office. We know nothing else about the duties and power of women chiefs.

Women in Politics: When Did It Pay?

Cross-culturally, women have seldom held public political office or been politically powerful in any formal sense. In many societies, of course, women are active in the "informal influence" sphere and within the kin group; these are harder to measure. In the societies in which women do or did hold public positions, it is frequent that, as among the Romans and the Saramacca, they held a position of power over women's affairs but had only informal influence on men's. For the Nama and the Montagnais, public offices were not always relevant because the societies were so dispersed and at such low densities; nonetheless (e.g., among the Montagnais), strong women could have considerable informal influence. Among the Nama and the Mbundu, women did not hold office—there was only one reported instance in each tribe of a woman's importance in the political sphere. Among the Marquesans, women "rarely" ascended to office. Creek, Bemba, and Ashanti women, though they held public power less often than men, appeared to be able to make the same decisions as men.

The societies in which women's decision-making capabilities centered on the same issues as men's were complex, matrilineal, or double-descent societies in which heritable resources existed.

Some societies not in the standard sample may prove instructive. The Tchambuli, or Chambri, in Papua New Guinea are well studied, but conflicting reports exist.[44] Margaret Mead originally suggested that women dominated men, and that this society represented a role reversal compared to Western ideas about sex and power. From this she inferred that our ideas of masculine and feminine roles in power issues, for example, were culturally determined and culture specific. Since this was a polygynous and patrilinial society, reports of great female power were unexpected (but reasonable if power were a fluke of culture and had no ecological payoffs).

Later studies found that women and men essentially operated in separate spheres. Men were indeed thought by the Chambri to be more aggressive than women, and issues of female dominance existed only with relation to women's matters—they never crossed the sex boundary. Thus, among the Chambri as among the Saramacca, women might dominate other women about women's matters only.

Concern over Galton's Problem, the possibility of bias introduced because neighboring societies may share practices, raises the issue of whether there might be an "African" bias in these results. The association between African location and overt political power for women, though statistically insignificant, suggests further exploration could be rewarding. Other African societies may be instructive. Among the southern Bantu, there existed an institution called the "female husband." Economic and political power were the con-comitants of female husbandship for women in Bantu society.[45] Among the southern Bantu, in which women could hold political office, female political leaders were expected to become female husbands. Women gained socially from becoming autonomous female husbands, but the impact on their reproduction is far from clear. If powerful "sons" from female husbandship returned gains to her genetic kin, there may have been indirect reproductive gains as well— but that is not discernible from the existing data. Among the Shilluk and the Nyoro, women who inherit or achieve high political status are forbidden to marry—a clear and direct reproductive cost.[46]

The most important relationship with women's overt political power is the descent system. I think this is related to the evolutionary ecology of resource gain and reproductive success. In the few societies in which women wield substantial public power, as opposed to informal influence, there is no evidence of clear, direct reproductive gain. In fact, in some examples, there is a conflict between political and direct reproductive gain for women. In some descent systems, women's power may accrue to their sons, who then reap a reproductive benefit;[47] in such systems, *women gain by increasing the number of their grandchildren.* In other systems, women's direct gains are not clear, but when power gives access to substantial resources, women in power similarly can make gains through their sons, or nepotistically.[48]

Once again, the sexual dimorphism in usefulness of resources and power in reproduction is critical. Men appear to seek overt political power for direct reproductive gain (wives, for owed reciprocity), while women seek resources for themselves and their offspring. Sometimes this is accomplished through indirect or informal influence and nepotistic gain. Most commonly, the amount of resources controlled by women is sufficient to support their family, but sometimes, as we have seen, particularly in matrilineal societies, women may exert public power to the gain of their families.

In societies in which women have little public power, they may nevertheless have considerable influence over resources within the kin group, with substantial reproductive impact. While women hold relatively few high-level public positions, they are major consumers; the relevant cross-cultural codes are controlling the "fruits of own, both, and men's labor." There is real point to the old joke about the man who tells his friend, "I leave all that day-to-day minor stuff to my wife: what we'll have for supper, what kind of car to buy. I make all the important decisions: whether to bomb Iraq, or recognize the Baltic States."

Women have evolved to use resources differently than men in reproductive matters, and this has had an impact on their political strategies and degree of influence. But this is surely not a justification for the rarity of women in public life today. It may be true that women in Western societies have comparatively great overt power, but it is also likely that for both men and women, the exercise of formal power may well have less impact on reproduction than in smaller societies. Thus, it is interesting that during west European history, as societies moved toward greater monogamy (reducing the potential rewards of power for men), politically powerful women seem to become more numerous (e.g., Mary Tudor, Elizabeth I, Mary of William and Mary, Queen Anne). Men may still reap reproductive advantage from political power, compared to women, through remarriage and children born to mistresses.[49]

Further, it seems clear that laws in both preindustrial and modern societies are far from neutral with regard to reproductive competition, and conflicts of interest exist between men and women. The accession of more women to positions of real public political power has the potential to change the balance of power on such issues.

13.

Sex, Resources, and Early Warfare

[These bloody feuds] are for the sake of the women.
—Jivaro informant, cited by R. Karsten, 1923

THE PRUSSIAN miliary strategist Carol von Clausewitz said war was simply "the continuation of state policy [politics] by other means." Although warfare gives rise to some of the strongest, most tightly knit and potent international coalitions in modern times, that is surely not its evolutionary context, for states are a relatively modern phenomenon, while organized intergroup conflict among competing coalitions is as old as, or older than, humanity itself. Conflicts of interest, if not coalitions in open aggression, are universal among living things, and certainly lethal conflict exists in many species. Thus it makes sense to begin our inquiry by focusing on simple conflict, asking: Over evolutionary time, what have been the ecological contexts of conflict and killing? Why and when are there sex differences? What were the costs and benefits to the individuals involved?

From our perspective, two phenomena matter: the reproductive impacts for individuals of fighting and killing (including formal war), and the potential conflicts of interest among different individuals involved in conflict. It is obvious from earlier chapters that conflict, which is risky, will center on items of real reproductive importance: mates, or status and resources when these lead to mates. Risky conflict is unlikely over trivial resources, and warfare in traditional societies *is* risky. It is far more lethal and pervasive than scholars had thought; it is war "reduced to its essentials." In contrast to most countries fighting in modern wars, a very high proportion of men fought in early societies, and mortality rates could be extreme.[1]

Lethal conflict looks, at first glance, as though it should decrease reproduction—certainly it's hard for dead individuals to reproduce. In fact, lethal conflict is like infanticide and delayed reproduction: in specific environments, for some individuals, risky conflict is a high-stakes gamble in which there is some probability of winning (greatly increased reproduction) and losing (death or disability). Were this not true, these behaviors would remain rare. Evolutionary novelty (see chapter 2) is an important complication. Certainly modern war is characterized by rapid change in technology, and it is likely that the original driving force of reproductive reward no longer exists.

Because of this complexity, we need to ask the following three questions: (1) Was there previously a reproductive advantage to engaging in lethal conflict? (2) Is there currently reproductive advantage to engaging in lethal conflict? and (3) Are the proximate triggers leading to conflict "unhooked" from any previous selective advantage?

RESOURCES AND CONFLICT

Risky, damaging aggression is surprisingly widespread in many species; fights occur both over direct reproductive resources (territory, mates, etc.) and indirect reproductive advantages (elimination of competitors or competitors' offspring). Typically, deaths arise from fights over mates or resources for getting mates, from infanticide, and from cannibalism.[2] The ultimate costs and benefits are reproductive, although the level of risk varies with the kind of aggression. The risk to an infanticidal killer, often a male taking over a harem and killing the offspring of his reproductive predecessor, is probably small. Not surprisingly, infanticide is more common than the killing of adults.

Lethal fighting among individuals of similar age and status is typically a male endeavor and occurs primarily among adults of the same population rather than between groups from different populations.[3] In a game-theoretic approach, combatants assess each other's resource-holding power: "The stake played for is infliction of loss of resource-holding power, and is determined by the fitness budgets of the opponents. . . . This defines a critical probability of winning . . . for each combatant, above which escalation (fighting) is the favorable strategy . . . and below which withdrawal is favorable.

Escalation, then, should occur only when the absolute probability of winning minus the critical probability of winning is positive for each combatant. Someone always loses; if information were perfect, the loser-to-be would never attack, and would always withdraw if attacked. The loss can be costly; in red deer, deaths from fights over matings represent 13 to 29% of all adult male mortality.[4]

Throughout military history, bluff has been a major way of keeping an opponent's information imperfect; think of Toussant L'Overture marching the same company around the block several times to fool his opponent Leclerc into thinking he had a large army. There is a long prehuman history to bluff; in many species, displays involve deception and bluff that make a potential combatant seem bigger and stronger, advertising that an attack might prove costly.[5] Conflict is most likely between potential combatants who are similar in status or power; in such a situation, the exact probabilities are hard to "calculate." In red deer, for example, subordinate stags are unlikely to escalate a confrontation; the risks of serious injury are too high. Depending on the costs, the stronger combatant may press an attack.

In some primate species, with the elaboration of coalitions among individuals, we see intergroup conflict: groups of varying size may separate from the main population on foraging trips, and if they encounter smaller groups or lone individuals from another population, they attack, exploiting the uneven balance of power. Chimpanzee raids are like this; male chimpanzees negotiate alliances within groups (chapter 12), and use those alliances to coordinate attacks against males of other groups. Intergroup, rather than interindividual, conflict is likely only in long-lived, social species; such group aggression is not qualitatively different (the context, benefits, and costs are similar) than individual fights, but is more complex, with cooperation among individuals. The rewards are reproductive. In many species, females change groups as spoils of war for the winners.[6]

Potentially lethal conflict is only likely when the possible reproductive rewards—mates, status, resources for mates—are high. Mating effort is the context of potentially lethal conflict, not parental effort: if the offspring is dependent on its parent, a live parent is crucial. Thus, within mammals, males will more often be in a position to gain than females from risky, possibly lethal fights (see chapters 3, 4). Sexual selection in competition over mates and kin selection in infanticide and intergroup conflicts are the context.

Why Women Warriors Are Rare

The United States is unusual in not having a major "warrior queen" icon in either its history or its mythology.[7] From Britain's Boudica (or Boadicea), said to have risen up against invading Romans, to Joan of Arc, to Queen Jinga of Angola fighting off Portuguese invaders, many cultures have at least a myth of important women warriors. In truth, of course, as the historian Antonia Fraser has recounted, women warriors are rare, and royal women, though often called upon to carry symbolic weapons, rarely or never fought like warrior kings. And women in the armed forces of nation states today are still fighting—internally—for the right to bear arms. Why?

Because the two sexes have different payoffs for risk taking, aggressive coalitions tend to be sexually dimorphic in mammals. Compared to females, males tend to form coalitions that are riskier, more fluid, more aggressive, and more often among nonrelatives (chapters 4, 11). Females' conflicts typically center on getting food or parental resources, or on infanticide, while males' conflicts are likely to focus on getting mates. This means that the reproductive impact of fighting may be many times greater for male mammals than for females. Further, mating effort conflicts, usually among males in mammals, are more likely to escalate to lethal proportions than conflicts over food, for example. This difference in risks and returns, of course, is what prompted Darwin to treat sexual selection differently from "ordinary" natural selection, even though functionally it is identical.[8]

In humans, too, men's reproductive success varies more than women's in most cultures. So it is not surprising that aggressiveness is one of the most consistent sex differences across cultures; for example, most homicides are committed by men. Women's politics and conflicts over resources tend to be at the familial and neighborhood level, while men's conflicts tend to have broader scope; this too is hardly surprising. Cross-culturally, men can often make enormous direct reproductive gains when they acquire power, status, and resources (chapters 4, 7, 11, 12). In bridewealth societies, rich and powerful men use resources to buy more wives, or to buy younger wives (who have higher reproductive value and are more expensive) than other men.

In the few societies in which women play high-stakes power

games or go to war (rather than simply wield informal influence), they show no immediate, clear reproductive gain.[9] Throughout evolutionary history, men have been able to gain reproductively by risky warfare; heroes gain status and access to women. Women seldom have been able to gain. Nonetheless, women have, although rarely, fought with or accompanied men to war. From at least the time of Alexander, women traveled and sometimes fought with their men; in Alexander's time, when campaigns were long and women might have children before the campaign was over, the children were legitimized after the soldier completed his duties. During the seventeenth, eighteenth, and nineteenth centuries, women occasionally passed themselves off as men and fought in the ranks of infantry and cavalry regiments.[10]

It isn't that men are bigger and stronger than women. In primate species, and in human societies, social complexities so outweigh the impact of physical size that size alone is a poor predictor of success.[11] Similarly, it is not that women are bound by the constraints of pregnancy, nursing, and child care. If that were true, sterile women and postmenopausal women might broadly be expected to engage in intergroup conflict.

As in politics, the critical factor behind our sex differences is that resources and power have different reproductive utility for men and women. Across cultures, three common conditions—patriliny (inheritance by sons) combined with patrilocality (sons stay home) and exogamy ("marrying out")—foster men's, but not women's, confluences of interests in war. Related men, but not women, tend to live together.[12]

WAR: RUNAWAY SEXUAL SELECTION?

Human war can become more complex and varied than intergroup aggression in other species, largely as a result of the development of technology (which itself is probably a product of intelligence). But the roots of lethal conflict clearly lie in sexual and kin selection. In view of these roots, the development of technology to today's superlethal levels raises an important question: Is war an example of runaway sexual selection?

Sir Ronald Fisher, a mathematician who made significant contributions to biology, noted that "remarkable consequences" follow if females exert a strong preference for particular traits in males. As Darwin noted, two influences are important in sexual selection: initial, sometimes considerable, advantages not due to female preference (e.g., the advantage of large antlers in combat for red deer); and any additional advantage conferred by female preference. This second advantage can appear "whimsical," but female preference centers on male advertisements that are expensive and hard to fake.[13] Females are forcing males to show their condition.

The intensity of female preference will continue to increase through sexual selection so long as the sons of females exerting the preference have any advantage over other males.[14] Fisher noted: "The importance of this situation lies in the fact that the further development [of the favored trait] will still proceed, by reason of the advantage gained in sexual selection, *even after it has passed the point in development at which its advantage in Natural Selection has ceased*" (italics added).[15]

Thus, when immediate reproductive gains are so great that they outstrip the countering pressure of ordinary natural selection for survival, lethal traits leading to extinction can arise in sexual selection. When, as among the Yanomamö, men with great warring skill have more wives and more children than others, sexual selection can be very powerful.[16] Even in modern industrialized societies, in which participation in wars (and other risk-taking behaviors) may be "unhooked" from the advantages given by sexual preference, if sexual preference still exists for "war heroes" or if there are other proximate rewards, previously linked to selective advantage, the risky behavior may still be common.

OTHER BIOLOGICAL APPROACHES
TO UNDERSTANDING WAR

One of the most influential biological approaches to understanding war was Konrad Lorenz's book *On Aggression*. Lorenz argued that aggression is an "instinctive" drive favored by selection. Although he specified that this did not make warfare unavoidable, others have inferred some sort of genetic basis (indeed, "instinctive" does rather imply a genetic basis), rather than a flexible response to ecological conditions, in which genetically identical individuals might act differently, depending on circumstances. Lorenz also argued that because humans lack lethal weapons in their simple physical makeup and rely on tools, they also failed to evolve reliable inhibitions against killing one another.[17] But this predicts neither the occurrence of aggressive behavior—what conditions (e.g., reproductively important conflicts of interest) are most likely to precipitate aggression—nor the constraints (e.g., individual costs and benefits).

Hopeful Assertions

Perhaps in response to just such arguments, well-known and well-respected scientists, in a statement (May 16, 1986) for the International Society for Research on Aggression, argued from the fact that there is no evidence for a specific allele for aggression, that warfare was "biologically possible, but . . . not inevitable, as evidenced by its variation in occurrence and nature over time and space." Here lies an assumption: if there were an allele for aggression (as has been found among males of one family recently), warfare would be inevitable—that any genetic basis might dictate something.[18] This view moves from patently true statements about the nonevidence of any special alleles for "warring behavior" to generalizations that "biology does not condemn humanity to war, and that humanity can be freed from the bondage of biological pessimism and empowered with confidence. . . . Just as 'wars begin in the minds of men,' peace also begins in our minds. . . . The responsibility lies with each of us."

Such an approach fails to come to grips with the ecology of war, the circumstances in which aggression profits the individual or lineage genetically. And, as we learn more about the complex interplay of genes, ontogenies, and environments, today it seems naive to imagine that anyone would postulate "an allele for aggression."[19] This approach is hopeful but remains insufficiently specific or predictive.

Echoing Lorenz, one argument posits that humans and chimpanzees have particular "Darwinian algorithms" that govern coalition formation and predispose both species to warfare.[20] These psychologically imposed structures have certain characteristics: cheaters must be identified and excluded or punished; participants are rewarded or punished in proportion to the risks they take and in proportion to their contribution to success. Each coalition member has impact on the coalition by regulating his own participation in the coalition and by the actions he takes to enforce the contract on other members. However, all of these specifications seem likely to be true for numerous other coalitions (hunting dogs, female lions, etc.), so the argument is not very convincing that humans and chimpanzees are unique in having the appropriate "algorithms" for warfare. In fact, we cannot eliminate the possibility that any evolved psychological mechanisms are the *result*, not the cause, of the patterns we see.[21] Finally, this argument is not a true alternative to the ecological argument. If, as I argue, lethal conflict exists because individuals and families have profited from assuming the risks of lethal conflict under specific conditions, over evolutionary time, then we expect proximate psychological mechanisms related to aggression. But we have yet to demonstrate any specific ones.

The Ecology of War: Uncertainty and Payoffs

Warfare, like a number of other human social patterns, appears to be related to environmental unpredictability: cross-culturally, warfare is more common when people in a society perceive unpredictability in the environment, when particular sorts of resources unreliably become (or threaten to become) limiting.[22] A strong predictor is the threat of weather or pest disasters.

An important empirical study raises another consideration of ecological influences. The anthropologist Napoleon Chagnon, working with the Yanomamö (see chapter 7) has found that the frequency of warfare differs among villages in an important way. In the rich lowlands, where the potential payoffs are high, villages are relatively large (300–400 people), warfare is highly developed and chronic, and men are aggressive both personally and in raids. In the foothills and mountains, plant productivity is lower, villages are smaller, and warfare is far less frequent; there seems to be little point to risking one's life.[23]

The behavioral ecological hypothesis predicts strong sex differences, and reproductive payoffs to male warriors; the environmental uncertainty hypothesis and Napoleon Chagnon's work suggest that the frequency of warfare should vary with ecological conditions. Societies with more warfare also encourage boys to be more aggressive, tougher, and show more fortitude, perpetuating the pattern. This is just the way we would expect sexual selection to affect warfare.

The Demography of War: Too Many Males

There is also some evidence that a male-biased operational sex ratio can predispose groups to war. This, too, makes ecological and evolutionary sense. Remember that in many societies, capturing women from other groups is a main purpose of war. Remember also that young males are societies' most violent members. We have sex-ratio data for few societies, but a trend is clear: societies with lots of young adult males are likely to see strife.[24]

INTERGROUP CONFLICT IN OTHER SPECIES

Red deer stags, elephant seal bulls, and mountain sheep rams all fight individually over females and other reproductively important resources. But there are true coalitional conflicts in other species—both within and between groups—that are more similar to human conflicts. Three social carnivores show intergroup aggression. In wolves, family-based packs (both male and female) occasionally invade neighboring packs' territories, attacking residents; intraspecific conflict accounted for 43 percent of wolf deaths not caused by humans in one study. Among spotted hyenas (which, like wolves, live in territory-holding family groups), intruders into a clan's territory are likely to be attacked and killed; smaller clan subgroups patrol the territory boundaries, confronting other "patrols." In lions, which also live in groups (prides) based on a group of related females and one or more associated males, interpride encounters occur, but lethal injury is rare. When invading males are attempting to take over a pride, there may be lethal injuries, though once a resident male gives up his reproductive rights, aggression typically stops. Males who have just taken over a pride are likely to commit infanticide, as in several other species.[25]

Aggression occurs widely in primates, including baboons, new world monkeys, lesser apes, and group-living prosimians. In many primates, a behavioral challenge (e.g., territorial incursion, conflicts over a specific resource, including females) generates a defensive response, and groups of individuals from different local populations fight.

Among apes, male-male coalitions may approach the complexity of human politics (see chapters 4, 11). And in primate intergroup aggression, we see for the first time a level of complexity and coordination that approaches the warfare patterns of pre-industrial societies. The complexity of group raids means that intelligence is important.[26] In chimpanzees, humans, and perhaps in gorillas, there are regular cooperative raids by breeding adults against adults of neighboring groups.[27]

In chimpanzees, such raids are probably low cost because groups vary in size—one can hide when big parties are encountered, and can attack smaller parties with little risk. Adult male chimpanzees make aggressive forays into the ranges of neighboring groups, sometimes fatally injuring defending males. Females are semisolitary, using a core area but often traveling outside that area. Adult and subadult males are more gregarious and travel more widely. Total community size ranges from 20 to 110 individuals, but temporary groups range from 1 to 20 animals; group size fluctuates, and it is not predictable how many conspecifics a group might encounter. Further, in chimpanzees it is more common for females to transfer from their natal group, while males are likely to remain among their relatives—and the costs and benefits of risky fights are different if one fights among relatives rather than nonrelatives.[28]

So males in traveling groups may profit from attacking smaller groups when they encounter them and capturing females when they can. Although male-male cooperation and the benefits of risk-taking are enhanced by groups of related males living together,[29] it is not a requirement. In gorillas, both sexes may leave the natal group. In lions, males leave the natal group while female relatives remain—yet lion males engage in intergroup lethal conflict.

Females join males in potentially lethal intergroup aggression in a few species: some baboons, some monkeys, wild dogs, wolves, and mongooses.[30] Each of these last three species has a monogamous, extended-family structure in which male and female costs and benefits are more similar than in polygynous species. These mixed-sex battles also tend to be at family territorial borders; females do not participate in mating-effort "raids" of the sort described above for male chimpanzees.

In nonhuman vertebrates, then, most aggression, both intra- and intergroup, has a reproductive cause. Among primates, groups of males may fight in ways that resemble ambush attacks reported in preindustrial human societies. Male-male coalitions are frequent among relatives, but also occur among nonrelatives. Males frequently come into open conflict over access to females, and over control of resources useful in attracting females. While most of the examples involve males, there are female aggressive encounters in a number of species. Within their group, females may work in related coalitions to attack reproductive competitors or the offspring of reproductive competitors.[31] And in a few species, females may join males in territorial disputes.

Peace-Making in Other Species

In most species, visibly giving up causes an attacker to stop. In chimpanzees, an attacker may continue in the face of submissive gestures. Remember that in the Arnhem Zoo study (chapters 4, 11), at least one male coalition had lethal consequences. Of all nonhuman species, it is among chimpanzees that we see the most elaborate peace-making. Frans de Waal suggests that the clear-cut dominance hierarchy provides a ritual format for reconciliation, which often follows a behavioral confirmation of formal status.

We humans have devised elaborate mechanisms, even institutions, for ending conflict. Even in the simplest societies there are important roles for third parties. Intergroup conflicts appear to have been both frequent and selectively important in our evolutionary history—and to resolve them we mastered new complexities, involving groups of individuals who may not only have disparate interests and incomplete information (or misinformation) about one another, but who may not know one another or be able to predict reliability or probability of default.

CONFLICT IN PREINDUSTRIAL SOCIETIES

What we call warfare in preindustrial societies is indistinguishable in context and function from much intergroup aggression seen in other species; it differs only in scope. Cross-culturally, 60 percent of the societies for which data exist engage in warfare at least yearly. Most attacks in traditional societies are ambush attacks, often well coordinated to take advantage of the element of surprise, and often with numerical superiority. A description of such attacks would differ little from the description of chimpanzee raids.[32]

Individuals in preindustrial societies travel and work in small parties of varying sizes, like chimpanzees, and attacks and escalations by larger groups can be fairly low-cost.[33] Further, patrilocal and patrilineal societies are more common than other types—and this means that groups of related males live and work together, influencing men's costs and benefits in ambushes and warfare. These patterns are repeated in otherwise divergent societies around the world.

Among the Yanomamö of South America (fig. 13.1; see chapter 7 and above), the reproductive context of men's warfare is clear. Men who participate in revenge raids and ambushes have more wives and more children than others, and men who avoid warfare suffer reproductively. Yanomamö men who have killed on a war party are accorded the title of *unokai,* and a man's performance in war parties affects him reproductively. War parties are small, from two to twenty men, and tend to comprise related men. It is true, nonetheless, that ecological conditions change men's costs and benefits; only in the rich lowlands is aggression rampant.

Although there are mystical aspects, most war parties arise from disputes about reproductive matters. Men may choose to avoid joining any particular warring party, and war parties may turn back, often as the result of a prophetic dream. Nevertheless, if a man avoids several possible opportunities, or behaves in ways perceived as cowardly on the raids, he becomes the butt of jokes, and other men may begin to make sexual overtures to his wife. Once a man establishes himself as *unokai,* he is likely to have more wives and more children than non-*unokai* (fig. 13.2). Thus there are clear reproductive advantages among the Yanomamö for men who participate in war parties, and particularly for men who kill.

Among the Jivaro, as among the Yanomamö, there is no stratification. During times of peace there's no chieftanship. When wars erupt, older experienced men who have killed many men and captured many heads are chosen as war chiefs. No Jivaro can be chosen if he has not killed. Bloody feuds, reported as functioning to obtain women, are frequent and follow familial lines.[34]

In North America, the Blackfoot Indians were known throughout the nineteenth century as formidable, aggressive warriors. Blackfoot warfare centered on capturing horses (for bridewealth) from neighboring tribes. Most parties comprised fewer than a dozen men for reasons of stealth, though raids of up to fifty men could occur. Many of the most active raiders were men from poor families "ambitious to better their lot." Even sons of reasonably well-off families needed

Figure 13.1. Men from two Yanomamö groups form a temporary alliance to raid a third village. (Photo by Napoleon Chagnon.)

more horses than their fathers could give them if they were to marry and set up their own household. Horse-raiding parties were led by experienced men judged to have a good war record and good judgment. Participants were volunteers, and though the leader might be a mature man in his thirties, most of his followers were in their late teens or early twenties. Occasionally, a childless woman would accompany her husband on a raid. Unless prior arrangements had been made (e.g., for equal distribution), each man could claim the horses he had led out of camp, or the range stock he had captured. Bitter arguments could occur over ownership, and it was the leader's job to settle these. Some leaders gave horses they themselves had captured to men who could claim no horses. A leader's generosity helped him maintain a popular reputation, and helped him recruit future followers easily. Successful raiders either paid their horses for a bride, or gave them to relatives, most commonly to their fathers-in-law or brothers-in-law.[35]

Among the Meru of Kenya, livestock were used for bridewealth; a man, to marry, needed to accumulate sufficient livestock (preferably cattle) to purchase a wife. Men fought to gain livestock and status, and the military cycle followed the seasonal pastoral cycle.

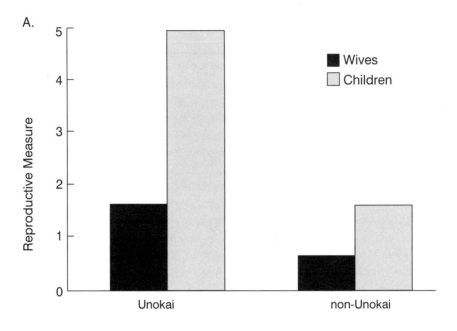

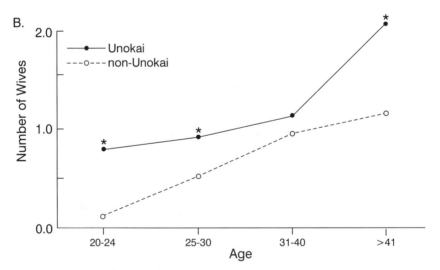

Figure 13.2. Among the Yanomamö of northern South America, men who achieve the status of *unokai* (roughly, revenge killer), (A) have more wives and children, and (B) do so earlier in life (* indicates p < .05).

Negotiation of bridewealth was done by the families, principally male relatives of the bride and groom, and interfamilial alliance was considered an important function.[36] The father of the warrior transferred five specified items—a cow, a bull, a ram, an ewe, and a gourd of honey—to the father of the bride. Additional gifts of other items were negotiable, to sweeten the deal, but could not replace the basic five. Further symbolic gestures of alliance were involved (e.g., the warrior's father would send beer ahead of the actual brideprice, then the bride's father would initiate a general beer-drinking fest to which the warrior's father was invited); however, a main point of the transaction was clearly economic. A portion of the bridewealth was often kept back by the warrior's family, often for years. War was the principal method of gaining livestock, and warriors were expected to earn bridewealth as well as increase familial wealth and status.

Rules and traditions of warfare among the Meru facilitated cattle stealing. In individual conflict between warriors, for example, a warrior could save his life by declaring that his opponent could take his cattle; this was accepted as a declaration of surrender. All captives, female and warrior, were redeemable for livestock. Married female captives could be kept as concubines or wives; unmarried and uncircumcised females were taken as "daughters" and later traded for brideprice. Among the Meru, as among the Yanomamö, men clearly gained reproductively by establishing themselves as successful warriors.

The broad cross-cultural data suggest these societies are fairly representative: reproductive matters lie at the root of war in most traditional societies.[37] Women (abductions, failure to deliver a bride) were causes of warfare in 45 percent of societies in one major study of seventy-five traditional societies. Material resources specified as useful in obtaining a bride were causal in another 39 percent, and in about a third of these, ethnographies specified that richer men obtained more wives than poorer men. In only twelve of seventy-five societies was there no immediately obvious connection between warfare and men's direct reproductive striving. Other studies have found that 75 to 80 percent of wars involved land (clearly useful in establishing a family), and that adultery and wife stealing were major sources of conflict. Similarly, in the societies of the Standard Cross-Cultural Sample, women were captured in 66 of 158 societies; in the vast majority of these cases, women were married or kept as concubines by their captors. In societies, like the Maori, in which there is no direct association between warfare and women or the resources directly used to acquire women, the warfare patterns still appear to reflect conflict over resources useful for the family line.[38]

It is easy to see a link between small-scale societies' warfare and reproductive and familial payoffs. Even in the Judeo-Christian heritage, women were a valued profit from warfare. John Hartung, in a cogent review, has reviewed the centrality of rape and forced marriage in biblical warfare.[39] Consider Moses: "Now, therefore, kill every male among the little ones, and kill every woman who has known man by lying with him. But all the young girls who have not known man by lying with them, keep alive for yourselves." Or the biblical injunction:

> When you go forth to war against your enemies, and the LORD your GOD gives them into your hands, and you take them captive, and see among the captives a beautiful woman, and you have desire for her and would take her for your wife, then you shall bring her home to your house, and she shall shave her head and pare her nails. And she shall put off her captive's garb, and shall remain in your house and bewail her father and her mother a full month; after that you may go in to her, and be her husband, and she shall be your wife. Then, if you have no delight in her, you shall let her go where she will; but you shall not sell her for money, you shall not treat her as a slave, since you have humiliated her. (Deuteronomy 21:10–14)

Interesting here, of course, is not only the proposition that forced marriage with a female captive is sanctioned, but also that the captor must wait a month—sufficient time to establish that his captive is not pregnant by another—before he can "go in to her."

The warriors themselves in traditional societies often have a very clear and straightforward perception of these costs and benefits. Napoleon Chagnon reported this conversation with a Yanomamö friend:

> "Who did you raid?" he asked.
> "Germany-teri."
> "Did you go on the raid?"
> "No, but my father did."
> "How many of the enemy did he kill?"
> "None."

"Did any of your kinsmen get killed by the enemy?"

"No."

"You probably raided because of women theft, didn't you?"

"No." (Chagnon reports that this answer disturbed his friend.)

"Was it because of witchcraft?" he then asked.

"No," I replied again.

"Ah! Someone stole cultivated food from the other!" he exclaimed, citing confidently the only other incident that is deemed serious enough to provoke men to wage war.[40]

14.

Societal Complexity and the Ecology of War

Four things greater than all things are,—
Women and Horses and Power and War.
—Rudyard Kipling

Is it possible to conceive of life without force?
—Charles de Gaulle

I T IS NOT SURPRISING that the functional relationships of warfare are clearer in smaller, simple societies than in large politically complex ones. The transition from preindustrial warfare to the complex multinational warfare discussed in treatises on military history may seem almost unfathomable, but we must explore it if we are to understand whether modern warfare is functionally different from tribal warfare.

The military historian John Keegan's description of Alexander the Great suggests that even in large hierarchical armies, as during Philip's and Alexander's rule in Macedonia, personal characteristics, kin-group size, and ability to inspire loyalty and reciprocity still were crucial to a man's success in warfare.[1] The Macedonian kingship was elective; Alexander had claim to the succession as the eldest son of the king's acknowledged wife, but if he had not been bold and eager for battle, he would have found it hard to press his claim.

Macedonia was an imperial power, and the Macedonian army was large, diverse, and hierarchical. It included cavalry and light cavalry, light infantry, and specialized troops—archers, siege artillerymen, engineers, surveyors, and supply/transport specialists. Soldiers were neither a tribal war band, as in the Yanomamö, nor were they conscripts; they were recruited from a variety of social classes. Nonetheless, there was a central, inner core of warriors, the Foot Companions, whose relationship to the leader was close, often a blood relationship. It was important that Alexander consistently led his men, fighting by their sides, performing dramatic feats of courage and leadership. While courage has probably always mattered to men's fates in war, in Macedonian times heroism became a prerequisite for leadership of large, complex organizations.

Similarly, Prithvinarayan Shah's ability to forge the modern state of Nepal owed much to his strategy of rewarding individual men's interests and thus commanding their loyalty. What is now Nepal comprised several small kingdoms; geographical constraints divided the area into small, self-contained units. Peasants' lives were hard. All land belonged to the state; those who worked the land typically paid half of the harvest to the state, as well as giving compulsory unpaid labor—even though few of the valleys were very fertile. Gaining freehold land was the only escape. With only perhaps eight thousand to ten thousand men, and serious logistical problems, Prithvinarayan Shah drew men of the hill tribes, notoriously pragmatic and unlikely to follow others' dreams, into his ambitions, offering land grants for services as a way for followers to break out of the cycle of agricultural poverty.

Prithvinarayan Shah was a real leader. He spelled out in concrete terms the advantages for his soldiers individually, and showed them the advantages of a farther-reaching group goal. In addition, throughout the long campaign, in negotiating with his enemies he offered substantial rewards to anyone converting to his views, and was usually successful. No other individual had been able to overcome fragmentary relationships and form a modern nation from the region's tribes.[2]

Leaders in war were likely to become leaders in peace. Indeed, war has been proposed as a mechanism involved in the very formation of states. Warfare has been argued to be a necessary, if not sufficient, condition for the formation of states, and there are clear relationships between political complexity and warfare patterns.[3]

War in traditional societies requires individual striving and centers on reproductively important resources. At all levels above the simplest ambushes, warfare involves organization and opportunities for gain through leadership; successful war leaders are likely to be good manipulators of others, and they accumulate an armed following. The path from war leadership to political leadership may be short but profitable: the risks of political leadership are sometimes less lethal than those of war leadership. I think that recognizing the importance of reproductive interests in lethal conflict can resolve some apparent discrepancies among earlier models of warfare and the rise of states.[4] Reproductive competition is a major evolutionary selective force underlying lethal conflict; warfare is a principal mechanism and may be waged in the name of women, bridewealth, revenge, agricultural lands, new territory, or any devised reason.

Not surprisingly, as societies become more complex, so does the scope of the problem.[5] In warfare involving hierarchies of power (i.e., rank and specialization; probably all but tribal ambush warfare), risk is correlated with prior status and/or rank. Since at least the Middle Ages in Europe, disenfranchised or low-status males have gone to war in positions of greatest risk. Sons of Portuguese nobles, for example, would take three-week crusades to nearby, relatively safe locations, while sons of poor families went to Jerusalem, often dying there.

In North America, among the Cheyenne, men who controlled more resources and had greater familial networks became peace chiefs; such men tended not to assume the risks of war. Instead, men who had no relatives (often orphans) became war chiefs; they could gain the proximate rewards of status but could not turn status into reproductive gain because they had to remain unmarried. The Cheyenne case appears to be an example of intragroup conflict of interest, between peace chiefs and war chiefs, with kinship networks and power on the side of the former. It would be extremely interesting to learn the history of the Cheyenne condition. Were the first peace chiefs (political leaders) originally successful war chiefs who then discovered how to avoid the risks of leading war parties without losing status? Taking extraordinary risks with no hope of reward should be rare, seen only when an individual's other options are severely constrained, or if he becomes convinced of some overwhelming benefit.[6]

If we put together basic sex differences in aggression, and ecological influences on "payoff" for aggression, some predictions follow. Truly lethal endeavors, such as the suicide missions of some Japanese in the Second World War and the extreme risks taken in jihads in the Middle East, should correlate either with very low status of the men, or with promises of gain otherwise unmatched. And this pattern is important: in developing nations, young males are increasing by twenty-five million and will continue to increase as a proportion of population; meanwhile, in the developed world, young males are expected to decrease by four million.[7] As I noted in chapter 13, as the proportion of young males increases in resource-limited environments, the potential for violent solutions in both local and larger conflicts seems likely to increase.[8]

There is a persuasive argument that even today kin selection still influences much conflict.[9] When societies become very large and are comprised of diverse groups, intragroup or ethnic conflict once again becomes important. If, as I have argued above, lethal conflict arose in the evolutionary context of reproductive striving, international wars are likely to be simply epiphenomena, although national leaders may use tactics of referring to proximate cues that are important in evolutionary time, such as familial terms to promote patriotism. Strife over local resources seems most likely to escalate into lethal conflict; thus it is not surprising that even today civil war and ethnic strife are far more common than international war.

GREEK HOPLITES: EARLY "WESTERN" WARRIORS?

The Greeks may have provided the first real shift (in the West, at least) away from the small-band, guerrilla ambush attacks that were the usual pattern in preindustrial societies. Indeed, the Greek situation shares some characteristics with modern warfare, but it also shows some important and perhaps underappreciated differences. The rise of the hoplite—a heavily armed and armored infantryman—in Greek warfare during the second half of the seventh and the sixth century B.C., had important political as well as military consequences.[10]

Hoplites fought in phalanxes, and each man's life depended on the behavior of his neighbor in the battle line. For the most part, they were small farmers. As with the Merus, Greek warfare followed the agricultural seasons. The constraints of vineyards, olive groves, and grainfields meant that battles were fought during the summers by local farmers to gain or protect local property. In most city-states except Sparta, there was little combat specialization and very limited drill. Men were vulnerable to the draft in any summer from their eighteenth to their sixtieth year, and in any battle the majority of men were likely to be over thirty. Men of rank fought as ordinary Hoplites among their less noble neighbors, and men fought with others of their tribal affiliation. When columns were decimated, they were not immediately reconstituted, but someone simply took the place of a man killed—usually a friend, relative, or neighbor. Men fought with their kin and their neighbors, with whom they held many interdependent relationships, and survivors of a battle were likely to know all of those killed. Men testified to the numbers of battles others fought, a sort of mutual recommendation system that fostered intensive mutual interest, strengthening coalitions beyond the battlefield.

Thus, while these conflicts involved hierarchy and trained warriors in organized combat rather than tribal ambushes, individual men were still fighting to protect or gain their own land resources, and they fought with their families, tribes, and neighbors. Commanders fought in the front lines with their men, suffering almost certain death if defeated, and a relatively high proportion of victorious generals died. In contrast, in most modern wars, commanders have moved farther from the front lines and share fewer risks with their troops, with some notable exceptions in the American Civil War.[11]

THE ECOLOGY OF RENAISSANCE WAR

European warfare in the Middle Ages, up until the mid-fifteenth century, involved many small-scale territorial wars and local, powerful men; it involved reproductive and familial gains and losses—a sort of violent housekeeping. Until perhaps 1450, knights fought largely as individuals and in small coalitions. Fights often started over revivals of old family claims to previously lost estates: "Political Europe was like an estate map, and war was a socially acceptable form of property acquisition." While particulars differed, familial economic interests were important. Noblemen, the landed gentry, had strong vested interests in waging war, not only for territory defense or acquisition, but for the spoils and riches to be gained. Later-born sons, with less access to resources and titles than their elder brothers, tended to end up in high-risk warfare. In Europe, from Roman times until Charles the Bold's military ordinance of 1473, differentiation by insignia in larger armies reflected social status.[12]

Just as among the Yanomamö, kinship mimicry was used to manipulate others: men called important people by kin terms to create or emphasize alliances. Military adventurers often became *fratres jurati*, or sworn brothers. When William the Conqueror invaded England, Robert de Oily and Roger de Ivery—unrelated men—were *fratres jurati*. After receiving honors for battle, Robert gave one to Roger as his sworn brother. Fifteenth-century *compagnies d'ordonnance* engaged the monarch's personal interest and were a route to high administrative office, confirmed disputed titles, and paved the way for handsome plunder of ransom profits. Successful warriors could gain higher status in peaceful times.

The introduction of guns permanently changed the nature of war, its conduct and conditions. At some level, of course, it is fair to say this of any technological development, any new war tool: horses, crossbows, and so forth. But gunpowder may have had an unforeseen impact that fed back into the nature of selection on men at war. Gunpowder led to larger and more costly armies, with more support personnel (e.g., masons to make balls of stone if the iron ones ran out). Once, conflicts were limited by an individual's ability to exert sustained physical effort; now they lasted longer. Sieges became more common.

Guns changed the nature of confrontation: guns could be poked through holes in fortifications. Perhaps most important, guns changed the requirements for soldiers: it took far less skill and fewer resources to fire a musket than to train as a longbowman or cavalryman. This separated the risk bearers from the profit makers in warfare. A new group emerged: weapons makers, with complex skills and political power, who could profit from lethal conflict without the risk of engaging in it.[13] The military-industrial complex was born.

Increased costs of the new technologies meant higher taxes. Although field armies may never have drawn on more than 5 percent of the population, recruiting was voracious and taxes became an important issue in all but the most local wars.[14] Mercenaries became prominent. Efforts to recruit the landed gentry were not too successful, for predictable reasons: because percentage quotas, rather than individuals, were drafted (except in the case of personal indentures), anyone who could buy or litigate his way out of the draft passed the burden on to someone poorer.

Early in the period, substitutes were likely to be younger sons of nobility, but later they became the unskilled, poor and hungry, and even those on the run from criminal proceedings. In 1572 the Venetian commander Giulio Savorgnan reflected on why his troops had enlisted: "To escape from being craftsmen, working in a shop; to avoid a criminal sentence; to see new things; to pursue honour—but these are very few. The rest join in the hope of having enough to live on and a bit over for shoes or some other trifle that will make life supportable." Similarly, in 1600, Thomas Wilson assessed English forces as comprising chiefly cottagers and copyholders, but also those who are "poore, and lyve cheefly upon labor, workeing by the day for meat and drinke and some small wages."[15]

Renaissance wars, then, came to be fought by mercenaries, not those defending or seeking lineage resources. The standing army came into being after the Renaissance, and power passed largely to others, not those who fought. Further, mercenaries had impact on civilians during peacetime: unlike knights, who were likely to return home, or paid soldiers, responsible to their employers, mercenaries during peacetime were "the responsibility of no man and they consequently became bandits."[16]

Over time, then, the potential individual reproductive gain of warfare fostered technological innovation, and then the technology of war changed the nature of warfare profits. The balance of benefits seems to have shifted, for most actors, from a familial, resource- or status-building strategy, as in the Yanomamö, the Meru, and Europe of the early Middle Ages, to a more stratified situation in which the poor and disenfranchised began to shoulder the risks and costs of warfare—as in Portugal among the nobles, and even in Vietnam. A shift occurred, in which more expertise was required for the manufacture than the use of common weapons, further separating the risk and the costs for warriors. Even today it would be difficult to disprove an accusation that individual fortunes and family empires are sometimes built from war—but typically by manufacturing entrepreneurs, not warriors. Of course, it's hardly so simple; rewards and booty remained important resource advantages for some men.

THE BEHAVIORAL ECOLOGY OF MODERN WAR

It is almost certainly true that past correlations between warriors' behavior and reproductive success are weak or nonexistent today, though we lack the necessary data to know. But aspects of men's behavior in modern wars, and of the organization of fighting forces, suggest that (1) proximate correlates of reproductive success due to risky and aggressive behavior still exist in modern wars, and (2) successful leaders organize field units in ways that play on the past kinship structure of warring groups.

Training

Many features of men's training for warfare mimic the proximate cues of both kin groups and close reciprocity. New recruits in many armies undergo forced transformation to uniformity (GI haircuts, uniforms). Training emphasizes communal values, often by using kinship terms; training is aimed not only at obvious skills, but also at ensuring cohesion, inciting hostility, enforcing obedience, and suppressing mutiny.[17] Recruits are likely to be called "son," "boy," or "lad." This paternalistic language goes far beyond basic training, and has done so for centuries. The "sworn brothers" fighting together in the Middle Ages engaged in reciprocity bolstered by kinship mimicry.[18]

David Hackworth argued that much U.S. combat failure in Vietnam arose directly from a few causes, including poor training that broke these important patterns.[19] Instruction was given by returned short-timers who had not wanted to go to Vietnam in the first place; instructors' dissatisfaction often led to war stories rather than to training when supervision was absent. Most significant, men were not trained and put into combat in units, with their mimic of familial structure and their strongly developed reciprocity. Instead, they were sent individually into combat in a strange land (the finishing preparation for this jungle war was often done on snow-covered fields). Without the required skills (e.g., none of Hackworth's trainees knew what to do when a gun jammed) and with their dependence on strangers, many died.

Pre-Battle Exhortation

Successful commanders from time immemorial have played upon the major themes of individual gain: wealth, social reciprocity, kinship gains, and sexual success. Thus Du Guesclin, a French knight of the mid-fourteenth century, mixed penitential motives with profit when he urged his recruits: "If we search our hearts, we have done enough to damn our souls. . . . For God's sake, let us march on the pagans! . . . I will make you all rich if you [follow me]!" Interestingly, Hugh of Caveley, an English knight fighting with him, responded by invoking kinship and reciprocity: "Bertrand, fair brother and comrade, mirror of chivalry, because of your loyalty and your valor, I am yours, I and all these here."[20]

Shakespeare captured the essence of this strategy when Henry V exhorted his rag-tag collection of men before Agincourt, calling them brothers even while highlighting their class diversity:

> We few, we happy few, we band of brothers;
> For he to-day that sheds his blood with me,
> Shall be my brother; be he ne'er so vile,
> This day shall gentle his condition:
> And gentlemen in England, now a-bed,
> Shall think themselves accurs'd, they were not here;
> And hold their manhoods cheap, whiles any speaks,
> That fought with us upon saint Crispin's day.
>
> (Henry V)

Henry invoked sexual selection, status-seeking, kinship mimicry— all to stir his men to do their utmost. The folks at home are sometimes similarly exhorted even today. In the news headlines of the Middle East conflict, headlines have reinforced kinship images: "Shipmates Become Like Brothers in Gulf Pressure-Cooker."[21] And men who serve under extreme conditions can come to value their comrades literally as kin. Stephen Ambrose, in *Band of Brothers*, the story of E Company, a rifle company exposed to face-to-face combat for long periods of time, recounts both the valor and the sacrifices men make under these conditions.

Resistance to Interrogation

Group structure and within-group reciprocity are crucial in fostering resistance to interrogation.[22] During the Korean War, for example, the most successful North Korean interrogations followed the breakdown of the prisoners' group structure. Reliable group structure—the presence of dependable comrades-in-arms—contributes not only to stalwart behavior in battle, but increased ability to resist interrogation after capture.

Insurgency and Terrorism

Technological advances have fostered newly dangerous ethnic conflicts. As the lethality of weapons increases, a small number of people can threaten the political stability and lives of large numbers of people. Major powers today shy away from escalation of international conflicts likely to lead to nuclear engagement. As local, usually ethnic, conflicts become more important, the danger posed by small groups increases: old conflicts, new lethality.[23]

Reviews of contemporary terrorism and low-intensity conflict suggest that despite exceptions like the Red Brigade, most conflicts in fact originate as local ethnic or religious conflicts of interest—e.g., Northern Ireland, the West Bank, Bosnia. Conflicts like these probably are intensified by the sorts of pressures prevalent throughout the evolutionary history of warfare: genetic lineages in conflict, expanded to become ethnic regional conflicts.

It is probably not irrelevant that in many successful (long-lasting) terrorist groups, the leader assumes a paternalistic role, and the group structure mimics that of families. Even a review of larger-scale conflicts suggests that ethnic and racial components are still important after World War II—far more important than ideology or nationalism. The current tensions in the Middle East and central Europe speak clearly to this point. The real danger is probably that small, originally local conflicts, because of technological advances, can wreak international havoc, and major powers can then be drawn into the fray. For example, U.S. funding for Special Operations Forces, typically involved in low-intensity interventions in local conflicts, increased 100 percent during the 1980s.[24] There is real potential for major powers to be drawn into confrontation through (originally) local conflicts, and balancing such conflicts can impose high costs and real risks—such as with NATO in Bosnia today.

Deception and Warfare

In other species, aggression is frequently accompanied by advertisements that exaggerate an individual's prowess (last chapter); some primates also deceive by using "neutral" postures to get close to a target. In human conflicts, there are probably not qualitative differences, but it may be that there are important quantitative differences in the importance of deception, compared to other species. We are capable of subtle deception; this may make conflict more, not less, likely—because conflict is more likely when at least one contestant's information is faulty.[25]

In human arms races, secrecy, deception, and misinformation are perhaps strategically useful, even if bluff doesn't always work. As a result, policymakers and government officials must often oppose citizen concerns.[26] This situation is complicated by the fact that leaders and their advisers play a double game—laterally, with other national leaders, and vertically, with the public—and the influence of the public on leaders' policies is maximized only periodically (in the United States, every four years), and the public receives imperfect, manipulated information.

Disadvantaged Men in War

From the engagement of relatively few, typically related, men in combat over resources that are directly related to their lineage's success, armies have grown and become hierarchical, with increasing divergence of the actors' interests. From related men who might squabble over this bride or those cows, we now have administrative groups sending others to fight. Maintaining discipline and loyalty in the face of unequal payoffs can be tricky. This problem lies at the very root of the transformation of societies from often highly polygynous states to (at least prescriptively) monogamous states.[27] In polygynous societies engaged in large-army warfare and conquest, formal reproductive and resource distribution schemes are common. As warfare technology changed and armies became larger, status differentials increased; no longer, as among the Yanomamö or even Alexander's armies, were the spoils of war a relatively simple reflection of individual courage and skill. More high-status men opted out of conflict, and more low-status men were recruited or drafted. The latter, from at least medieval times to the Vietnam War, suffered higher casualties than their richer competitors.[28]

Yet the fate of disadvantaged men is not so simple; if it were, the only question would be why they ever serve. Just as the costs and benefits for any individual of living in a group must be weighed against the costs and benefits of living alone or in another kind of group, the question must be asked: Did these men, on average, fare better or worse than if they had remained civilians? No data exist. For those who survived, it is possible that they did, in fact, fare better than if they had not served, through the status of wearing a uniform, just as in older high-risk, high-gain endeavors.[29]

WAR AND REPRODUCTIVE SUCCESS TODAY

In societies like the Yanomamö and the Meru, a man's lifetime reproductive success was likely to be closely correlated with his performance in war. These two examples are not atypical; in many societies, reproductive rewards for valor were standard.[30] Both direct reproductive rewards and heightened status or privilege before the law were common rewards for leadership and success in war. As armies became larger and more stratified, and as direct formal reproductive rewards for performance disappeared, the situation became less clear. For many, there were obvious risks—but no longer obvious reproductive rewards. For disenfranchised men today, with little or no chance of success in peacetime environments, there is possibly a correlation, but it is hard to measure.

Has the reproductive nature of lethal conflict really disappeared? Rape is a well-known but little-studied and seldom-quantified concomitant of warfare. In the Bosnian conflict, for example, there were reports of rapes of Muslim women by Bosnian soldiers: "We are seeing the same pattern repeatedly, of Chetniks telling women, 'It is better to give birth to Chetniks than to Muslim filth,'" said Mahir Zisko, executive director of the war-crimes commission. "Another statement that recurs is: 'When we let you go home you'll have to give birth to a Chetnik. We won't let you go while you can have an abortion.'"[31]

PROXIMATE AND ULTIMATE CAUSES OF WAR:
EVOLUTIONARY NOVELTY

In evolutionary terms, warfare should become and remain common only in circumstances in which the net fitness of warriors is enhanced. Throughout the history of conflict, in humans as well as in other species, reproductive profits have been associated with the risks of lethal conflict. With the elaboration of war and the increased pace of weapons development, selective outcomes have become less tied to individual actions and characteristics. Those with the most to gain from warfare came to suffer lower risks than those with little to gain.

We may well have unhooked the reproductive rewards from the behavior, so that lethal conflict is now counterselective for actual warriors, and driven only by proximate cues. Now, perhaps (though there are no data), war may not profit anyone directly involved in the conflict—yet the driving cues remain. With this in mind, consider briefly some of the proximate causes of warfare assigned in modern conflicts. This subject is far too broad for a detailed analysis here; specific causes are multifarious, and others deal with it in more detail. There are, however, some major patterns that are worth reviewing briefly. Oft-cited causes and correlates of war include the number of contiguous neighbors, economic conflicts (comprising approximately 29 percent of wars from 1820 to 1949), territory disputes, ideological differences, and misperception or distortion of information (e.g., in wars in this century). Pacifying influences include shared ethnicity, common government, recent alliance, and extended deterrence (when military strength is sufficient in the short term). Leaders' personalities can have a potent influence on events. Even in the huge number of specific causes of wars in modern times, the "ecological" categories are still rather limited: conflicts arise over resources (economic or territorial), and are less likely the longer and deeper are common bonds (kinship or reciprocity); open conflict is often precipitated by faulty information.[32]

There is, however, a behavioral ecology of war in our evolutionary past. The multiplicity of proximate correlates in modern warfare does not mute the importance of the kin, ethnic, and male competitive forces. There have been at least eighty small-scale wars since 1945, resulting in fifteen to thirty million deaths. The vast majority of these are between local groups; few are international. A review of recent atlases of war—or perusal of this morning's paper—reinforces the importance of essentially tribal conflicts of interest.[33]

Can Evolutionary Theory Help Avert Arms Races?

If potentially lethal conflict has very old evolutionary and ecological roots, can learning about these roots help us avert arms races? It is difficult to tell. Understanding the evolutionary background of lethal conflict and arms races may simply help us to understand that costs and benefits, deception and misperception, may involve currencies other than the immediate and obvious ones. Whether this could help us mitigate modern arms races, I don't know.[34]

In a major work on the kin selection roots of warfare, after eloquent analysis, the authors are reduced to calling for "some form of world government, some management force that might stabilize the most immediate threat to humanity—nuclear destruction."[35] The entire work, however, is an acknowledgment that the power of in-group amity and out-group enmity would likely force any such world government to be a conquest state, a chilling prospect.

There are hints, however, that even in the absence of power hierarchies or hegemony certain costs and benefits may promote cooperation.[36] The reproductive profits may be largely gone, but the proximate drivers remain. Writing of just such problems, biologist Richard Alexander found that "it seems to place me in a camp of those who see mutual deterrence as the basis for peace, even though I doubt that either self-extinction or massive destruction can be prevented indefinitely by deterrence alone."[37] He then suggested that our finely honed social and predictive intellect might provide several partial brakes on arms races—for example, if sufficiently numerous and powerful individuals and groups perceive that no matter who wins the confrontation, we all will lose, their power in internal social/political coalitions may allow them to force some solutions.

The difficulty is that the brakes are weakened by the dilemma a biologist would call the "levels of selection" problem (chapter 9): because natural selection fosters genetically selfish ways, long-term costs and benefits to groups are discounted compared to immediate, short-term costs and benefits to the individual. The larger and less related the group, or the farther in the future costs and benefits must be calculated, the greater will be the discount. Thus, given a short-term gain in status or tax base for a local constituency versus an unspecifiable risk of nuclear warfare some time in the future, we do not predict restraint. Of course, modern war is not the only issue for which our large-scale activities are problematic (see chapter 15).

Any successful attempt to foster peaceful behavior must change individual costs and benefits in the proximate sense; currently many social institutions and rewards mediate strong status pressure on individuals to enter into wars.[38] This sounds simple enough and has parallels in economic approaches, but there are serious problems: (1) there are no longer simply a few easily identified, powerful groups or coalitions with focused costs and benefits, so it is hard to figure out how to direct any such proposed manipulations; and (2) now, to an extent previously unknown, small groups can control highly destructive devices. When these individuals are unaffected by the sorts of costs and rewards humans have evolved to recognize, we call them terrorists. Our technology is sufficiently advanced that even such small groups can wreak havoc, killing great numbers of people.

15.

Wealth, Fertility, and the Environment in Future Tense

Desire is the very essence of man
—Spinoza

No species has ever been able to multiply without limit. There are two biological checks upon a rapid increase in number—a high mortality and low fertility. Unlike other biological organisms, man can choose which of these checks shall be applied, but one of them must be.
—Harold Dorn

I T IS EASY to imagine that our evolutionary past is remote, unconnected to our lives today, and of interest only when we think of traditional societies or ancient history. Yet, as we saw in the last few chapters, our evolved tendencies interact with today's novel environments. Today's cities, no less than yesterday's rain forests and savannas, *are* our environments. Just as soldiers in modern warfare display remnants of behaviors from past times, so do we all in our daily lives, as we live, work, and raise our families. Some physical and social aspects of our current environments are evolutionarily novel, largely the result of our own actions; and we may well have disconnected many behaviors and tendencies from their ancient results. Even so, our evolved tendencies interact with our current environments—and our future depends in part on the results of those interactions.

Complex as they are, the problems of modern warfare interacting with the remnants of our evolutionary history are more obvious than the paths that multiply our evolved resource acquisition tendencies into regional and even global problems of population and environment.

Three strands of argument are central to this book; they contribute, I think, to the curious resource situation in which human societies find themselves today. First, in past environments, those who strived successfully for resources were those who had children. It is clear from earlier chapters that in most species, males who get control of unusually rich resources have more offspring than others; females who fail to acquire some threshold amount, like unsuccessful males, fail to reproduce at all. In human evolutionary pasts, at least as reflected by the demography of traditional and demographic transition societies, the same pattern is clear (chapters 7, 8).[1] Even when monogamy was the rule and people married late, as in the demographic transition in nineteenth-century Sweden, men with rich resources were more fertile than others. In the evolutionary history of all living things including humans, then, getting "more" has always been reproductively more profitable.

Second, our continued creation of novel technologies and environments has resulted in a phenomenal spreading and interchange—we affect others across time and space in truly new ways. (In songwriter Tom Lehrer's words, "The breakfast garbage that you throw into the Bay, They drink for lunch in San Jose.") This brings the *levels of selection* problems of chapter 9 into stark relief. Our individual and corporate actions can affect populations far away, but it is not for their sakes that we do what we do.[2]

Finally, for humans in particular, an additional complication exists: we have created and intensified the importance of heritable wealth. Sheer fertility—number of offspring—is no longer a good predictor of lineage success in many environments today. We evolved to be fertile, to help our offspring survive, and to enhance their reproductive success. There are obvious trade-offs involved here. Do absolutely or relatively wealthier people have more children than others? Can highly consumptive parental expenditure make children's survival or reproduction more certain?[3]

Competition from other humans has always been a key selective pressure on us, and when intense this may lead to an advantage not to *more* babies, but to *better-provisioned*—and thus more consumptive—babies. That is, if only well-invested children become reproductive, the trade-off between numbers versus success of offspring becomes acute. In other species, this is what demographic transitions are all about: if only better-invested offspring succeed reproductively, and if making better-invested offspring means that fewer offspring are possible, fertility will fall. For many species in competitive or densely populated environments, the most successful reproductive tactic in many environments is not to make the maximum number of offspring, but to make fewer, better-invested offspring.[4] Sometimes the investment is in schooling, as in Thailand, sometimes in market training—the particulars vary, but I think the general pattern will become clear.

These three strands suggest to me that we will face new problems as growing, and increasingly consumptive, human populations interact with environmental quality and stability. The interactions of the strands will, I fear, affect our very ability to continue to live as we wish. We already have hints in the news headlines: global warming, increases in atmospheric CO_2 levels, destruction of the ozone layer, acid rain, precipitous decline of ocean fisheries on which some national economies depend, and so on. I suggest that while the proximate causes of all of these differ greatly, at a deep level, we have created the difficulties simply by doing well what we have evolved to do: garner, consume, be fertile, give to our children, and not look too far ahead.

FERTILITY, CONSUMPTION, AND SUSTAINABILITY: WEAVING THE STRANDS

In 1992, the "Northern" (developed) and "Southern" (developing) nations squared off at the Rio Conference. Each group accused the other of making the earth unable to sustain human life as we know it: "your fertility is too high" versus "we could afford our fertility if you were not so profligate with world resources." Later, the same issues were reraised, with little progress and no resolution.

At the heart of the debate lay conflicts over resources, fertility, and power—the issues of this book—multiplied up from individual to population levels. Human population is at an all-time high, per capita consumption grows daily, and wealth and power are unevenly distributed. Uneven resource control and power are older than humanity, of course, but the combination of high population and high consumption is relatively new.

The stage is set for real-world impacts of our evolutionary history. Suppose lowered fertility in some environments produces greater lineage success through fewer, better-invested children. What is likely? Per capita consumption will rise and costs are likely to be externalized; both have typically been effective competitive strategies.[5]

In a scenario such as this (arguably the First World versus Third World situation), *lowered fertility alone produces no solution to the population-consumption dilemma.*[6] A population's total impact on its environment arises not simply from the number of people, but includes consumption rates and the impact of consumption technology. The relative importance of each part of this relationship differs for developing and developed nations. Although the particulars vary, ignoring either population numbers or per capita consumption will mislead us.

WEALTH, FERTILITY, AND CONSUMPTION TODAY: EMPIRICAL DATA

Analyzing fertility relationships today is complicated for a number of reasons. First, because so many conditions are evolutionarily novel, there is a possibility that whatever patterns we see will be emergent phenomena, influenced by proximate cues once—but no longer—related to selective pressures. Second, societies are large, containing quite diverse subcultures whose members may face different pressures. Third, "wealth" and "status" today can mean a multitude of things (and those meanings can differ for men versus women, or people in different subgroups of a large and heterogeneous population). It can be difficult to be certain that relationships we think we see are not simply varying along with other factors. Perceived status is (as we would expect from our small-society evolutionary history) based on those we know, not large statistical aggregates—so broad patterns of wealth and fertility might no longer exist, even though we are doing what we evolved to do.

What do we actually know about resource consumption and fertility around the world? We know that average fertility is lower in developed nations than in the developing world, and that the age structure of developed nations is older than the developing world; both facts mean that population growth will be greater in the developing world. But we may be putting the pieces together in the wrong order.

Consider Nancy Birdsall's study on population growth in developing nations.[7] She found a slight negative trend between Gross National Product per capita and the Total Fertility Rate in developing countries (all of which have higher fertility rates than Europe and North America). Such data are widely used, for example by policymakers, to argue that high income correlates with lower fertility; thus the prescription to industrialize and to raise per capita GNP, and all will be well.

Though studies like this one come to mind when people speak about wealth, fertility, and consumption, they are inappropriate. They only tell us that the resource ecology of developed and developing countries is vastly different. The resource-fertility question can only be reasonably asked about individuals within the same population: do wealthier families have more or fewer children than poorer families? Do they consume more or less? Birdsall's *within-society* comparisons show a linear positive relationship between wealth and fertility in India, and a curvilinear relationship in the other three countries she examined. In no case did the wealthiest fifth of families have fewer children than the poorest. More to the point, the per-child wealth—a reflection of likely per capita consumption—in each country was vastly greater for the wealthiest families compared to others.[8]

So we need within-society comparisons. Even then we may have the wrong form of data; we usually have aggregate data, and these are often inadequate to answer these questions. Remember the Swedish patterns of chapter 8.[9] Aggregate data for the nineteenth century in Sweden show no significant relationship between status and fertility, but the individual-based data show considerable, complex impact of resources and status on men's total fertility and women's age-specific fertility. Individually based studies in several societies have analyzed the effects of cohort (when you were born, reflecting historical factors), wealth, and markets on human fertility. In the best empirical studies so far, results are mixed. Several show continued relationships between some measure of status and lineage continuance; in others, once-evident wealth-fertility differentials have disappeared in younger generations.[10]

Both theory and empirical data suggest a complex and locally varying picture (as in chapter 8). It appears that (1) children are an economic cost that is less-defrayed today than in some agricultural societies; (2) resources (including heritable resources) matter to the competitive success of children in most modern societies, and unless a family's wealth is boundless, fertility must decline as per capita investment goes up.[11]

Demographers, like behavioral ecologists and evolutionary anthropologists, are wrestling with these questions of wealth and fertility. Nobel laureate Gary Becker treated families as a neoclassical economic problem; Richard Easterlin suggested that a man's perception of his relative wealth would predict fertility, and introduced shifting preferences into the models. Demographer John Caldwell noted that in many societies, children can defray at least some of their economic costs. However, some assumptions that might seem obvious to the behavioral ecologists among us are lacking among at least some economically oriented scholars: a recent review came to the conclusion that "in short, there is no explanation for why Americans still want children."[12]

Despite our human preoccupation with preferences and proximate cues, despite all the complexity, some generalizations emerge. Fifteen of twenty-two recent tests of the Easterlin hypothesis (relative wealth promotes fertility), using micro- and macro-data, and in a variety of cultures, support the hypothesis. A person's marital status and number of existing children make a difference in people's deliberate fertility decisions. It also matters whether a government offers pronatalist tax breaks or maternity payments, or taxes fertility. Work to increase wealth represents harsher trade-offs for women than for men (more below). Finally, some additional complexity is introduced by simple preferences. For example, individuals who value the social capital (new relationships) created by having children, or whose religious beliefs prohibit birth control, may be likelier than others to have children.[13] No wonder patterns are complex! People who perceive themselves as well-off tend to want more children than their competitors; but these perceptions (at least so far) do not translate into strong positive wealth-fertility patterns across large heterogeneous groups.

WEALTH, WOMEN'S AGE-SPECIFIC FERTILITY, AND WOMEN'S LIFE PATHS TODAY

One of the best predictors of slowed population growth is delayed fertility. The strongest correlates of delayed fertility, in both developed and developing nations, are women's participation in the workforce and women's education. Women face harsher trade-offs than men when we consider resource-garnering activities compared to the production of children and the dispersal of investment to them, and fertility declines may arise in part from women's solutions of these trade-offs.

What trade-offs do work and schooling represent for women? Time and opportunity costs? Wealth versus fertility?[14] There may be multiple and complex reasons for earlier versus later fertility, and to understand what relationship these patterns have (if any) to evolved patterns requires that we can rely on the accuracy of the data.

Many widespread preconceived notions ("welfare payments increase women's fertility") are simply not true.[15] In fact, "wealth" is not tidily related to women's fertility in the United States: it looks as though it affects total fertility differently from the timing of fertility.[16] But, worse yet, what is wealth? Men's earnings? Women's? Both men and women today work to gain wealth. How does men's income versus women's income affect fertility? Work generates income, work costs time, and lost time affects men's and women's fertility differently. One study of current fertility of Swedish women, and white and non-white U.S. women, sorts out covariates like time versus income (though not across generations). Some major evolved patterns are apparent, although considerable complexity exists in transitory (and possibly nonselective) economic and fertility preferences.[17] Women's fertility trade-offs are as real in industrial nations today as in traditional societies (chapter 7): when women's work experience is controlled for, women's wages have several positive effects on fertility (first births for Swedish and non-white U.S. women; second births for white U.S. women). From hunter-gatherers to women today, lost time may decrease fertility, but wealth, even women's income, tends to increase it. Education and work have time and opportunity costs for women, while income has a positive impact on fertility—but it is very difficult to separate the three.[18]

Other dynamics (e.g., the per capita investment in children) clearly differ for wealthy professional women and poor single mothers—but both face time-money-fertility trade-offs. Thus women who earn their own income are not likely to be comparable to middle-income families in which household income arises only from the man's work. As a result, in Western industrialized nations today, in which women are a large part of the labor force, and in which divorce is prevalent, the wealth-fertility correlation for women may not be linear.[19]

This discussion has centered on developed nations, for which we thought we understood the dynamics of fertility. We know even less about the dynamics of fertility in developing nations; so policy prescriptions of "women's work and women's education to lower fertility" may or may not be effective, depending on the relationships.

AN EVOLUTIONARY PERSPECTIVE: REDUCING BOTH FERTILITY AND CONSUMPTION IS NOVEL

Let us return to my third strand of interest. With economic development, fertility is likely to decline; within populations, wealth and fertility may be positively, negatively, or not correlated. But with development, per capita consumption increases: estimates are that the resources used per child raised in the developed world are fifteen to twenty times the levels of the developing world. Reduced fertility does not mean reduced consumption.

These are issues of population, consumption, and long-term sustainability. Population patterns—birth, reproduction, and death—are the sum of what individual men and women do: they mate and marry, have children, and die, consuming resources along the way. Meanwhile, policy is being made and implemented. If, as most ecologists think, we face resource and population limitations in the near future, problems of wealth, health, and fertility for men and women are not just of interest to a few ivory-tower academics.

Estimating both consumption and population trends is fraught with difficulty; what will happen depends not just on numbers, but on age and sex ratio of populations, early versus late fertility, per capita consumption, and impacts of resource extraction and production. Despite the complexities, current estimates are that the world population will grow to perhaps 11.2 billion—more than twice its current level—by the year 2100; most growth will be in developing nations. The "carrying capacity" of the earth—the human population that can be sustained—depends on one's assumptions, but most scholars think it likely that people in developing nations will continue to want improved standards of living. After all, if you live in a hovel but can watch *Life Styles of the Rich and Famous,* you may soon yearn for more. And most of us in the developed world are unlikely to want to live at much lower levels.

How many people can the earth support? Sustainability is some function of (population) × (per capita consumption) × (technological impact of that consumption). These terms will depend on how people want to live—how many resources one wants to consume (what standard of living), how many children, and how evenly resources and children are spread among families. This is clearly a problem of political planning, institutional effectiveness, and prevailing ideas about social equity and economic restructuring. I'm no expert in any of these—but it doesn't take an expert to imagine that one cannot increase both numbers and per capita consumption endlessly if resources are not infinite. Many experts are increasingly uneasy about the level of realism and comprehensiveness in current approaches to the issues of population and environment in the twenty-first century.[20]

WHAT'S MISSING IN CURRENT STRATEGIES?

We saw in the last chapter that the behavior of men in war shows remnants of past social environments. Calling recruits by familiar terms, exhorting soldiers to bravery before battles, probably worked well in small groups of men facing risk (and potential individual gain). Today, because of the context of military organization, these may still work well in fostering the development of dedicated soldiers.

We also show remnants of our past in our resource and fertility behavior, and I am far from sanguine about current approaches to large-scale ecological problems: the contexts are very different, and none considers fertility and consumption together. Let us examine very briefly some current approaches, asking "what's missing" along the way. I can think of five: the "noble ecological savage," reduced fertility, reduced consumption, ecofeminism, and technological fixes.

Noble Savage

The concept of the noble savage, in the sense of the morally superior human uncorrupted by civilization, was strong throughout the sixteenth to nineteenth centuries. This concept has been expanded from moral to ecological arenas, most notably by Kent Redford. A typical environmental argument runs that we have, in important ways, "lost touch" with ecological constraints as we have developed technological insulation against ecological scarcity—traditional peoples were more conserving and respectful of those resources than we, and likelier to be willing to sacrifice personal benefit for the good of the group when conditions demanded it. It's all technology's fault; hunter-gatherers lived in a "balanced and harmonious" state, altering nothing.[21]

Perhaps the most cogent response to this view was offered by Marcos Terena, president of the Union of Indigenous Nations (UNI), in Belem, Belize, in 1990 at an international seminar on ecological problems in Amazonia:

> Why do you white people expect us Indians to agree on how to use our forests? You don't agree among yourselves about how to protect your environment. Neither do we. We are people just like you. Some of us view nature with a great sense of stewardship whereas others must perforce destroy some of it to obtain what they need to eat and pay for expensive medical treatment and legal counsel.[22]

Reduce Fertility

As a nation's wealth increases, family size typically decreases. Around the world, fertility is highly uneven, with most population growth in Africa and the poorest countries of Asia and South America. But falling fertility only means that the rate of population growth slows down. Even if fertility rates fell dramatically today, these populations would continue to grow for some time, because their age structure is such that increasing numbers of children will move into reproductive ages in the next twenty years. Even if all African countries today—instantly—assumed only "replacement" fertility of two children per couple, because the number of ten-year-olds is greater than the number of twenty-five-year-olds the total number of reproductive women, and population, would increase for some years. Reduced fertility may help but isn't enough.

Don't forget: "a nation's wealth" is an excellent reflection of per capita consumption. Many successful family-planning programs around the world have clear messages that "fewer children means wealthier families." Of course, wealthier families consume more resources, and if one asks about the interplay of population and consumption, fewer-but-more-consumptive families do not lead to decreased or stabilized resource consumption.

Reduce Consumption

"Live simply, that others may simply live" reads a bumper sticker on cars around my home. Can restraint by the wealthy promote sustainability? Certainly it's true that the developing nations were right: if we in the developed world used fewer resources per capita, more resources would be available for others. And I have friends and colleagues who work hard at creating lifestyles that have ever-lower consumption. But they are in the minority, and their actions have no direct effect on people in developing nations. If one thing is clear from earlier chapters, it is that reduced per capita resource consumption has not been, for humans or other species, a strategy that got an individual's genes into the next generation better than competitors under most conditions. Striving for what is rare and difficult to attain, showing off, has worked but is consumptive.

Like many other reasonable-sounding strategies, this one must become the most common strategy and it must be continually reinvented to work; it is also highly vulnerable to successful cheating. It is not an evolutionarily stable strategy (ESS); it is an argument that would only work if everyone acted identically, often against obvious self-interest. Could we, using evolutionary insight, substitute non-consumptive status competition for resource competition? For example, can we shift competition to center on hard-to-get (and thus status-enhancing) ends that are nonetheless less consumptive than many current status icons? I hope so, but am not ready to bet my next paycheck on it.

Ecofeminism

This proposition is the obverse of Professor Higgins: the world would fare better if only men were more like women: nurturing, caring, giving. But many of these approaches are almost stereotypic in their portrayal of sex differences. Consider: when males and females (in any species) succeed through different strategies, of course they behave differently. So elephant seal females are maternal and caring (at least to their own offspring), while males galumph about, sometimes squashing their own pups (whom they do not recognize) as they pursue further matings. The male loses less than the female if an infant dies. But as males' and females' successful strategies converge, so will their behavior. No matter that my female colleagues and I might like this idea. We look around and see that Norwegian prime minister Gro Bruntland is a good environmentalist (and so is Al Gore); Margaret Thatcher was not (and neither was George Bush). This approach, in most of its forms I have encountered so far, is exhortation, not analysis, advocacy rather than logic. There are excellent feminist approaches to problems of behavior and evolution; if these authors begin to tackle the population-environment dilemma, I would hope for some progress.[23]

Technology to the Rescue

Julian Simon was a major advocate of this position: in all of human history, there has been a correlation between an increased standard of living and increased numbers of people.[24] Every time we have faced an apparent "crunch," someone has invented something clever to solve it (and has usually done pretty well by doing so). I think our evolutionary background predisposes us to be attracted to this argument. It has never, in our entire evolutionary history, paid us to worry much about things too far in the future—we couldn't control it. As a result, we tend to use heuristic operating rules much like those of other species: do what feels good (à la Hemingway), and what makes us feel successful and admired. In our past, that has led to reproductive success, and to descendants over time.

Why worry? Biologists do worry; we are the Cassandras on this issue. As one of the epigraphs to this chapter notes, no other species has ever grown in numbers without limit. And while we humans are extremely clever, unless we figure out how to circumvent the Second Law of Thermodynamics, or colonize other planets or systems, we too will eventually be limited by the earth's resources.

This is a longer-term view than we have evolved to hold,[25] and I wouldn't want to be foolhardy enough to predict much about the outcome, or things like prices of goods along the way. But biologists, both evolutionary and ecological specialists, suspect human populations will eventually be limited: the only questions, they argue, are how pleasant the process and the result will be. Will extinction rates of other species continue to rise? Will human infant mortality rates rise as sanitation declines? Will a "sustainable" human population be able to live at the level enjoyed today in the developed nations, or will Bangladesh set the standard?

CAN NEW STRATEGIES AND TACTICS HELP?

If the approaches above strike me as incomplete or misguided, what, then, do I think are the components of a successful strategy? Can understanding the issues of earlier chapters be of any use? How can we foster strong normative conservation ethics, if we wish to? Throughout the earlier chapters I have argued that individuals (even nonhumans) act as if they could "calculate" some kinds of costs and benefits arising from their actions; those costs and benefits are current ones, not in the far future, and local, not global; and the costs and benefits were not, and need not be, monetary. Our costs and benefits as a social primate are older than the invention of barter and money, though not older than family structure and reciprocity. We evolved as a highly social species, and reciprocity is a powerful force, one we have probably underestimated in our attempts to encourage ourselves to act sustainably.

If these things are true, some solutions may come more easily with social levers rather than solely economic levers. Potentially important rewards include advertising one's status as a good cooperator. Warning: this is not identical to the strategy, "Exhort ourselves to be altruistic"; I am talking about social costs and benefits in a local and immediate sense. Recent information about human decision making suggests that, faced with complex decisions (as most social and environmental decisions are), we have unconscious biases, and that "knowing" overt factual information doesn't help as much as we wish.[26] I suspect, though we have no data yet, that one strong set of biases involves the costs and benefits of earlier chapters, and that the only strategies likely to work consistently are those that manipulate individual, familial, and reciprocal costs and benefits.

Whatever approach one favors, successful tactics boil down to a few: information, social incentives and disincentives (social norms) like exhortation and shame, economic incentives and disincentives (taxes, credits, discounts), and governmental regulations; and, slightly apart from these, "think globally, act locally" and "local control" for indigenous people. Each has both promise and problems; there are few data that allow us to analyze their relative effectiveness.[27] In sum, we face complex problems on temporal and spatial scales that are unique in our evolutionary history; we have made some small progress in tackling them, through ever-increasing information, manipulating our own and others' costs and benefits, and social norms that exhort ourselves and (especially) others to "do right."

AN EVOLUTIONARY BOTTOM LINE

If, as many scholars agree, current trajectories of population growth and resource consumption continue, few of the likely outcomes are pleasant.[28] Darwin's "Hostile Forces" are virtually identical to the Four Horsemen of the Apocalypse. We live in an evolutionarily novel world that we have created: can we manage it?

I wish I had an evolutionary panacea to propose; I know it is customary to make one's last chapter an exhortation: only learn or do what I propose, and life will improve. Yet it is not clear, when we are talking about large heterogeneous societies, that any small set of strategies, evolutionarily informed or not, will be quickly or widely effective. What is possible, I think, is far more modest. We almost certainly will muddle through, as ever, tinkering here and there, changing significantly only rarely, when we can get principals to agree. Analyzing the conditions we face thoughtfully, and putting more effort into those strategies that are consistent with our evolutionary past, seems to make sense. Tactics to reduce consumption that play on our perceived short-term self-interest seems more likely to persist than simple exhortation.

I suggested above that we have created these problems by doing what we have evolved to do—but that gives us only modest guidance about what to do next. We have, I think, far to go.

Notes

PREFACE

1. The relative importance of natural selection versus historical accident in evolution (e.g., E. O. Wilson 1975, 1978 versus Stephen J. Gould and Richard Lewontin 1979): an issue much debated between behavioral ecologists and paleontologists, for example. Scholars' opinions seem to reflect their backgrounds: paleontologists' records are obviously much affected by history, while behavioral ecologists look closely at strategies and their local selective outcomes, and thus see selective pressure at work; see Alcock 1998a,b and references therein for a concise statement of the behavioral ecologist's view. The importance of mechanisms: the biologist Stephen Rose (1997) and many evolutionary psychologists (e.g., Cosmides and Tooby 1992) argue that mechanisms are crucial to understanding behavior at all; most behavioral ecologists and evolutionary anthropologists suggest that mechanisms may help us understand the historical particulars (e.g., why daylength is a cue for migration rather than highly variable daily temperature) but that they are not required to understand function (why migration has evolved in this species in this location). As we learn more about mechanisms, they may enrich our understanding; but until then, I will proceed as an ecologist, looking at patterns in environments and organisms. Today's novelty: in general, behavioral ecologists begin by asking whether particular behaviors—predicted to be reproductively profitable—are so in modern environments; evolutionary psychologists begin by assuming the behaviors are not profitable now, but are Pleistocene remnants. See Sherman and Reeve (1997) and E. A. Smith (1999) for nice comparisons of these approaches, and Hammerstein (1996), Marrow et al. (1996) for new directions. My own thought is that if we assume selective irrelevance and do not seek to deduce, our failure arises from lack of effort, not failure of method.

CHAPTER 1. INTRODUCTION

1. See, e.g., E. A. Smith and Winterhalder 1992, Ridley 1993, 1996, Cronk 1991a, Borgerhoff Mulder 1991. The evolutionary anthropologist E. A. Smith (1999) has an excellent review of three current approaches to human behavior:

behavioral ecology, evolutionary psychology, and dual cultural-genetic inheritance.

2. Behavioral ecology and its intellectual relatives: human behavioral ecology is one of a number of quite similar evolutionary approaches to human behavior, with a variety of names—evolutionary ecology, biosocial science, human ethology, sociobiology, socioecology, evolutionary biological anthropology, and others. Depending on who is reporting, these may vary subtly, or they be different names for the same thing. Sometimes professionals in these fields have strikingly similar models of how human behavior evolved and functions; often they differ, sometimes wildly and incompatibly. But one common thread seems to be emerging from several of these approaches: very simple rules can generate emergent complex behavior—perhaps even the complex array of human behavior. Genetic influences: Hamer and Copeland 1998; historicity: Williams 1992a.

3. Barber 1988.

4. Heinrich 1989: 44.

5. Heinrich 1989: 302–313.

6. See Grafen 1991: 6. The phenotypic gambit is common in behavioral ecology; it assumes only two things: a strategy set, and a rule for determining the success of a strategy. It implies that strategies we see are at least as successful as any nonoccurring strategy would be if it existed in small numbers. Applying the gambit to a particular strategy set and payoff rule is a powerful way of testing the joint hypothesis that the strategy set and payoff function have been correctly identified, and that the gambit is true. Endler (1986) (chapter 8) has a nice discussion of how this behavioral ecological "equilibrium" approach compares to other views on the relative strength of selection, random events like mutation, and constraints.

7. E.g., Cronk 1995, Cancian 1973, Deutscher 1973, Harpending et al. 1987.

8. These two can be broken down further when it is profitable; see Holekamp and Sherman 1989. For a nice review of the issues of "proximate-ultimate" see Holekamp and Sherman 1989, Sherman and Reeve 1997, and Alcock 1998b.

9. See, e.g., Boyd and Richerson 1985: 157–166, Richerson and Boyd 1992, Hirschleifer 1977. As Richerson and Boyd 1992: 90 put it, "Conceiving of Darwinian models of (substantively) 'un-Darwinian' hypotheses, analyzing simplified versions of them, and considering the implications of the results clarifies complex long-disputed issues such as the possibility of group functions and the role of historical explanations." Specific approaches vary: Alan Rogers (1990, 1991, 1995) and Marcus Feldman and colleagues (1994), for example, use formal mathematical models to study such problems as wealth conservation and the interplay of cultural and genetic factors in sex-ratio selection, but to do so they must often make unrealistic assumptions such as haplodiploidy, or single-locus determination of the trait. In contrast, E. O. Wilson, Richard Alexander, and Richard Dawkins construct less formal, verbal, hypotheses. John Holland and Robert Axelrod, and Josh Epstein and Robert Axtell, in the agent-based com-

puter model Sugarscape, simulate what will happen when simple "cellular automata" are turned loose with simple rules in specified environments. Each group has its favorite approach, but no single approach promises perfect understanding. These approaches follow the *modus operandi* of the "hard" sciences, rarely applied in the past to human behavior. Each group makes simplifying assumptions; each can produce novel insights from which surprisingly powerful new predictions follow, just as in Euclidean geometry.

10. Retold by Alexander 1988.

11. E.g., Lenski and Travisano 1994, Travisano et al. 1995, Travisano and Lenski 1996. See also Rose and Lauder 1996. We are learning about the genetics of more human traits (including behaviors) daily, through the work of behavioral geneticists; see, e.g., Hamer and Copeland 1998 for an overview.

12. A caveat belongs here: the phenotypic gambit works best in relatively stable selective environments, when there has been time for strategies to appear, compete, and come to equilibrium—something we can not be sanguine about in human evolution. But simply assuming that the relevant portions of the environment have changed, without testing, seems illogical.

13. What we know comes mainly from two sources: studies of twins (Segal 1999), and identification of abnormalities linked to single-locus genetic changes. Mostly we describe the relevant features of an environment and can see that some strategies should work better than others in that environment—Alan Grafen's phenotypic gambit (above): we examine the evolutionary basis of a character (phenotypic trait) as if a simple genetic system controlled it, as if there were simple "payoff" rules for how many offspring each allele would generate, and as if enough time had passed and mutations had occurred to allow alternative strategies the opportunity to invade. This gambit obviously makes assumptions—that the strategies we see are the successful ones in the particular environment, and at least as successful as any non-occurring strategy. Even this argument involves a leap of faith that can lack justification (Grafen 1991: 7).

14. E.g., Richerson and Boyd 1992. This is a simple, parsimonious argument, an application of Occam's razor using the phenotypic gambit: let us begin with the simplest hypothesis, and add complications to it only when it no longer makes accurate predictions. We can make predictions and test them as we wait for the genetics to be sorted out.

15. The first was E. O. Wilson's 1975 *Sociobiology*. Cogent groundbreaking summaries are by G. Williams 1966, 1992a, W. D. Hamilton 1964, and R. Trivers 1972.

16. "Society" is not even a peculiarly human phenomenon. We may have more complex social rules than other species, but other organisms exhibit complex and multilayered behaviors such as alliance formation in intragroup and intergroup conflict (an older, but classic, discussion is found in Bonner 1980), and can have reciprocal and rather complex "political" behavior. Humans have developed many such phenomena to new extremes, but it's rare to find truly unique human phenomena with no predecessor in other species. It is often hard to find a point at which to draw the line. Other species, for example, exhibit ter-

ritoriality; we go further and define property rights, in which the holdings of absent owners are supported and protected by third parties. Our task is to find out when the functional significance changes—when does the actual reason for the behavior change?

17. Here is an important contribution of such approaches. If, as in traditional economics, we must "assume rational (economic) behavior" and make no other assumptions, we can get in great difficulties; only if we understand the problem thoroughly, have all the currencies in mind, and people really are largely rational, will we have a hope of coming close to an answer (e.g., see Arthur 1991, 1994; Smith 1997). Genetic algorithms and behavioral ecology may more quickly get to testable propositions.

18. See Belovsky 1987. Early attempts to adopt optimal foraging models for hunter-gatherer societies, based on classic foraging theory (Stephens and Krebs 1986) applied to humans, were disappointing. Gary Belovsky pointed out that foraging patterns were optimal, only if one calculated family, not individual adult, patterns of acquisition and consumption. Thus, ignoring the social complexity of providing for one's children meant that one could not figure out how the system worked. Exciting new work by anthropologists Kim Hill, Magdalena Hurtado, Hillard Kaplan, and Jane Lancaster (all at University of New Mexico), forthcoming, promises to enrich this considerably.

19. Dobzhansky 1961. Although for obvious reasons we began analyses (as in every beginning biology class) by acting as if one gene coded for one trait, and no other trait of the organism mattered, that's limited for all the reasons I want to explore here. Here is a simple example: there is an allele that raises the risk of atherosclerosis (coronary artery disease). And that is the advice physicians give. But is the advice correct? Sometimes, but not always: the "dangerous" allele, which typically raises the risk of atherosclerosis, in combination with certain other alleles, actually lowers the risk (see Sing et al. 1992, 1995, Templeton 1995). So what seemed simple at first, an apparently one gene–one response condition, actually involves epistatic interaction among genes.

For complex systems in particular, I suggest that it is more profitable to leave behind the specifics of coding, and concentrate on how the traits interact with the environment.

20. Dawkins 1986: 296.

21. E.g., Plomin et al. 1990: 81.

22. Plomin et al. 1990: 401.

23. Evolutionary psychologists, for example, frequently argue that we can only study things *known* to be adaptations: traits we can prove to have evolved specifically in response to a given pressure, and for which we can identify the resulting particular structure. Typical arguments also assume adaptations formed in "the EEA," the environment of evolutionary adaptedness (usually assumed to be hunter-gatherer social systems in the Pleistocene); see Irons 1998 for a critique. As a behavioral ecologist, I suspect humans are like other wide-

ranging species in showing some basic conformity, with considerable elabora-tion and diversity in traits. See E. A. Smith 1999 for an excellent overview com-paring evolutionary approaches to human behavior.

24. Lessells 1991 has a nice discussion of this problem.

25. For recent good discussions of the concepts of adaptation and current util-ity, see Williams 1992a, Sherman and Reeve 1997, and Seger and Stubblefield 1996.

CHAPTER 2. RACING THE RED QUEEN

1. Van Valen 1973, Ridley 1993. Others (e.g., Alexander 1987) use the term "arms race" for the evolution and counterevolution of competitors, predators and prey, parasites and hosts, for example.

2. Darwin 1859, 1871. Darwin's original formulation of the theory of natural selection began with the observation that organisms seemed well suited to their habitats. Even though he wrote before the birth of genetics, he said things re-peatedly that suggest he was well aware that what we now would call "genetic self-interest" was paramount in creating the "fit" between organism traits and environmental pressures (Cronin 1991). Selfish gene: Dawkins 1989.

3. It has always amazed me that optimization approaches work as well as they do, for "success" for any allele is just producing a better-surviving or bet-ter-reproducing organism, not the best possible (see, e.g., Grafen 1991, Kirk-patrick 1996). Rolling about on fitness landscapes is not, by design, a very effi-cient process (one must always travel "uphill" and so can get stuck on a minor fitness foothill, whenever to get to the Mount Everest of fitness would require first going downhill). Optimization approaches simply ask: what will work best (what can the highest peak be?) and take no note of closer, lower hills.

4. E.g., Dawkins 1982, 1986, 1989; Daly and Wilson 1983; Alcock 1998b; Trivers 1985; Krebs and Davies 1991, 1997; and specific studies cited therein.

5. Empirical tests of selection and historical accident: Travisano et al. 1995, Lenski and Travisano 1994.

6. This is the crux of reproductive strategies in other species, and perhaps of the "demographic transition" in humans (chapters 13, 14). That is, *ceteris paribus*, more resources → more offspring → more genetic persistence. But if the envi-ronment is highly competitive, more resources per offspring may be required for successful offspring. The result is likely to be fewer offspring but equal or greater resource consumption per capita. High consumption, as it relates to per-sistence, is seldom important in other species, but may become an issue for humans.

7. E.g., Hausfater and Hrdy 1984; see also chapter 6.

8. An additional complication we face in asking about human behaviors such as infanticide is that the behaviors themselves may be repugnant to some of us.

Behavioral ecology has nothing whatever to say, so far as I can tell, about the moral or normative "rightness" or wrongness" of behaviors; human individuals and societies make those determinations. So, when I observe that infanticide tends to occur in humans and in other species under specific environmental conditions, that does not mean that it is somehow "right" (the "naturalistic fallacy"), nor does it mean that infanticide is "wrong".

9. E.g., nonreproductive helpers at the nest: Woolfenden and Fitzpatrick 1984; sterile honeybee workers: Seeley 1985.

10. Darwin 1859: 236. See Cronin 1991.

11. Hamilton 1964. Actually $rb > c$ is Grafen's 1991 recasting of Hamilton's formula, which was $b/c > l/r.$ b = recipient benefit, c = donor cost, r = degree of relatedness.

12. Haldane 1955. See also Dunbar et al. 1995, Johnson and Johnson 1991, and McCullough and Barton 1991 for historical examples of how benefits, costs, and relatedness interact.

13. Notice that we have already begun to talk about different costs and benefits: energetic (calories) and reproductive (genes). Implicit in this and following arguments is the idea that organisms routinely spend calories, or take risks, to get genes.

14. Hamilton 1964: II, 19. See Alexander 1974 for a discussion of kin selection and parental manipulation in social insects; see also Grafen 1991 and Dawkins 1979 for a cogent analysis of misunderstandings of kin selection.

15. As I point out in chapter 9, some authors like to treat kin selection as though it were *kin-group* selection (e.g., D. S. Wilson 1980, Wilson and Sober 1994, Sober and Wilson, 1998, Dugatkin and Reeve 1994). The functional focus is different, and even Dugatkin (1997: 44) recognizes the utility of separating the approaches for most analyses. Here I agree with biologist George Williams (pers. comm.) that the appropriate interpretation of kin selection is *not* kin-group selection, but "individual selection for the adaptive use of genealogy."

16. E.g., Belding's ground squirrels: Sherman 1977, 1981; sphecid wasps: West-Eberhard 1975, 1978; wolves: Mech 1977; Florida Scrub Jays: Woolfenden and Fitzpatrick 1984.

17. Of course, calculating actual relationships can be fraught with difficulty; e.g., Dawkins 1979, Grafen 1984, 1991.

18. Risk and relatives in crisis: McCullough and Barton 1991, Diamond 1992; conflict and cooperation in Icelandic (and Vikings in other countries) and English history: Dunbar et al. 1995, S. Johnson and Johnson 1991.

19. Krebs and Davies 1997, Dawkins 1989. For an excellent treatment of selection in a broader way, see Bell 1997a. Readable introductory material includes Alcock 1998b, Krebs and Davies 1993, and Bell 1997b.

20. More in chapters 8 and 15. A debate currently rages among "evolutionary psychologists" and "evolutionary anthropologists" on just this issue. Both

groups are interested in the question, "Why do we see some traits rather than others today?" Psychologists typically use a "forward" method (Sherman and Reeve 1997, Buss 1999), arguing "what we see now is here because it was adaptive in the Environment of Evolutionary Adaptedness (EEA, roughly hunter-gatherers in the Pleistocene)" (e.g., Barkow et al. 1992). Evolutionary anthropologists, keenly aware of the variation in human ecologies (from the Arctic to tropical rain forests to Australian deserts), tend to use the "backward" method, asking "what would be useful in this environment as described?" and argue that one might be better off characterizing the adaptively/selectively relevant features of environments (e.g., Irons 1998), current or Pleistocene.

21. Hamilton (1996: 14) does not attribute the anecdote to the bishop's wife. Whether apocryphal or true, the response to ideas linking natural selection and humans is nonetheless typical.

22. Williams (1989) called natural selection "immoral," but it has existed since the beginning of life, while moral systems came along much later; it may, in fact, give us a way to think about ethics—it has nothing whatever to say about what "ought" to be, but it might help us think analytically about systems of ethics. Thus, I wince at the cartoon in which the defendant, before the judge's bench, pleads, "Not guilty, Your Honor, by reason of genetic determinism."

23. Onerousness of invoking selection: Williams 1966.

24. Male chimp: Goodall 1986; Great Tits: Krebs and Davies 1993; gull population density: Kadlec and Drury 1968, Drury 1973, Podolsky 1985.

25. The most frequent reason that animals become rare or endangered today is that their habitat is altered by humans (e.g., Ehrlich and Ehrlich 1981; see also Ehrlich 1997, and chapter 15).

26. The Pill as a modern Wonder of the World: *The Economist,* December 25, 1993; sexual access of wealthy men: Pérusse 1993, 1994.

27. The accusation of "tautology" has been ably answered by others, both for the principles of natural selection (Maynard Smith 1969, Stebbins 1977, Alexander 1979, Dawkins 1986), and the concept of fitness (Dawkins 1982, ch. 10).

28. Cronin 1991.

CHAPTER 3. THE ECOLOGY OF SEX DIFFERENCES

1. The issue of sexual versus asexual reproduction is itself an interesting and incompletely resolved question: Why is sexual reproduction so widespread? For recent and readable treatments, see Wuethrich 1998, Barton and Charlesworth 1998, Michod and Levin 1988, Bell 1982. For Hamilton's parasite argument, see Hamilton et al. 1981, Hamilton and Zuk 1982, Ebert and Hamilton 1996.

2. Size affecting the sexes differently: Roff 1992, Stearns 1992.

3. E.g., Warner et al. 1975.

4. Mating and parental subsets of reproductive effort: Williams 1966, Low 1978, Alexander and Borgia 1978, E. A. Smith and Winterhalder 1992; definitions of fitness: Dawkins 1982, ch. 10.

5. Questions of why and when there is sex in the first place are broad and well treated by others: e.g., Bell 1982, 1997a,b, Ghiselin 1969, Williams 1975, Maynard Smith 1978, Michod and Levin 1988. A fun popular treatment of human sexuality is Diamond 1997.

6. This suggests that the parthenogenesis is a derived condition.

7. E.g., some rotifers, aphids; see Williams 1975.

8. Parthenogenetic and hermaphroditic species are relatively rare and found in particular circumstances (see Williams 1975, Ghiselin 1969). Ryan (1998) reviews ways in which sexual selection creates sex differences.

9. In fact, this aspect of sexuality is so nearly universal that the critical arguments come from thought-experiments and modeling exercises, for not enough variation remains in the real world to do empirical tests. See, e.g., Parker et al. 1972. Arguments about the evolution of anisogamy have been expanded and elaborated by Bell 1978, 1982, Charlesworth 1978, and Maynard Smith 1978. Additionally, genomic conflict (discussed below, Hurst 1995, 1996; also Partridge and Hurst 1998) will lead to anisogamy.

10. See Bell (1982: 21–23) on biologists' use of the terms "gender" and "sex."

11. Sometimes the two sexes are physically so different that early zoologists, concerned with describing species, described males and females (of what we now know are the same species) as different species.

12. Today humans are so slightly constrained by our environment that we seldom imagine that past constraints can make much difference to us; for example, a recent (March 14, 1995) *New York Times* headline touted, "Evolution of Humans May at Last Be Faltering." Our discussions of sex differences often begin with assumptions that observed differences are, by and large, "simply cultural." Here I hope to challenge that assumption.

13. E.g., see Ghiselin 1969.

14. Some species have more than two sexes—like the multispecies slime molds. Laurence Hurst (1991, 1992) predicted the critical difference we find.

15. The well-known vaginal infection *Chlamydomonas* engages in a spectacular war of attrition between the cytoplasmic materials of its *plus* and *minus* (rather than *male* or *female*) gametes that wipes out 95 percent of the chloroplasts (cytoplasmic instruction carriers). And it is not an even destruction: only some of the *plus* parent's chloroplasts are destroyed, but *all* of the *minus* parent's are.

16. See Hoekstra 1987 and Hurst and Hamilton 1992. Ridley (1993: 97 et seq.) has a lively and readable account.

17. Low 1978.

18. Such offspring-specific expenditure is true parental investment (PI): Trivers 1972; parental effort: Low 1978. In any time period, $\Sigma \, PI = PE$.

19. Thus, any extrapolation from seasonal success to lifetime success is risky; Clutton-Brock 1988, 1991; also Clutton-Brock et al. 1986.

20. Elephant seals: Le Boeuf and Reiter 1988; butterflies: Elgar and Pierce 1988. See also Strier 1996, reviewing this problem in New World primates. Variance in male lifetimes: Clutton-Brock 1988: 474.

21. Red deer: Clutton-Brock et al. 1982, 1986; elephant seals: Le Boeuf and Reiter 1988.

22. Gaulin and Fitzgerald 1986.

23. E.g., Gaulin and Hoffman 1988, Maccoby and Jacklin 1974, Moir and Jessel 1991, Silverman and Eals 1992, 1998, Eals and Silverman 1994, James and Kimura 1997, McBurney et al. 1997; see also Hyde 1996. The general pattern of sex differences suggests a history of sexual selection; because differences arise at puberty, sex hormones seem likely to be the important proximate mediators, both organizational (e.g., C. Williams and Meck 1991) and activational (e.g., Hampson and Kimura 1988, Silverman and Phillips 1998).

24. Orians 1969, Emlen and Oring 1977, Borgia 1979, Davies 1991.

25. Elephant seals: Le Boeuf and Reiter 1988; red deer: Clutton-Brock et al. 1982.

26. See, e.g., Buss 1985, 1989, 1994, 1999. It is worth noting that Buss found that among the characteristics women seek in men across at least thirty-seven cultures today, resource holding and resource potential were always near the top of the list.

27. E.g., Betzig 1986.

28. Rattray 1923: 88–89. Laura Betzig (1986) gives other equally dramatic examples, as well as analysis.

29. In some primates, this behavior continues as the male expends some parental effort as well as mating effort; Smuts (1985) has called these more complex and lasting relationships "friendships."

30. Burley (1979) suggested that natural selection conceals ovulation to counter "a human or prehuman conscious tendency among females to avoid conception through intercourse near ovulation." Alexander (1979: 135) makes a suggestion similar to Burley's about the concealment of ovulation from the female herself. Hrdy (1981) suggests that concealment allows for "paternity confusion"; in primates in which females are receptive for a period longer than the ovum is fertilizable, and in which males may give some paternal care to infants, female receptivity-without-certain-fertility might allow females to garner care for infants from more males than only the one who is the father. Alexander and Noonan (1979) argue that concealed ovulation may have enabled females to force desirable males into consort relationships long enough to preclude the male's access to other matings, and simultaneously raised his confidence of paternity, making paternal investment more profitable. Strassmann (1991) makes the further point that concealment of ovulation might favor subordinate males,

inferior in direct physical competition for females, but willing and able with regard to parental effort.

31. Höglund and Alatalo 1995.

32. Result of skewed female choice: Fiske et al. 1998; lek paradox: Taylor and
Williams 1982, Williams 1992a.

33. Hamilton and Zuk 1982, Ebert and Hamilton 1996.

34. As Daly and Wilson (1985, 1987, 1988) note, spousal and child abuse cases
are in some cases appropriately viewed as violent mate-guarding techniques by
males with little to offer.

35. E.g., Murdock 1967, 1981.

36. Darwin 1871, I: 319.

37. "Expenditure" was not defined by Fisher (1958: 159), who first used the
term parental expenditure and presumably meant caloric cost; Low (1978) defined parental "effort" as the sum of parental "investment" (Trivers 1972) over
time.

38. Roff 1992, Stearns 1992.

39. S. Frank 1986.

40. I want to be clear here that I am not arguing that male and female humans
"must" do anything, or that only "genetically determined" mechanisms dictate
behavior. I am talking about the relative ecological costs and benefits of different
behaviors. As I will argue repeatedly, genes and environment interact, and no attempt can succeed which either ignores genetic components or seeks to explain
behavior simply as a series of genetically programmed events. We know that
members of families are more alike than strangers, and the wealth of twins-
raised-apart studies (e.g., reviewed by Segal 1999, L. Wright 1997) makes it clear
that many very subtle traits have strong genetic components. But we also know
that intrafamilial correlations of traits such as cognitive ability differ among populations and there are strong sex-by-generation and ethnic group-by-generation
interactions (DeFries et al. 1982). That is, intrafamilial similarities arise both
through genetic and environmental (family environment) influences, and are
further modified by external (cultural environmental) factors. Yet if we have a
grasp of the commonalities and variations among mammals in general, we may
reach a better understanding of the bias that exists. Then, if we wish to follow patterns different from our evolutionary history, we are better equipped to
do so.

41. These patterns have different impacts on mating systems. Continuous
two-parent care inclines to monogomy (geese); alternate "relief" care (sandpipers) tends toward multiple-clutch monogamy, and polygynous, polyandrous,
or polygynandrous desertion and remating. See Emlen 1982a,b, 1995, 1997.

42. Social monogamy is common, but biologically certain paternity is rarer
than we thought (see Morrell 1998). Male confidence of paternity and male
parental care: Alexander 1974, Hartung 1983, 1997, Kurland 1979, Kurland and
Gaulin 1984. Mating competition as shaping male paternal investment in birds:

Birkhead and Møller 1992, Davies 1992, Møller 1994; in mammals: Smuts 1985, Whitten 1987, Smuts and Gubernick 1992; among human foragers: Hawkes 1990, 1993, Hawkes et al. 1995, Hawkes, O'Connell, and Blurton Jones 1997. Frequency-dependence of trade-offs: Hawkes et al. 1995.

43. Social scientists use the term differently for human marriage systems; see chapter 4.

44. Intensity: Wade, 1979, Wade and Arnold 1980; high variance in monogamy: Clutton-Brock 1983.

CHAPTER 4. SEX, STATUS, AND REPRODUCTION AMONG THE APES

1. Bottom line: Ellis 1993, Dewsbury 1982, table 1, table 3, Silk 1987, Cowlishaw and Dunbar 1991; trade-offs: Ellis 1993, Packer et al. 1995.

2. Smuts 1985, 1987a,b, de Waal 1986, McGrew et al. 1996, Rodseth et al. 1991.

3. The phenomenon of "male mating-season mobility" is well documented, but the reproductive consequences are difficult to establish for particular males, since we typically have data on mating, but not on actual paternity.

Female choice: a caveat is important here. Since female mate choice (e.g., Andersson 1994) can be constrained by male behavior, including sexual coercion (Smuts and Smuts 1993, Clutton-Brock and Parker 1995a,b, Hrdy 1997a,b, Hooks and Green, 1993), in a number of primate species (e.g., Watts 1989) mate *choice* in primates may not be the simple reflection of mate *preference* that we usually assume it to be.

4. Rank and copulations in macaques: D. Hill 1987, Paul et al. 1993; copulations not the issue: Rowell 1988, Fedigan 1983. In fact, in some species, females appear to trade copulations for a male's contribution to group defense. A male is more likely to take risks in defense of a female with whom he has copulated; females in some species actively seek copulations with subordinate males when they are not fertile (e.g., see Kano 1996, Inoue et al. 1991, Shively and Smith 1985). Hrdy (1997a,b) suggests that female solicitation of multiple copulations (either simultaneously or sequentially, depending on the breeding system) is characteristic of prehominid females and a tactical response to male efforts to control the timing of female reproduction (see also Hooks and Green 1993).

5. E.g., Cowlishaw and Dunbar 1991, Strier 1996. Studies from the wild measuring paternity and showing a positive rank-success correlation: Pope 1990, deRuiter et al. 1994. Studies of captive populations (e.g., Berard et al. 1993, Inoue et al. 1991) sometimes fail to find a positive relationship. One potential explanation is that in these captive groups, the intensity of male-male competition is skewed; the groups are of unusual demographic composition (more males live in these groups than live in wild groups). Several relevant papers are found in volume 34 of the journal *Primates*.

6. De Waal 1982, 1986.

7. Complexity of interactions, including female-male: Noë et al. 1980; male negotiations: de Waal 1986, 1996a,b. De Waal suggested that, for chimpanzees, these sharings are essentially mechanisms to reduce aggression, to appease and to reassure. Males form coalitions that change over time and show little correlation with social bonds, as measured by association. These coalitions are reciprocal and shifting, and if a male defects, the coalition fails. These guys may be keeping track of who has helped and who has harmed them: sex, power, and politics among the apes. No wonder de Waal titled one book *Chimpanzee Politics*.

8. Interpopulation variation: Takahata et al. 1996; females seeking furtive matings: Gagneaux et al. 1997. More reproductive access by high-rank males: Nishida and Hosaka 1996, Tutin 1979, Hasegawa and Hiraiwa-Hasegawa 1983, Nishida 1983; no clear relationship between rank and reproductive success in the one population: Tutin 1979. In the Gombe population, while the dominant male had the highest reproductive success, he did not father two infants for whom paternity is known (Morin et al. 1994). In the Bossoi population, which has had a single male since 1985, one of four infants born since that time was not fathered by this male (Sugiyama et al. 1993).

9. Bonobo female alliances: Parish 1996; rank and mating, female choice: Kano 1996; mildness of male aggression: Nishida and Hosaka 1996, Kano 1996.

10. See reviews by Ross 1998 and by Strier 1996, who also examines the ecology of mating systems in New World primates.

11. Reviewed by Silk 1987, Dewsbury 1982, Fedigan 1983; gelada baboons: Dunbar and Dunbar 1977; chimpanzee females: Pusey et al. 1997. Silk's review has extensive references. See also Boyd and Silk 1997, chap. 7, for a recent overview.

12. Packer et al. 1995.

13. Food defense and female bonding: Wrangham 1980, 1987, van Schaik 1989, Barton et al. 1996; predation: Cheney and Wrangham 1987.

14. Two-parent care: P. Wright 1984, Kurland and Gaulin 1984; male inability to monopolize multiple females: Mitani 1984; infanticide protection: van Schaik and Dunbar 1990, van Schaik 1996.

15. Terborgh and Goldizen 1985, Goldizen 1987.

16. Body size, testis size, and canine size comparisons: Harvey and Harcourt 1984. Note that large body size may confer additional advantages in species like gorillas and orangutans, in which males coerce females into copulations (Watts 1989, Wrangham and Peterson 1996).

17. This general statement is true for other primates, but with an important difference: in nonhuman primates, resource distribution influences female distribution and group size, and males follow. In humans, biparental care and male control of important resources appear to lead to a situation in which males comprise the primary groupings (e.g., patrilocality is common), and females follows.

18. E.g., Freedman 1974, Kagan 1981, Monroe and Monroe 1975.

19. For example, females with extra X chromosomes (XXX is most common, but up to XXXXX exist) have an extra Barr body for each extra X chromosome, and these females are at risk for retardation (Plomin et al. 1990). Males with an extra X (XXY; Kleinfelter's syndrome) are sterile, and have very small testes after puberty; they represent about 1 percent of individuals institutionalized, about ten times their representation in the general population. Males with an extra Y chromosome (XYY) were for some time thought to be more aggressive than others; this is not true (see review by Plomin et al. 1990), but such males are taller, tend to show delayed language development, have learning problems at school, and may have a slightly raised incidence of mild retardation. So there are a few clear observable sex differences attributable to the sex chromosomes.

20. See Skuse et al. 1997, Henn et al. 1997. The X chromosomes from mother and father are thus "imprinted." One kind of genetic imprinting, reviewed in chapter 6, arises from conflict between alleles from Dad and alleles from Mom (e.g., Haig 1992, 1993). The conflict theory does not explain this imprinting case well (Pagel 1999). Rather a form of dosage compensation appears to operate: because females have two X chromosomes and males have only one, females can receive twice the dose of gene products found on the X. This could be disruptive to other genes shared by males and females, and so females have evolved to switch off genes on one of their X chromosomes (Iwasa 1998). If, as seems likely, for whatever reason, selection favors better social skills in females than in males (or, socially clumsy males are not at such a disadvantage as inept females), then that information must be contributed by the paternal X (which always goes to daughters, while a maternal X can contribute to a daughter or a son). Thus we see the apparently counterintuitive result that Turner's syndrome daughters, who have only one X, from mother, are socially less adept than normal XX daughters (see also Thornhill and Burgoyne 1993 for a related effect in mice). The dosage compensation and the conflict theories are appropriate for explaining imprinting in different evolutionary settings.

21. Betzig 1986; Turkmen: Irons 1979a,b; Kipsigis: Borgerhoff Mulder 1987, 1988a,b, 1990, 1995; Ifaluk: Turke and Betzig 1985; Yanomamö: Chagnon 1979, 1982, 1988; Ache: K. Hill and Kaplan 1988a,b, Hill and Hurtado 1996; Bushmen: R. Lee 1979a,b. In another dozen societies reviewed in less detail by J. Hill 1984, resource control enhanced reproductive success in 10/12, and Mueller 1993, Mueller and Mazur 1996, 1998 discuss military men today.

22. Keyfitz 1985: 142–161.

23. Atlas: Murdock 1981. Age, brideprice: Borgerhoff Mulder 1988b, 1995.

24. You might also say that women with high reproductive value are free to choose men with greater resources, although direct female choice is difficult to demonstrate in many societies. This situation holds not only in marriage markets, but in remarriage. In most societies, widows and divorcees remarry less frequently than their counterparts, and those women who do remarry are younger (of higher reproductive value) than the women who do not (e.g.,

Bideau 1980, Åkerman 1981, Cabourdin 1981, Imhof 1981, Bideau and Perrenoud 1981, Corsini 1981, Knodel 1981, Wolf 1981, Glick and Lin 1986, Griffith 1980). Demographers, concerned only with economic aspects of mating markets, have been at a loss to explain this widespread pattern (e.g., Knodel and Lynch 1985), but it appears obvious to behavioral ecologists.

25. Sweden: Low 1991; Norway: Drake 1969; England: Hughes 1986.

26. Women's status and reproduction: Mealey 1985, Mealey and Mackey 1990, Essock-Vitale and Maguire 1988; women, resources and politics: Low 1992. See also chapter 12.

27. For the wives of such polygynous men in traditional societies, fertility may be uneven; in a number of societies, second and subsequent wives, for whatever reasons, have fewer children than first wives (e.g., Dorjahn 1958, J. Smith and Kunz 1976, Daly and Wilson 1983, Bean and Minneau 1986, Garenne and van de Walle 1989, Pebley and Mbugua 1989, Sichona 1993). The fertility effects are sometimes small (e.g., Timæus and Reynar 1998, and sometimes ephemeral in terms of number of grandchildren (Josephson 1993). There are many covariants (e.g., women who remain unmarried until their reproductive value is low are unlikely to become first wives, and also likely to have few children in their remaining lifetimes).

28. Uneven fertility: Daly and Wilson 1983: 283; general female reproductive ecology: Low 1992, 1993a,b. A. Campbell (1995) explores the conditions under which young women do engage in physical aggression, usually triggered by three issues related to reproductive fitness: management of sexual reputation, competition over access to resource-rich men, and protection of heterosexual relationships against rival women.

29. Packer et al. 1995, Hawkes, O'Connell, and Rogers 1997; see also chap. 7.

30. See Wickler and Seibt 1983.

31. Such societies are what Alexander et al. 1979 termed "socially imposed monogamous" systems: socially "monogamous" but biologically polygynous.

32. Usually measured as percentage of men and women polygynously married.

33. Low 1990b,c. White and Burton (1988; see critique by Low 1990b) suggested polygyny is "favored by homogeneous and high-quality environments," perhaps reasoning that rich environments are easier to exploit. Unfortunately, the real relationships are difficult to determine, because very different climate zones are lumped, as if the tropical rain forest were comparable to seasonal, high-rainfall areas. Even brief reflection suggests that it is far too simplistic ecologically to assert that there is a single measure of environmental quality one that decreases with cold or aridity, so that dry polar regions are lowest on environmental quality, and moist tropical regions highest. This not only confounds extremeness, range of variation, and predictability, but ignores, for example, the fact that pathogen stress, hardly a contributor to "high environmental quality"—but a factor promoting polygyny (see above)—is highest in

moist, tropical regions, and that protein availability, probably a requirement for "high" environmental quality, can be quite high in cold and dry regions.

34. This argument is, of course, a major focus for work on the evolution of sexual reproduction itself (see chapter 3). The difficulty lies in defining the sort of uncertainty that will favor the advantages of sex, or variable offspring, to a degree sufficient to compensate for the loss of genetic representation. W. D. Hamilton and his colleagues (Hamilton 1980, Hamilton et al. 1981, Hamilton and Zuk 1982, Ebert and Hamilton 1996) have argued cogently that pathogen stress is one of the few, perhaps the only, environmental uncertainty that will meet the criteria.

35. Polygyny threshold: Orians 1969; ecological correlates of polygyny in humans: Low 1988b, 1990b,c.

36. E.g., Emlen and Oring 1977, Borgia 1979, Alexander and Borgia 1978.

37. This is a highly significant relationship (p = 0.00001); Flinn and Low 1986.

38. Borgerhoff Mulder's (1988a,b, 1995) work is the most detailed; she found that younger (higher reproductive value) women commanded higher bride price (see chapter 7).

39. Kanuri: R. Cohen 1967; Yomut Turkmen: Irons 1975; Tsembaga-Maring: Rappaport 1968.

40. E.g., Native American societies of western North American: Jorgensen 1980: 167.

41. Low 1988b.

42. Distribution of dowry: Gaulin and Boster 1990. See also Gaulin and Boster 1997.

43. Rao 1993a,b. Further, domestic violence and spousal abuse have correlated with these increases in dowry worth; although alcohol is one significant factor, "insufficient dowry" is another (Rao 1997).

44. Aswad 1971, Barth 1956, Alexander 1979; see review by Flinn and Low 1986.

45. Alexander 1979, but see Flinn 1981.

46. Fathauer 1961, Flinn 1981, Flinn and Low 1986.

47. K. Hill and Hurtado 1996, Nordenskiold 1949.

48. E.g., Wade 1979, Wade and Arnold 1980.

49. Fig wasps: Hamilton 1979.

50. Bullfrog RS: Howard 1979; conditional strategies: Thornhill and Alcock 1983: 287.

51. Heritable class: Murdock 1967, 1981, White 1988.

52. Cf. Bateman 1948.

53. E.g., Parker 1974.

54. Sunfish: Dominey 1980; Ruffs (*Philomachus pugnax*): Rhijn 1973, Högland and Alatalo 1995.

55. Sade 1967.

56. See also Trail 1985.

57. This measure is a direct descendant of Crow's (1958) widely used intensity of selection measure in which the variance in gain (fitness) is standardized by the square of the mean gains. Professor James Crow devised the following modification, which is simpler than our first attempts. Let x be reproductive success. Its variance is

$$V(x) = Sp_i(\bar{x}_i - \bar{x})^2 + Sp_{ij}(x_{ij} - \bar{x}_i)^2 = V(b) + Sp_iV(w_i),$$

where p_{ij} is the proportion of individuals of behavioral class j in stratum i, p_i is the proportion in stratum i, x_j is the mean success in the ith stratum, $V(b)$ is the variance among stratum (e.g., size class) means, and $V(w_i)$ is the behavioral variance within the ith stratum. Then the index of total opportunity for reproductive gains is

$$\begin{aligned}
I_t &= V(b)/x^2 + Sp_iV(w_i)/\bar{x}^2 \\
&= I(b) + Sp_i(\bar{x}_i/\bar{x}^2)(V[w_i]/\bar{x}^2) \\
&= I(b) + Sp_iR_i^2I(w_i)
\end{aligned}$$

where $I(b) = V(b)/\bar{x}^2$ is the index from stratum mean differences, $I(w_i)/\bar{x}_i^2$ is the index from differences within the ith stratum, and $R_i = \bar{x}_i/\bar{x}$ is the ratio of average within-stratum success to overall mean success. When there are only two strata, and reproductive success occurs in only one, this reduces to Crow's formula. If the trait is completely heritable, $I_t/2$ is the proportion by which mean reproductive success increases through the sex being considered. This procedure apportions the intensity index to differences within and between strata; it further describes the relationship between the commonly used notion of total variance in male reproductive success when considering male-male competition, and the components of variance relevant for predicting behavior.

Steve Frank (personal communication) notes that this set of ideas has been little developed in the literature, although there is a widespread sense that they are logical. Frank (1996) reflects on Haldane's (1932) appendix, in which Haldane considered the conditons under which selection favors high variance rather than high mean response, and goes on to connect Haldane's discussion with models of the evolution of sex, such as Williams (1975).

58. E.g., data summarized by White 1988; see Low 1988a, 1990b. Data are available on variance and heritability of resources (extent of stratification) for a number of societies, although individual status-and-success data for stratified societies are rarely published (see chapters 7 and 8).

59. E.g., see Hames 1996 and discussion of the Yanomamö in chapter 7.

60. Costs to women: e.g., Chisholm and Burbank 1991, Strassmann 1997, 1999. Relative fertility of first and other wives: e.g., Dorjahn 1958, J. Smith and Kunz 1976, Daly and Wilson 1983, Bean and Minneau 1986, Garenne and van de Walle 1989, Pebley and Mbugua 1989, Sichona 1993. However, in at least some societies, these differences disappear after a generation: Josephson 1993.

61. Children' survival: Dorjahn 1958; divorce: Betzig 1989, 1996, Strassman 1997, 1999.

62. Borgerhoff Mulder (1992, 1997) for reviews this and other influences.

63. E.g., Dyson-Hudson and Smith 1978. Alexander et al. (1979) called this kind of monogamy, the only kind seen in other species, "ecological monogamy." See also Emlen 1995, 1997.

64. Remember that biologists define monogamy as a system in which the relative variance in reproductive success is equal between the sexes ecological monogamy. Anthropologists and sociologists define a monogamous system as one in which an individual may be married to only one spouse (at a time). If families break up, and people remarry, we see a dilemma in these definitions. If men remarry more often than women, and if they have children from second and subsequent unions more often than women, the system is socially monogamous but biologically polygynous (male success will vary more than female success).

65. E. A. Smith 1998a calculates the fitness effects, and analyzes Tibetan polyandry as a Member-Joiner game; he also includes a perceptive analysis of the tangled literature on the functional significance of polyandry.

66. Gorer 1967.

67. Peter 1963: 453.

68. Tambiah 1966.

69. E.g., Beall and Goldstein 1981, Goldstein 1976, Hiatt 1981, Peter 1963, Alexander 1974, Tambiah 1966, Crook and Crook 1988, Yalman 1967, Durham 1991. Quote: E. A. Smith 1998a.

CHAPTER 5. SEX, RESOURCES, APPEARANCE, AND MATE CHOICE

1. Expensive displays: Zahavi 1975 (overview in Zahavi and Zahavi 1997); heterozygosity: Brown 1997; European Barn Swallows: Møller 1994; grasshopper choice of good foragers: Belovsky et al. 1996; cockroach discrimination: Clark et al. 1997; elephant seal choice of beaches: Le Boeuf and Reiter 1988; Red-Winged Blackbird choice of marshes: Langston et al. 1997.

2. Pérusse 1994, Buss 1994, 1999.

3. Neoteny in female faces: D. Jones 1995, D. Jones and K. Hill 1993; hips, breasts, and buttocks, fluctuating assymetry: Gangestad et al. 1994, Thornhill and Gangestad 1994, Thornhill et al. 1995. See also Low 1979, Alexander 1971, Low et al. 1987.

4. The dowry societies (chapter 4) reinforce this point. See Darwin 1871. Trivers 1972, Low 1990b, and Flinn and Low 1986 review the evidence on human mating patterns.

5. Dugatkin 1996, Godin and Dugatkin 1996.

6. Iban illness and haircut: Roth 1892.

7. Ethnographies from the ninety-three odd-numbered societies of the Standard Cross-Cultural Sample (see Murdock and White 1969) turned up more comments made by men assessing women (76) than by women assessing men (19). About half of the comments (52/95) were about clear fitness signals such as clear and/or unwrinkled skin, firm breasts, symmetrical features and breasts; 20/95 concerned wealth or skills, 6 concerned traits that are profitable only in specific environments (fat or steatopygy in harsh environments); and 17 comments were recorded about largely or entirely culturally influenced preferences (light skin, long fingers, etc.; this includes the comment about disliking short hair when the cultural practice was to cut hair during illness). Thus in this preliminary sample, direct fitness comments were reported significantly more than other sorts of traits ($n = 95$, $\chi^2 = 49$, $p < 0.005$).

8. Singh 1993a,b, 1994; Singh and Luis 1995. Yu and Shepard (1998) note one exception.

9. Fat and deception: Low et al. 1987, 1988; lactational difficulties: Niefert et al. 1985.

10. Fat preference: review in Low 1990a,c, also R. Smuts 1992, Darwin 1871, II: 345; fat and deception: Low et al. 1987, 1988, Low, 1990d.

11. Definition: Fisher 1958: 27–30; mate choice: Buss 1994, Pérusse 1994; female status and mate choice: Gallup 1982, 1986, Strassman 1991, chapter 15. See Buss 1999 for differences in long-term versus short-term preferences and strategies.

12. Knodel et al. 1997.

13. Buss 1994, 1999.

14. See Jones 1995, 1996, Jones and Hill 1993; conflicts of interest between the sexes are pervasive, both in ancient and modern environments (e.g., Kenrick et al. 1996, Smuts 1996, Malamuth 1996).

15. R. Smuts 1992.

16. Fisher 1958: 152.

17. Zahavi 1975, Zahavi and Zahavi 1997. See Pomiankowski 1987a,b for the formal proof.

18. Møller 1994. Not only do females choose longer-tailed males, they make up their minds about such males early: the premating delay is shorter for longer-tailed males than for others. Females prefer males with symmetrical tails, apparently because symmetry reflects a life history of good health—freedom from disease and parasites. See also Møller 1987, 1989, Andersson 1994.

19. Bower birds: Borgia 1985.

20. Murdock 1981.

21. E.g., Lewis, 1959, 1970.

22. Brown 1997.

23. Low 1979, 1990a.

24. Alexander et al. 1979.

25. Whyte 1979, Low 1979, 1990a.

26. Low 1990a, 1994a.

27. Whyte 1978, 1979.

28. Borgerhoff Mulder 1987, 1988a,b, 1990, 1995.

29. Schlegel and Barry 1991.

CHAPTER 6. SEX, RESOURCES, AND HUMAN LIFETIMES

1. Jared Diamond (1992) called humans the "third chimpanzee," recognizing both that we face the same ecological problems as other primates, and that we are most closely related to common chimpanzees (*Pan troglodytes*) and pygmy chimpanzees, or bonobos (*Pan paniscus*).

Life histories—the lifelong patterns of maturation, courtship, reproduction, and death in any species—are the outcome of competing costs and benefits of different activities at any point in the life cycle. Many life history patterns arise from allometric (size-related) relationships. For example, the timing of many life history events (e.g., early or late reproduction) is strongly driven by age-specific mortality patterns—and small organisms die earlier than large ones; they simply tend to be more at risk (e.g., Charlesworth 1980, Charnov 1991, 1993, Stearns 1992, Roff 1992). One can compare the "relative value" of a number of life history traits in the primates, including humans, controlling for these allometric effects (Harvey et al. 1986). The relative value describes the variable after the effects of size have been removed, so that we can ask: Does this trait follow the pattern we would expect from size alone? If not, what other factors are important?

Life history theory links behavior, natural selection, and historical and phylogenetic accident to explore variation in maturation, birth, death, and behavior patterns. It lies at the heart of understanding diversity, precisely because it deals with natural selection, adaptation, and constraint (Charnov 1994, Roff 1992, Stearns 1992). It is a subset of natural selection theory, and shares the same logic: it argues that the characteristics we see represent trade-offs in allocation of effort (energy and risk) between survival and current reproduction; between current versus future reproduction; and, within current reproduction, among offspring of different sex, size, and number. As in any zero-sum game, an organism's effort spent in one endeavor cannot be spent in another. A few life history traits are central to any analysis: size at birth; growth pattern; age and size at maturity; allocation of reproductive effort; age schedules of birth and death; number and sex ratio of offspring. The trade-offs among these traits lead to a variety of patterns in, for example, mating, parental care, and senescence. These are patterns we examine in humans, usually without asking how human patterns compare to those in other species.

2. Leutenegger 1979, Stearns 1992: 87, Harvey et al. 1986: table 16–5.

3. Altricial descriptions: e.g., Ricklefs 1983, Nice 1962, Case 1978; altricial growth rates: Ricklefs 1983; possible advantages of altriciality: Ricklefs 1983, Dienske 1986, Alexander 1990.

4. Harvey et al. 1986. Several scholars (e.g., Humphrey 1976, Alexander 1990) have argued that large brains have been favored by the importance of intelligence in social evolution—that the uncertainties and risks of social life require more complex intellectual skills than, say, simply avoiding predation and finding food (see chapter 9). Certainly, human brains are relatively expensive metabolically; in a resting human, brain tissue consumes 20 percent of the energy budget (compared to 9 percent for a resting chimpanzee, or 2 percent for a typical marsupial).

5. Growth constraints: Martin 1983; metabolic expense: Hofman 1983, review by B. H. Smith 1990, 1992; postnatal growth rate: Martin 1983, Jolly 1985, Harvey et al. 1986.

6. General prediction of early maturation: Roff 1992: 347; subfecundity: Lancaster 1986. See also Hawkes et al. 1998, who, using Charnov's 1993 dimensionless comparative approach, predict later maturity for humans.

7. E.g., Gray 1983.

8. Hunter-gatherer IBIs: Harvey et al. 1986; Sweden: Low 1991; Germany: Knodel 1988: 322; chimpanzees: Harvey et al. 1986; dental maturation: B. H. Smith 1991, 1992; death risk before sexual maturation: Stearns 1992, Roff 1992, Charnov 1991, Promislow and Harvey 1990. B. H. Smith 1991, 1992 notes other oddities in the pattern of human developmental life history events, compared to other primates. Comparing *Pan, Australopithecus,* and *Homo* (from *H. habilis* to modern humans), the length of infancy to the eruption of the first permanent tooth has increased by three years, time to the last permanent tooth has increased by ten years, and life span has increased by thirty years. But we are weaned at ages 40 percent younger than would be expected from size alone (fig. 6.1). In other primates, weaning is closely associated with the eruption of molars; if human mothers waited to wean until children's molars erupted, they wouldn't wean them until they were six to eight years old!

9. Hawkes, O'Connell, and Blurton Jones 1997, Hawkes, O'Connell et al. 1998 and 1999.

10. See Dennett 1991, Humphrey 1983, Alexander 1979, 1987, Cavalli-Svorza and Feldman 1981, Lumsden and Wilson 1981, Boyd and Richerson 1985, 1992, 1996, Richerson and Boyd 1992, Durham 1991.

11. Parent-offspring conflict: Trivers 1974; romantic conceptions: Haig 1993.

12. E.g., Haig 1992, Moore and Haig 1991; see also Hurst and McVean 1997. Genomic imprinting, the term for this maternal-paternal conflict of interest within the fetus, may even affect brain (especially cortex) patterns in mammalian evolution (Keverne et al. 1996). There are several theories of genomic imprinting, and while they share commonalities they are not identical; see Hurst and McVean 1997.

13. E.g., Haig (1992, 1993) and Hurst (1992) argue that paternal genes in the fetus may function to manipulate maternal physiology for the fetus's benefit even if serious cost to the mother may result. Hurst and McVean (1997) did not find strong support. See also Reik and Surani 1997.

14. Trivers 1972.

15. Maternal physiological responses: Peacock 1990, 1991; nutrition in traditional societies: Bailey et al. 1992; high-status women: Kasarda et al. 1986.

16. !Kung IBIs: Blurton Jones and Sibley 1978; backload model: Blurton Jones 1986, 1987a. !Kung patterns in compounds: Pennington and Harpending 1988. See also Blurton Jones 1997. Anderies 1996 has extended the model; inclusion of maternal mortality further predicts optimal ages at first and last birth.

17. See Downhower and Charnov 1998, and chapter 15.

18. Ache and Hiwi: Hurtado et al. 1985, Hurtado et al. 1992; Ye'kwana: Hames 1988a.

19. Turke 1988.

20. Hrdy 1992.

21. Like the high-status baboons in chapter 4, these women experienced unanticipated costs. Here, the very high-fertility women "endured a range of problems ranging from chronic anemia to prolapsed uteruses" (Hrdy 1992).

22. Harem takeovers in langurs: Hrdy 1974, 1978, 1979; in lions: Packer and Pusey 1983, 1984; in gorillas: Watts 1989. See also Struhsaker and Layland 1987, table 8.4. Human abuse: Daly and Wilson 1984, 1985, 1987. Stepfathers, like langurs taking over a harem, reduce not only their costs for a nonrelative, but reduce the potentially competitive investment by the mother; and women may be less able to prevent a spouse's abuse than a man.

23. E.g. Hughes 1988, Daly and Wilson 1988. Hill and Ball 1996 suggest that cultural "ill omens," as well as biological cues, may be used. Note that "adaptive" means here only "reproductively profitable in these circumstances" not "an evolved adaptation." We can measure something about the former, but the latter I think would be a difficult question.

24. Overview: Daly and Wilson 1984, 1988, Hrdy 1992. Daly and Wilson (1984) compared sixty cultures in the Human Relations Area files to identify causal circumstances; Hrdy (1992, table 1) compared infanticide rates in a set of traditional societies for which there are relatively good data. Ill omens: C. Hill and Ball 1996; maternal circumstances: Bugos and McCarthy 1984.

25. France: Fuchs 1984; Spain: Sherwood 1988; Russia: Ransel 1988.

26. Boswell 1990.

27. Abortion patterns: Hill and Low 1991, Torres and Forrest 1988; attitudes: Betzig and Lombardo 1991.

28. Women's attitudes: Betzig and Lombardo 1991; "issue evolution" in U.S. politics: G. Adams 1997.

29. General model: Sibley and Calow 1986; human model: Hill and Low 1991.

30. Fisher 1958.

31. E.g., Whyte 1978, 1979, Low 1990a, 1992, 1994a.

32. Chagnon 1979, 1982, 1988, 1997.

33. Coalitions in other male and female mammals also follow these patterns (chapters 11, 12). Some of these differences arise simply from being mammals, captive by the differences in mating effort versus parental effort return curves. Other things being equal, male mammals achieve maximum reproductive success through expending their reproductive effort as mating rather than parental effort, and by expending generalizable parental effort rather than true offspring-specific parental investment. Female mammals, equipped to nurse their young, do best by producing healthy, viable offspring, optimally apportioning effort to specific offspring (fig. 3.2, chapter 3).

34. Trivers and Willard 1973. See also Charnov 1982, Leimar 1996.

35. That is, young men have usually only the promise of resources, while young women have current reproductive value; successful older men have real current resources while older women have declining reproductive value.

36. Fisher 1958: 159.

37. Trivers and Willard 1973, Leimar 1996.

38. See Clutton-Brock 1991.

39. Leimar 1996, Low 1991, Williams 1966. Note, however, that for small clutch and litter sizes, there can be a "packaging problem"—one very large or two very small offspring? See Charnov 1993, Downhower and Charnov 1998.

40. Gorillas: Mace 1990; Sweden: Low 1991; parental sex-preference: Knodel 1988.

41. E.g., J. Hill 1984, Betzig 1986, Flinn and Low 1986, Low 1990e.

42. Traditional polygyny and sex bias: Hartung 1982. Cowlishaw and Mace 1996 have reanalyzed these data using a phylogenetic approach, enriching the inferences we can make. Cultural shifts to polygyny are most commonly associated with male-biased inheritance, while changes to monogamy are most strongly associated with no inheritance bias.

43. Portuguese nobles: Boone 1986, 1988; Kipsigis: Borgerhoff Mulder 1998a; Gabbra pastoralists: Mace 1996.

44. Swedish siblings: Low 1991. Further, men who inherited land had more children than their landless brothers (Low 1990e).

45. Germany: Voland 1984; U.S.: Vinovskis 1972.

46. Tennessee: Abernethy and Yip 1990; U.S. generally: Gaulin and Robbins 1991, Mealey and Mackey 1990. This well may be a broader phenomenon than only the modern U.S.; see Cronk 1991a,b,e, 1993.

47. Sex differences in bequests: Judge 1995, Hrdy and Judge 1993, Judge and Hrdy 1992. In one study (Judge and Hrdy 1992) decedents with two or more daughters treated them more equally than decedents with two or more sons. College student survey: Gaulin et al. 1997. The results were obtained through surveys about perceived care. Interestingly, bequests fit "rational" economic models rather than "altruism" models (Altonji et al. 1997).

48. Fisher 1958: 159. Laland et al. (1994) have shown that persistent sex-preferential infanticide may create selective forces. See also Das Gupta and Mari Bhat 1997 and chapter 10.

49. Inuit: E. A. Smith and S. A. Smith 1994, E. A. Smith 1995; hypergynous societies: Dickemann 1979, 1981; male-biased infanticide in high-status families: Parry 1979; fitness in stratified societies: Harpending et al. 1990.

50. E.g., Barry et al. 1957, 1976, Konner 1981, Whiting and Whiting 1975, Whiting and Edwards 1973, Blurton Jones and Konner 1973, Ember 1981, Rosenblatt and Cunningham 1976. Early analyses of sex differences in child rearing have not found any clear logical patterns in the existing variation. In part, this may have arisen from a failure to use the extraordinary advances in evolutionary theory of the past decade.

51. Barry et al. 1976, Low 1989b.

52. Because broad patterns in the intensity of training boys and girls generally covary, it is possible that the patterns in boys' training simply reflect general patterns in child training, and are not the result of sexual selection; to eliminate this possibility, we must eliminate the possibility that the traits of interest simply covary for boys and girls. There are twelve traits that show some pattern in the training of boys or girls, with the degree or intensity of polygyny, in stratified or nonstratified societies. Seven of these are exclusive to one sex; six are exclusive to males, one to females. As intensity (maximum harem size) of polygyny increases, boys, but not girls, are trained to show fortitude, competitiveness, sexual restraint, and obedience (nonstratified societies), or industriousness (stratified societies). As the degree of polygyny (percentage of men and women polygynously married) increases, boys, but not girls, are taught to show fortitude, be aggressive and industrious (nonstratified societies). In stratified societies, as the degree of polygyny increases, the boys are taught to show less sexual restraint and more self-reliance. Girls, but not boys, are trained to be responsible in nonstratified polygynous societies. For girls, training in responsibility and industriousness increases with both intensity and degree of polygyny in both stratified and nonstratified societies. No such general pattern is evident for boys. Thus, the observed patterns do not appear to be simply due to covariance in training of boys and girls.

53. Low 1990a.

54. E.g. Laland et al. 1994, Kumm and Feldman 1997.

55. Children quickly show sex differences in perceptions of dominance and aggression. By age three, boys play in groups and play more aggressive games than girls (e.g., Laland et al. 1994; also see Omark and Edelman 1975). By four, boys advertise themselves as toughest; by six, they have formed dominance hierarchies, perceive them accurately, and attempt to manipulate their position— while girls find the entire question irrelevant! Eckel and Grossman (1998) find consistent adult sex differences in dictator games.

56. E.g., Dweck 1975, Dweck et al. 1978, Dweck and Wortman 1982.

57. Rates of praise: Dweck and Goetz 1978, McCandless et al. 1972; type of praise: Dweck et al. 1978.

58. Human maximum lifespan = 115 years, versus 29 for macaques and 44 for chimpanzees; Stearns 1992: his table 8.3. Allometric calculation: Harvey et al. 1986.

59. Harvey et al. 1986 note Sacher's 1959 finding that bigger-brained primates live longer, although they point out that (1) other measures such as adrenal gland weight (Economos 1980) correlate even better than brain weight with longevity, and (2) correlational analyses do not suggest any appropriate causality.

60. E.g., Hill and Hurtado 1991, Early and Peters 1990, Howell 1979; see Hill and Hurtado 1991 also for a review.

61. Hutt 1972.

62. Death clears the way for others: Curtis 1963; selective irrelevance: Comfort 1956.

63. Williams 1957 and Hamilton 1966. Hamilton (1996: chapter 3) has precisely the right quote from James Joyce's *Portrait of the Artist as a Young Man*. Temple says: "The most profound sentence ever written is the sentence at the end of the zoology. Reproduction is the beginning of death." Hamilton had the wit most of us lack to see the profundity in it.

64. Female schedule of reproductive decline: e.g., Dunbar 1987, Gaulin 1980; percent function at sixty-five: e.g., Mildvan and Strehler 1960; life expectancy at first reproduction: e.g., Nishida et al. 1990, T. Smith and Polacheck 1981: 108–110.

65. See Hill and Hurtado (1991, fig. 1), which combines data from Wood 1990 and Mildvan and Strehler 1960. New evidence (Johnson and Kapsalis 1998) suggests that up to 10 percent of rhesus macaques experience the equivalent of menopause.

66. Whether this represents a shift from patterns in traditional societies is unclear. Traditional societies: Ravenholt and Chao 1984, Wood 1990, reviewed by Hill and Hurtado 1991; developed nations: Snowden et al. 1989.

67. Female elephants (Croze et al. 1981: 306) and perhaps horses show increasing length of interbirth interval with age. In toothed whales (Marsh and Kasuya 1986) and some strains of laboratory mice (Festing and Blackmore 1971, E. Jones 1975), female reproductive function appears to cease entirely well before the end of life, as in humans.

68. Reproductive shift: Williams 1957; other primates: Fairbanks and McGuire 1986, Cheney and Seyfarth 1990; traditional human societies: Hawkes et al. 1997, Hawkes, O'Connell, et al. 1998, 1999, Hurtado and Hill 1990, Lancaster and King 1992. Postmenopausal women's resource extraction rates: Hawkes, O'Connell, and Blurton Jones 1997. Contemporary patterns of parental support: Cooney and Uhlenberg 1992, who find that parental support declines

somewhat after children reach age thirty but is responsive to children's status and needs.

69. Using Charnov's (1993) "dimensionless" approach to life histories, comparisons of human patterns to those of other primates Hawkes, O'Connell, et al. (1998) suggest that this is a likely selective path; hypothesis also would account for human late age at maturity, small size at weaning, and high fertility. It does not explain, unless as a by-product, late male senescence.

70. Hawkes, et al. 1999.

71. Hill and Hurtado 1991, 1996. See also Rogers 1993, Kaplan 1997. I personally like the grandmother hypothesis. Why is this empirical test no help? I suspect it is a twofold issue. First, as Hawkes et al. suggest, it may be that reproduction has not been shortened, but other systems have been extended; the gain from this extension seems supported by the wide patterns of grandparent investment and the comparative life histories of humans versus other primates. Second, there may be issues of scale of analysis: once the systems have come to equilibrium, we don't expect to see longer-than-profitable reproductive lives. So within the remaining existing variation, we would still expect to see later menopause correlated with higher total fertility. And finally, evidence from rhesus macaques (Johnson and Kapsalis 1998) suggests that costly female reproductive failures may be simple senescence (although this does not explain the rate problem).

CHAPTER 7. SEX AND RESOURCE ECOLOGY IN TRADITIONAL

AND HISTORICAL CULTURES

1. See Ridley's 1993 nice discussion of this paradox.

2. Murdock and Provost 1973. The Standard Cross Cultural Sample of 186 societies, used here and elsewhere for comparisons, is chosen to represent high-quality data from all the world regions and representative language groups within regions.

3. E.g., Murdock and Provost 1973.

4. Of course, I am not saying anything whatsoever about women's work today. I do not think that past patterns dictate either current utility, or individual choice (see chapter 15).

5. Collective action and sharing: Hawkes 1992; similar data arise from Cashdan's 1985 studies in Africa. Overview: Winterhalder 1996a,b,c, 1997, Connor 1995a,b; tolerated theft: Blurton Jones 1987b, Bliege Bird and Bird 1997; showoff men: Hawkes 1991, 1992, 1993.

6. See Whyte 1979, Low 1990a. In societies in which women can control the fruits of men's labor and the fruits of joint labor, they are also likely to inherit property of some economic value. Societies in which women can inherit prop-

erty are also the societies in which women are likely to be active in community affairs. Although the ability to hold formal leadership positions in the kin group is not associated with the ability to control the fruits of labor or to inherit property, it is positively associated with the ability to be active in community affairs. The ability of women to hold political posts outside the kin group does not correlate with any other measure of resource control or power. Women's ability to gain political power appears to approach men's in only two societies. Women's resource control and influence showed no relationship to the risk of starvation or protein deficiency. Women's ability to control resources is not related to group size or mobility, suggesting that women do not have a greater voice in small, mobile, hunter-gatherer societies; or to the presence of community-wide male work groups, which appear to be organized around problems of male-male competition. Divale and Harris (1976) suggested that in areas of the world with protein insufficiency, male dominance characteristics that contribute to successful hunting, and male power, become overvalued, resulting in female powerlessness. The cross-cultural data do not support this hypothesis. Similarly, Leacock and Lee (1982) suggested that women in hunter-gatherer societies had more power, but the data do not support this hypothesis.

There are several problems in trying to assess the actual degree of control of resources by either sex, and they are frequently not resolvable from the ethnographies. For example, when women are reported as controlling the fruits of men's labor, there are several possibilities. A woman may control only a small proportion of the resources a man garners, and only because he allows her to distribute this portion after he has disposed of all he wishes. The man's and woman's interests may overlap completely so that either may dispose of resources without conflict. The woman may distribute significant proportions of the resources garnered by the man, by following the directions he gives for their distribution, and for an outsider this condition may be difficult to distinguish from true female control. Finally, she may control a significant proportion of the fruits of his labor, and make decisions without regard to his desires. The meanings of each of these conditions are quite different, and they are not easily discerned from codes or many ethnographies.

7. Chagnon 1979, 1982, 1988, 1997.

8. Chagnon 1997: 7.

9. See Hames 1996.

10. Hill and Hurtado 1996. Earlier ethnographic accounts include those of Clastres 1972a,b and Bertoni 1941.

11. Hill and Hurtado 1996: 316–317, Hurtado et al. 1985. See also Hawkes 1993, Winterhalder 1997.

12. Hill and Hurtado 1996: 316–317, Kaplan and Hill 1985.

13. Manners 1967.

14. Borgerhoff Mulder 1988a,b, 1995.

15. Borgerhoff Mulder reviews the few cases reported.

16. Borgerhoff Mulder 1997. Temporal changes: Borgerhoff Mulder 1995.

17. Cronk 1989, 1991b,c,d,e, 1993, 1999.

18. Interestingly, this income does not appear to make these women's brothers, for example, able to marry more. So favoring daughters is not a matter of local resource enhancement.

19. Irons 1975, 1979b, 1980.

20. Irons 1979b.

21. Lee and Campbell 1997, Lee et al. 1993, Lee and Guo 1994, Wang et al. 1995.

22. Wang et al. 1995, table 1; see also Lavely and Wong 1998. The European and Chinese patterns differed: marital fertility in Europe was high, but a large proportion of adults failed to marry (see chapter 8); among the Qing, marriage was far more common but marital fertility was lower.

23. Monogamous men who survived to age forty-five had, on average, 4.5 children, and polygynous Qing men had on average 6.5 children, compared to 8–10 children for monogamous and 16–20 for polygynous European men (Wang et al. 1995 and references therein).

24. See Lee and Guo 1994 and Wang et al. 1995.

25. Lee and Campbell 1997; see also chapter 6.

26. Lee and Campbell 1997: 138.

27. Lee and Campbell 1997, ch. 9.

CHAPTER 8. SEX, RESOURCES, AND FERTILITY IN TRANSITION

1. Demographic transition theory rests on the argument that environmental changes (in most arguments, linked to industrialization) caused family sizes to drop. Sometimes the arguments are made at the population level (e.g., Coale and Watkins 1986, Wrigley 1983a,b, Viazzo 1990), sometimes at the level of perceived value of children by their parents (Hammel et al. 1983). Simpler, more parsimonious arguments exist, which require no group selection (below; chapters 9, 15) and which are a subset of the well-tested and well-supported behavioral ecological explanations of changes in reproductive output of other species. If fertility is influenced by environment, as earlier chapters suggest, many patterns of starting, stopping, and spacing children are possible adaptive responses to environmental conditions, rather than not-optimum-because-not-maximum patterns.

2. In earlier chapters, it is apparent that the behaviors that become and remain common are those that produce reproductive profit for their performers. Even for nonhuman species, because the world is often far more complex than one might at first imagine, we have seen that "maximum" fertility is not always optimum. In fact, "most successful reproduction" does not necessarily mean producing the most offspring, or even the most surviving offspring (e.g., Lack

1947, 1966; Dawkins 1982, 1986, 1989; Williams 1966). No species produces off-spring at its maximum physiologically defined rate; reproductive success depends not only on production, but also on the investment required to produce viable, competitive, reproductive offspring. Thus producing fewer, better-invested offspring, compared to the maximum physiologically possible, can be, depending on environmental conditions, reproductively more efficient. What really influences the variation in success is almost always the failures—what statisticians call the "zero success" reproductive group (Falconer 1981). Costs and benefits may differ for male and female parents (chapter 3). This reiterates and extends the "quantity versus quality" dilemma first raised by Darwin (1871).

Such a view gives a new perspective on the demographic transition, the period in the late nineteenth century during which western European and North American family sizes fell dramatically, and on today's demographic transition in developing nations around the world. The competitive environment into which humans are born is extremely complex, in part due to the long lifetime over which individuals must respond to changing conditions that will affect themselves and their offspring. This, then, may be both a simpler and more general approach, compared to classical transition theory.

3. This is an extremely important point, to which I will return. These data, like the data on traditional societies in previous chapters, are comparable to the nonhuman animal data and unlike most modern analyses, which use "aggregate" data. Different forms of data can affect the answers we get (chapter 15, Low 1999). Consider the aggregate data of a census: these are summary measures that do not track individuals over complete lifetimes, and that lack any estimates of within-class variation. There are systematic biases in just which individuals are most likely to be censured, and which missed; these mean that aggregate data won't work to answer the questions I am asking here.

4. Late marriage was sometimes called the "European" pattern (Hajnal 1965). Spatial and temporal variation: it is difficult to elucidate relationships that vary both temporally and spatially; hence the analyses are structured to make comparisons by decade and by parish. Doing this allows us to analyze the relationship between resources and lifetime reproduction through the changing economic and social times of the demographic transition, and in parishes of very different base fertility levels. Thus we can follow sometimes subtle relationships through the changing times of the nineteenth century.

5. Proportion never marrying: Low 1989a, 1990e; Low and Clarke 1991, 1992; proto-industrialization: Mendels 1981; see also Flandrin 1979, Tilly 1978. In Sweden it is probably related to land enclosure and inheritance changes during the nineteenth century (Jörberg 1972, 1975). From 1686 to 1810, the nobility practiced *fideicommiss*, or male primogeniture, with the constraint that the eldest son must continue the practice (Malmström 1981, Inger 1980). Until 1845, sons in-

herited twice as much as daughters; after that date, daughters had equal inheritance rights, although in practice sons had first choice of the land and goods which were to be their inheritance, and sons could purchase their sisters' inheritance from them (Lo-Johansson 1981, Inger 1980). This meant that disputes occasionally arose over the value of the exchanged inheritance items; purchasing needed land from a sibling could prove economically onerous, but also siblings sometimes complained that they did not receive fair value (not uncommon elsewhere in Europe). Even after the shift from *fideicommiss,* and even after establishment of legally equal inheritance rules for both sons and daughters, inheritance biased by birth order was often evident (see Gaunt 1987, Low 1989a, 1990e, 1994b), and a bias toward the first son was perhaps more evident in the northern areas. Legal agreements in which a father ceded his land to one (usually the eldest) of his sons before his death, typically in return for room, food, and certain other rights, were common. But as Gaunt (1987) noted, during the nineteenth century the payments delivered to the retiring father increased in size, and receiving a farm became an economic burden. Indeed, default was common, and contemporary jokes abounded about arsenic as "retirement medicine" (Gaunt 1987, 1983). Thus, there probably existed some tension both within and between generations over resources.

6. Low and Clarke 1991.

7. Anders Brändström, Umeå University.

8. Geographic and land enclosure: Gerger and Hoppe 1980; demographic and economic data: Low 1989a, Low and Clark 1993. For an excellent and detailed study of marriage, fertility, and crop prices in England (where they also covaried), see Wrigley and Schofield 1981.

9. Low 1989a, Low and Clarke 1991.

10. See Knodel 1988 regarding the usual correlation between age at marriage and peak fertility.

11. Population and demographic behavior: Low and Clarke 1991, 1993; also see Jörberg 1972; Sundin 1976; iron foundries: Ostergren 1990, Sundin and Tedebrand 1981, 1984; industrialization: Norberg and Rolén 1979.

12. Cow's milk: Brändström 1984; infant survuval: Low 1991.

13. Low 1989a, Low and Clarke 1991.

14. Low 1990e, Low and Clarke 1991.

15. Sundin 1976.

16. Low 1989a, 1990e, Low and Clarke 1991, 1992; Low 1991; Clarke and Low 1992.

17. Geographic and temporal variation are summarized in Low and Clarke 1991, 1992. Making lifetime fertility comparisons within local populations at the same time is parallel to many nonhuman demographic analyses, but is seldom done for humans (see chapter 14).

18. Interestingly, in both wealthier and poorer families, daughters survived

better than sons and there was no evidence of significant undervaluing of daughters among wealthier or landed families (cf. Voland et al. 1990, who did find such a pattern in German villages).

19. Clarke and Low 1992 discuss the influence of skills and liquidity of resources on migration.

20. Low and Clarke 1992.

21. Low 1990, 1991. Also see Wrigley and Schofield's 1981 classic *Population History of England*.

22. Others also have found this generally to be true (e.g., also see Røskaft et al. 1992, Voland 1990; Turke 1989, 1990). Individual patterns in such important items as age of marriage typically vary with resources (e.g., Wall 1984, Sharpe 1990, Cain 1985, McInnis 1977, Pfister 1989a,b, Thompson and Britton 1980, Hayami 1980, Schultz 1982, Simon 1974). Depending on their own resource bases (e.g., Galloway 1986, Schultz 1985), families may respond quite differently to such influences as market shifts, treat their children quite differently (e.g., Georgallis and Wall 1992, Bailey and Chambers 1998, Merrigan and St.-Pierre 1998, Mitterauer and Sieder 1982: 110); even aggregate data tend to reflect resource influences, as individuals make decisions (e.g., Thomas 1941, Wrigley 1983a,b).

23. We humans invent these at a great rate: ability to purchase medical services, ability to will reproductively useful resources such as land or status, investment in education, etc. As parents' ability to influence their children's eventual success by *investment* increases, we expect *fertility* to decline and resources to be routed into investment, decreasing mortality and increasing success of children.

24. This problem of census techniques and reliable predictions remains an issue today—and not just for academic reasons. In 1998, the U.S. Congress brought suit against the U.S. Census Bureau for counts that miss individuals, mainly poor and homeless ones. However, the new requirement for total counts rather than improved statistical sampling will be expensive and may be impossible to do accurately.

25. Traditional taboos: Campbell and Wood 1988; wealth and copulation frequency: Pérusse 1993, 1994.

26. Exogenous survival: Tilly 1978; investment leading to success: MacArthur and Wilson 1967, Rogers 1990, 1991.

27. This ecological approach echoes some classic demographic models centered on "individual decision" and "proximate variables" (e.g., Becker 1981, Becker and Lewis 1974, Easterlin 1978, Tilly 1978, Bongaarts 1978, 1982, Lindert 1978, Simon 1974, Lesthaeghe and Wilson 1986), as well as Mosk's (1983) "leveraging" approach to fertility, and Easterlin and Crimmins's (Crimmins and Easterlin 1984, Easterlin and Crimmins 1985) models of the factors favoring a shift to conscious control of fertility. All of these models have two important charac-

teristics: individuals are not assumed to be uniform, and there is an explicit trade-off between quantity and "quality" (probable success) of children.

28. MacArthur and Wilson 1967, Rogers 1990, 1991. MacArthur and Wilson (1967: 145–150) argued that, when the density of conspecific competitors (in any species) was low, selection favored "productivity" and competitive efficiency of offspring was relatively unimportant to their eventual success; in more competitive environments, selection favored the production of more competitive (better nourished, better taught) offspring, at the cost of number of offspring; parents should shunt resources into offspring investment, even at the expense of offspring numbers—net lifetime reproduction was enhanced not by high fertility, but by lowered fertility—producing fewer but better-invested offspring. See also M. Smith et al. 1986 on inheritance as investment.

29. See review by Low et al. 1992.

30. More in chapter 15. Rank 1989 found that at "low" socioeconomic levels, women on welfare have fewer children (age-specific fertility is lower at all ages) than women not on welfare, and that these welfare recipients specifically cited the need for resources to invest in their existing children as the reason for avoiding further pregnancies.

31. Education: e.g., Knodel et al. 1990; labor markets as driving forces requiring more investment: Kaplan et al. 1995; raising the stakes: Turke 1989. Turke has also suggested that as kin networks are disrupted, fertility may decline, especially when older children and nondescendant relatives initially comprise a resource—nepotistic effort. When that resource declines, children pose an increased cost to their parents, a cost no longer defrayed by kin help. Thus, fertility will decline.

In sum, investment level required to produce successful offspring may vary with environment, and specifically with the threshold level of investment required for a child's success—often a correlate of competition, and in this sense, precisely analogous to the proper use of MacArthur and Wilson's biological concept (r- and K-selection). If poorer parents cannot substantially enhance their children's success, then we might expect larger families, concentration of resources in one or a few children, with others living with the family or leaving early (behavioral ecologists would call this an "alternate strategies" situation). Couples at the high end of the socioeconomic "ladder" might do better by investing more per child to allow them to be competitive with their peers (e.g., education, clothing, status acquisitions). The required investment may limit the number of children they can afford. Within subgroups, however, those with more than sufficient resources may be able to support additional children and still have all be adequately invested.

32. E.g., Easterlin 1978; Becker and Barro 1988; Turke 1988.

33. E.g., Hartung 1982, Low 1990e, Low et al. 1992.

34. Rogers 1990, 1991. One of the most consistent differentials in a child's abil-

ity to turn investment into successful reproduction, of course, is that of sex (chapter 6).

CHAPTER 9. NICE GUYS CAN WIN

1. Alexander (1987: 128) writes bluntly: "Modern society is filled with myths perpetuated because of their presumed value in regard to self-images: that scientists are humble and devoted truth-seekers (of course we are! Trust me!); that doctors dedicate their lives to alleviation of suffering; that teachers dedicate their lives to their students; that we are all basically law-abiding, kind, altruistic souls who place everyone's interest before our own; that our country (church, family) is always right and benevolent and has never done anything with malevolence as a part of the motivation; that our nation is on the side of right." Indeed, there are individuals who clearly behave in selfless ways (e.g., the teacher who died protecting children in a recent school shooting incident); the important questions are (1) Are these behaviors widespread? (2) Are these behaviors helpful to their doers—typical high-risk-high-gain strategies?

2. True, an individual's life (and by implication, what solutions will be effective) is influenced by his or her physical and social environment, particular genetic makeup, sex, and genetic relatedness to neighbors. Nevertheless, it is a simple but central biological fact that not all individuals are identical; very few species routinely produce genetically identical offspring.

3. Following Hamilton 1964, 1970, and Alexander 1974.

4. Celibacy may, in some conditions, be genetically profitable: see chapter 10.

5. Of course, one can sometimes turn true altruism to genetic profit after the fact, as one of my students, disturbed by the depressing nature of our discussion of altruism, gave away: "But I gave to charities for a year before I told my friends!" He, of course, meant that he was a true altruist; I am afraid I got a different message! Advertising our "altruism" (whether it was originally genetic altruism or not) may convince our friends to trust us more: "Look how noble and good I am; cooperate with me." This is Robert Frank's 1988 "commitment problem" (next chapter). I suspect we ask our friends, in subtle ways, to make just such declarations, and that we make it clear that it's more effective if the showoff altruism has real costs.

6. See especially Pusey and Packer 1997, who make this point and discuss several empirical examples. Consider grooming, often counted as reciprocity—is it really costly to the groomer? If it is directly profitable, we would call it pseudoreciprocity. Assuming that a behavior such as grooming costs the groomer is, in game theory terms, equivalent to making up the relationship between payoffs. That's fine in game explorations (what would we expect if these were the payoffs?) but circular and confusing when employed to explain observed behavior.

Several issues are easy to muddle in this complex problem. First there is the issue of whether our observations are correct: in the Disney films, for example, lemmings were being driven over cliffs and thrown out of boats, not sacrificing themselves to lower population numbers and relieve nonrelatives of competition. In the real world, striking migrations of large numbers of lemmings are observed; they tend to be adolescent individuals unable to establish themselves in the crowded natal habitat. So leaving is risky, but likely to be reproductively more profitable than staying and getting trounced by established adults (Chitty 1996). It is true that lemmings and their relatives do live in populations that fluctuate, but not that lemmings leap off cliffs routinely.

Next we must ask whether we are measuring all relevant currencies, and whether we're measuring them correctly. In the lion example, the females are sisters: nursing one's sisters' cubs costs energy, gains genes, and can be genetically profitable. Being relatives seems to matter in other lioness interactions, as well: when groups of (related) lionesses defend territories from intruders, some lioness must take the risky "point" position (Heinsohn and Packer 1995). Here you would expect Tit-for-Tat, sharing the risk. But in fact, some lionesses almost always lead, often looking back at laggard lionesses. There is no evidence of punishment for the laggards, or of reciprocation in some other currency. In lions, as in other species, laggards can get away with a lot, if they are close relatives. And, again, do we really know that grooming is costly to the groomer? There is no convincing evidence that it is (Dunbar 1988, Hemelrijk 1994). Once we have eliminated such sources of error, we gain much clarity. We can distinguish situations in which cooperation pays regardless of what the other guy does ("by-product mutualism") from situations in which it pays to discriminate. We may identify true genetic altruism: there are indeed a few Mother Teresas in the world, but I do not expect us to find it in dominant practice over generations. In fact, new evidence (Segal and Hershberger 1999) suggests that even monozygotic ("identical") twins may not routinely cooperate, although monozygotic twins cooperate more than dizygotic twins.

7. By-product mutualism: Connor 1995a, Brown 1983, West-Eberhard 1975; pseudoreciprocity: Connor 1986, Winterhalder 1997. Dugatkin (1997, 1999) has a broad survey of cooperation in a wide variety of species. Connor (1995b) and Reeve (1998) have good summaries. Reeve, like Dugatkin and Wilson, views the "new" group selection (treated later in the chapter) as simply an alternate view of all these mechanisms (see also Dugatkin and Reeve 1994). I prefer not to do this until I can find new predictions from the alternate views; although Dugatkin and Reeve (1994) suggest that, for example, group selection is a likelier route to good hypotheses about intragenomic conflict, I think this arises from an extremely restricted view of "individual" or gene selection.

8. Mutualism: West-Eberhard 1989; win-win from win-lose: e.g., mitochondria may have become part of the genomes of other species from a parasitic beginning; now they are an intrinsic part of the very cells of those species.

9. Axelrod and Hamilton 1981, Axelrod and Dion 1988, Axelrod 1984.

10. The game is called "Prisoner's Dilemma" because in early versions the two players were envisioned as two thieves who were interrogated by the police. If the two "cooperated" by not admitting anything, the police could prove nothing about the most serious charges associated with the theft, and the charges were minimal. If both "defected" by telling on the other, both got intermediate sentences. If one defected on his partner while the other held silent, the defector went free while the cooperator got the maximum sentence. The situation of varied payoffs to interactors is actually far broader than the "prison" metaphor suggests. Versions of the game are reviewed in, e.g., Binmore (1992) and Daly and Wilson (1983).

11. Prisoner's Dilemma is one of several "mixed-motive" or nonzero sum games (see also Binmore 1992, Dugatkin 1997, 1998). For 2-person, 2-strategy games, there are twenty-four ways to order the payoffs P, R, S, and T (this reduces to twelve because the players are symmetrical). Eight of these possess optimal pure strategies for both players and are thus uninteresting. So we have four major categories of non-zero-sum games: Prisoner's Dilemma ($T > R > P > S$), Battle of the Sexes ($S > T > R > P$), Leader ($T > S > R > P$), and Chicken ($T > R > S > P$). (Maynard Smith 1974, 1979, and Maynard Smith and Price 1973 recognized this years ago as the payoff probabilities for mate desertion in animals and applicable in many territorial disputes; there it is called Hawk-Dove.) The order of the payoffs affects the probability of cooperating; for example, in Battle of the Sexes, "Sucker" (cooperate with a defector) pays more than T (temptation to defect), so cooperation will be likelier than in other games. Dugatkin (1998) reviews animal examples, and Hofbauer and Sigmund (1998) treat evolutionary games broadly.

12. Repetition affecting R: Dugatkin et al. 1994. This problem is cost sensitive, and small cheap favors may become widespread; there is some evidence that the proximate (perceived) trigger is often emotional (self-esteem, friendship) (e.g., Nesse 1990, Nesse and Berridge 1997). For example, the economist Robert Frank has noted that most of us leave tips even when traveling, even in situations in which we are unlikely to see the server again. But the cost is 15 percent of the meal, and there is some possibility of a cost to self-esteem and of being observed (more about this in chapter 10) if one considers "defecting." A rule to expand cheap cooperation may have multiple effects. (Note, for example, that for large parties, for which the 15 percent tip may be a nontrivial amount, restaurants do not typically rely on their customer's observance of practice: the tip is automatically added into the bill.)

13. E.g., Emlen 1982a,b, 1995, Reeve and Nonacs 1992.

14. Strategy dependence: e.g., Binmore 1992; Axelrod's Tit-for-Tat and variations: Axelrod 1984; see also Axelrod and Hamilton 1981, and Axelrod and Dion 1988. TFT has perfect regularity, but "Generous Tit-for-Tat" (GTFT), discussed in a moment, can cooperate occasionally after experiencing a defection.

15. See especially Binmore 1992, Sigmund 1993, Maynard Smith 1974, 1979, 1982, Dugatkin and Reeve 1998, and Hofbauer and Sigmund 1998.

16. Trivers 1971, Connor 1986; see also Winterhalder 1996a–c. For optimal discrimination, see Reeve 1989.

17. Other species enforcing norms: e.g., de Waal 1996a: 89 gives an example of chimpanzees in a group punishing adolescents who were late for a feeding (the group had to wait until all individuals had arrived, to be fed). Nash outcomes in untrained human experiments: V. Smith 1982, McCabe et al. 1998.

18. Vampire bats: G. S. Wilkinson 1984, DeNault and McFarland 1995.

19. Pusey and Parker 1997. See also Axelrod (1986), who described the evolution of metanorms such as "punish those who fail to punish cheaters." But it also may be true that in some potentially complex relationships (e.g., soil microorganisms) some (perhaps vague and cryptic) mutualisms will turnout to be TFT.

20. See Axelrod 1986.

21. See Mesterton-Gibbons and Dugatkin 1992, Pusey and Packer 1997.

22. Not only humans, but chimpanzees, appear to use memory of past interactions to make decisions about current behavior (de Waal 1997). This strategy goes by the inappropriate name of "Pavlov," though it is the opposite of a reflex strategy (a highly readable review of this is in Nowak et al. 1995).

23. Remember, though, that all these games involve some highly restrictive assumptions: pairs of animals, making simultaneous decisions without knowing anything about what the opponent is about to do (see Nowak and Sigmund 1994 for alternating decisions). This pairwise aspect is important; Boyd and Richerson (1988) have shown that reciprocity over communal resources is unlikely if there are more than two individuals. However, in groups of >2, a common pattern I'll say more about is: watch, and not only refuse cooperation to defectors, punish them. Experimental economists are discovering a phenomenon they call "strong reciprocity," characterized in part by a "taste" for punishment.

In the real world, many interactions exist that don't quite match the definition of reciprocity. Mutualism, in which actors may profit without attendant costs, is probably far more common than we realize; as is pseudoreciprocity, in which A helps B at some cost to itself, then B helps A later in an interaction in which both parties profit—A pays, B doesn't. Analyses of the costs and benefits of such interactions are becoming more sophisticated, though we really have a long way to go before we produce convincing empirical measurements. For evidence on the robustness of generosity and contrition see Wu and Axelrod 1995.

24. Machiavellian intelligence: Humphrey 1983, Byrne and Whiten 1988, 1992, Alexander 1971, 1979, 1987, Wilson et al. 1996, Wilson 1998a.

25. These conditions may be simple, but are not equally easy to generate in varied societies; for example, widespread cheap-to-enforce norms are most likely inside small, homogeneous populations. Conditions for ESS: Boyd and Richerson 1992. Boyd and Richerson 1996 suggest that cumulative cultural

adaptation may be likely to be rare, for the conditions required to generate (as opposed to maintain) observational learning are rather stringent.

26. Interdemic selection: Wright 1945, 1949, 1977, D. S. Wilson 1975; culture/ gene interplay: Boyd and Richerson 1985, Durham 1991, Lumsden and Wilson 1981; within-group coalitions: Alexander 1987, Irons 1991; group benefit: Wynne-Edwards 1962.

27. See Wright 1945, Hamilton 1975, D. S. Wilson 1980, E. O. Wilson 1975.

28. At some level, interdemic selection, like kin selection, is a group structure that generates genetic correlates (although these are not required in the specification of interdemic selection, it is the outcome of viscosity). For problems in partitioning the variance in outcomes of actions, this kind of hierarchical approach may be analytically useful. However, evolving "Mother Teresa" altruism in either system is empirically unknown.

29. Quote: Boyd and Richerson 1985: 240. Group norms affecting individual costs and benefits: Boyd and Richerson 1992, Alexander 1979, 1987, Axelrod 1986, Frank 1990. Note also that humans are unlikely to have rigid heritable strategies, but rather to follow conditional ones (see, e.g., E. A. Smith 1998b).

These models are analytically interesting; they can be used to highlight the tension between individual good and group good. Alexander (1987) first noted the importance of suppressing costly competition within the group, and Steve Frank (1995, 1998) has shown how, once individuals are unable to leave the group, policing and suppression can occur even in very simple models such as replicators within a cell (see chapter 1). High degrees of relatedness lead to self-restraint in such models; low degrees of relatedness lead to mutual policing, the extent and intensity of which depend also on whether the cost is borne by individual actors or a group as a whole. Even in these extremely simple systems, then, the problems of "public goods" exist.

Finally, these models are a way of partitioning variance into within-group versus between-group variance. This can be useful for some questions. But the history of this approach highlights a real danger: some (perhaps mostly nonspecialists) have made what Binmore called the "fallacy of the twins" in reference to the Prisoner's Dilemma. That is, "Since groups of altruists will do better than mixed groups, perhaps we can generate altruism." As W. D. Hamilton (1975) noted: "But, of course, being able to point to a relevant and generally nonzero part of selective change is far from showing that group selection can override individual selection when the two are in conflict."

30. These approaches are not properly lumped with models like Wynne-Edwards's or (unless population structure plays a critical role in the outcomes) interdemic selection. Like Irons and Alexander, cited in this section, I consider these approaches obvious examples of individual profit-maximization strategies in conditions in which other humans are a strong selective force. Boyd and Richerson, also cited here, model the influence of others in a group; this they call "cultural group selection." In all of these, natural selection is assumed to operate in the usual way.

31. Coalitions and laws: Boyd and Richerson 1990, 1992, Alexander 1987, Irons 1991. Advertisement: R. Frank 1985, 1988. See Irons's 1996a,b reviews. Coalition-coercion models and culture-gene interaction models share much; I think the distinction is mostly an issue of the kind of approach—mathematical or verbal/heuristic.

32. Nonrandom copying: S. A. Frank 1990, Boyd and Richerson 1990, 1996; Ball and Eckel (1998) also find that we treat "high status" individuals preferentially—even when the signal (a star) is clearly meaningless. Mukogodo daughter-son real versus claimed preference: Cronk 1991d.

33. Thus, many species of long-lived social animals behave in ways that among humans would be termed "ethical egoism," and long-term reciprocation and assistance are common, though not randomly dispensed; see de Waal 1996a. Failure to share and "tolerated theft": see Blurton Jones 1984, 1987b, Winterhalder 1996a,b,c, Hawkes, Bleige Bird, and Bird 1998.

Cosmides and Tooby 1992 suggest that humans have mental modules that predispose them to share, for example, when hunting is boom-and-bust and one makes a really rich kill. Thus the sharing is not learned, but evoked. Similarly, in the Mukogodo example of the previous paragraph, because daughters can do better than sons, better care of daughters may be easily evoked, whether or not there is someone to imitate.

34. Boyd and Richerson 1985, 1990, Durham 1991, Laland et al. 1994.

35. Wynne-Edwards 1962.

36. Wynne-Edwards collected observations on social behavior of all sorts of vertebrates in an attempt to demonstrate that anything other than maximum fertility (smaller clutch or litter size, not reproducing in certain seasons, and so on) constitutes behavior that contributes to the well-being of the group, even though detrimental to the individual. In his earlier work, he argued that all other species were groups selected, but humans were not—and that's why other species' populations never outpaced their resources, but human populations would! Recently he argued that all species, including humans, are group selected (see Low 1993a and others for reviews).

Wynne-Edwards fell prey to the problem of not assigning cost correctly; he assumed that less-than-maximum fertility imposes an individual cost, but we know this need not be true (the winning strategy might be: make fewer, raise them all; see chapters 8 and 15). If lower fertility were really a cost, the immediately obvious problem is that highly fertile "cheaters" would win.

37. For example, even the classic *Population History of England* (Wrigley and Schofield 1981: 462) accepts Wynne-Edwardsian "population regulation." For a cogent commentary, see Hawkes and Charnov 1988.

38. Recently there has been a movement to lump varied models in order to explain the spread of true genetic altruism. It can be useful to look at selection in a hierarchical way, and clearly group norms and group sanctions affect individual costs and benefits. D. S. Wilson and colleagues have aimed to broaden the definition of "group selection until it is basically a non-conflicting view, and struc-

ture them as alternative fitness-accounting schemes." As Wilson noted, "It would be wrong to claim that groups invariably evolve into adaptive units . . . , but it would be equally wrong to claim that groups never evolve into adaptive units." But: try to figure out the group structure and frequency of group switching required to generate actual genetically altruistic behaviors reliably in any case (see S. Frank 1998). This is an area of considerable interest, but not a settled matter.

39. Even in this example, defining costs and benefits precisely is difficult.

40. E.g., see Dawkins 1989. Recently, D. S. Wilson 1998b, Sober and Wilson 1998, and Wilson and Sober 1994, in lumping almost all interactions as "group selection," conflate kin selection, clade selection, interdemic selection, and "co-operator selection," including reciprocity and coalitions. Indeed, it may be useful to partition the within-group and between-group variances, but such lumping, I think, obscures the functional aspects of genetic costs and benefits, and yields approximately the same analytic power as the *New York Times* crossword puzzle. See E. A. Smith 1998b, Hawkes, Bleige Bird, and Bird 1998, Harpending 1998, and Palmer et al. 1997.

41. E.g., Alexander 1979, Boyd and Richerson 1985, Richerson and Boyd 1992.

42. Williams 1966, Hamilton 1964, Dawkins 1982, 1989. For an overview, see Daly and Wilson 1983 and Krebs and Davies 1991. Williams 1992a is a current and complete analysis.

43. Bell 1997a: 520.

44. Even more vulnerable to free-riding exploitation than common-pool resources (which anyone in the specified group can access) are open-access resources, which cannot be protected from outsiders at all. In the first case, free-riders inside the group are problematic, and coalitions and mores may be of help; in open access resources, the free-riders are strangers (see Ostrom et al. 1999).

45. Hawkes (1992, 1993) and E. A. Smith and Boyd (1990) have nice discussions of this problem in traditional societies. Ostrom (1990, 1998a,b) and Ostrom et al. (1994) have comparative and experimental data on common-pool resources. Atran (1993) and Nkrumah (in press) have examples of how relatedness and village stability affect use practices. David Sloan Wilson (1998b) has proposed commons issues as another example of group selection; commentaries (e.g., Hawkes, Bleige Bird, and Bird 1998, E. A. Smith 1998b, Harpending 1998) make it clear that truly costly behaviors are theoretically possible but not shown to exist in any examples or applications.

CHAPTER 10. CONFLICTS, CULTURE, AND NATURAL SELECTION

1. Karlsen 1987: 51.

2. Wealthy women (estates valued at greater than £500): Karlsen 1987: 79 poor women and women with no male protectors: Karlsen 1987: 73.

3. Women of high reproductive value were unlikely to be accused as witches;

except in the Salem outbreak (which was somewhat anomalous in many regards), fewer than 35 percent of accusations were directed at women under age forty (Karlsen 1987: 65, 86). And if one considers only married women, those from poorer families (with estates less than £200) are overrepresented in convictions.

4. The different return curves of mating and parental effort suggest some biases in patterns of sex differences we might expect to see (e.g., see Wilson et al. 1996 on sex differences in Machiavellianism). Add human cultural and historical particulars, and (even though human culture, like other behavior, has basic operating rules) there is no way we can predict a priori all possible outcomes (see especially Boyd and Richerson 1992). This, of course, is part of the virtue of modeling approaches. But even knowledge of the basic rules may cast a new light on complex examples—may help us see a pattern where it was obscure before. The fact that in our evolutionary past a man's value as a mate was largely based on his ability to control resources, while a woman's value derived primarily from her ability to produce children and not the resources she brought to the marriage, seems perhaps at once simple and obvious, and irrelevant to our lives today, when we have so few children and many women work. Yet a number of social patterns, both cross-cultural and historical, seemingly unconnected to resource and reproductive value, make sense in this context. Similarly, we can make some testable predictions about the strength and enforceability of norms in small stable-membership groups versus large heterogeneous and mobile societies. Alexander (1987) cogently described moral systems as emergent phenomena resulting from these conflicts of interest. Irons (1991) and Cronk (1995) have made the point about manipulation of systems eloquently. Flinn (1997) has recently reviewed the important hypotheses about the role of social learning, as well as the difficulty of testing them.

5. Males homozygous for the t allele are either sterile or die before maturity: Lewontin 1970; cytoplasmic war: Hurst 1991, 1992, Hurst and Hamilton 1992; parliament of genes: Van Valen 1973.

6. E.g., Axelrod 1986, Boyd and Richerson 1992. Of course, some of the strongest reciprocal and coalitional relations exist to harm "outsiders" rather than to spread comfort within the group.

7. This conflict between "my genes, my lineage" and "all our fates," sometimes called the "levels of selection" problem, is crucial to understanding human resource use. Chapter 9 has suggested why it is so hard to convince ourselves to make really costly sacrifices for some unrelated group. More subtly, why are we eager to make the sacrifice of, say, a benefit dinner or concert, especially if our friends are aware of our sacrifice? Why is "nepotism" a dirty word in politics and administration? Why are some patterns widespread, repeated in culture after culture even if otherwise quite different? These become central issues as we face large-scale complex problems such as population growth and individual consumption, the temptation to externalize costs and let them be borne by others who do not share the profits, and so on.

8. "Moral thinking" may be literally an adaptation (contraints on harm of others within the group, solidarity for intergroup competition) that allows the formation of larger groups than would otherwise be possible, e.g., Alexander's 1987 *Biology of Moral Systems* (also Alexander 1974, 1977, 1979, 1988), Campbell 1975, 1979, 1983, Gibbard 1992, Irons 1991, 1996a,b, Ruse 1982, 1986, E. O. Wilson 1978, Cronk 1994, 1995. A readable recent synthesis is Matt Ridley's 1996 *The Origins of Virtue*.

9. See especially Alexander's 1987 *The Biology of Moral Systems,* Ridley's 1996 *The Origins of Virtue,* and references in note 8.

10. Kohlberg 1981, 1984. See Alexander's 1987: 131–140 review.

11. E.g., Alexander 1987, Hamilton 1975, Frank 1988, Irons 1996a,b.

12. E.g., Cronk 1994, 1995, Irons 1996a,b.

13. We might begin by asking waiters how much the tip percentage changes, depending on whether there is more than one person at the table.

14. Cavalli-Sforza and Feldman 1981.

15. Ritchie 1990, 1991.

16. Great variation may reside in very local cultural responses to environmental pressures. In the high desert regions of Peru, for example, Quechua mothers swaddle their infants tightly. This swaddling alters the microenvironment for the infant, making temperature higher and more stable, and humidity higher (Tronick et al. 1994). Such practices may, over time, select for infant characteristics that do well under the cultural practice.

17. Durham 1991 has done much to bring theory to a review of empirical data. Cronk 1995 has an excellent discussion of the issues; also see Boyd and Richerson 1985, Richerson and Boyd 1992, Cavalli-Sforza and Feldman 1981, Lumsden and Wilson 1981, Durham 1991, Alexander 1979, 1987, and Irons 1996a,b.

18. Pulliam and Dunford 1980: 66.

19. Boyd and Richerson 1985, 1990, 1996, Richerson and Boyd 1992.

20. Richerson and Boyd 1992. Boyd and Richerson (1996) note that the conditions for generating observational learning (as opposed to maintaining it) are stringent, and this is "why cultural transmission is common, but cultural evolution is rare."

21. Sign language: Aoki and Feldman 1994; hunter-gatherer skills: Hewlett and Cavalli-Sforza 1986; Stanford undergraduates: Cavalli-Sforza et al. 1982.

22. Cultural and reproductive success: Irons 1983a, Alexander 1979, 1987; cultural manipulation: Cronk 1995.

23. Durham 1991.

24. Guttentag and Secord 1983 review these data.

25. Chinese policy and birth sex ratios: Hull 1990, Johansson and Nygren 1991, Wen 1993, Yi et al. 1993, Li and Choe 1997.

26. Peng 1991.

27. Two-province, all-births sample: Wen 1993; China-wide one percent census: Yi et al. 1993. Formal models confirm that a culture-gene interaction can be

powerful, and indeed, the relative value of women in China, at least as brides, appears to be increasing (Kumm and Feldman 1997).

28. E.g., see review by Edgerton 1992.

29. Durham (1991: 194 et seq.), Boyd and Richerson (1985), Richerson and Boyd (1992) examine the conditions and limits under which cultural variations that do not enhance inclusive fitness can be maintained or spread. See also Logan and Qirko (1996) on the role of psychological mechanisms. "Gene-culture coevolution theory" (Laland et al. 1994), derived from Cavalli-Sforza and Feldman (1981) is the most open to nonadaptive interpretations, for, as Laland et al. (1994: 151) note, it treats culture as purely ideational; culture is a "pool of knowledge which is stored in the nervous systems of individuals."

30. Of course, religions are not immune from hierarchical treatment of members or risk taking among young men.

31. See, e.g., Palmer et al. 1995.

32. See Qirko 1999, who reviews other institutionalized celibacy examples as well.

33. In A.D. 556–561 the pope made priesthood candidates who were married with children sign a document prohibiting the children from inheriting church property. Pope Gregory tried to enforce clerical celibacy by declaring that sons of priests would be illegitimate and could not inherit any church property. Although his attempt failed, it is interesting that the debate arose over property. In 1022, Pope Benedict prohibited marriage and concubinage of all clergy; he declared all clerical offspring as serfs—clearly an attempt to deprive clerical offspring of resources that otherwise were likely to revert to the Church. This prohibition in many cases was not obeyed. In 1059, bishops tried to persuade priests to make their marriages "discreet," and there were often monetary penalties for priestly marriage, which nonetheless continued to increase (Heinemann 1990, Sipe 1990, Rice 1990, Sweeney 1992).

34. Anthropologist William Irons calls this the "spare tire" theory of parental investment: when resource richness is uncertain, starting more offspring than can be made independent can be a reasonable strategy. Among raptors, a simple version of this exists: parents start incubating the first egg before the second is laid, and continue. The resulting clutch has chicks of different ages. When food is plentiful, all chicks are reared, but when resources are constrained, the smaller chicks die.

35. Borgias: Mallett 1987: 109–110, 112.

36. Here is an example ripe for comparing predictions from two slightly different approaches. Conditions can change, and change the utility of any strategy, as in this example. This important phenomenon is a topic of dispute between at least some evolutionary anthropologists and evolutionary psychologists. Evolutionary anthropologists, focusing on the diversity of human behavior, are likely to ask about the "adaptively relevant environment" (Irons 1998): What are the costs and benefits of this behavior in this environment? This

leads to the sorts of comparisons biologists and behavioral ecologists are likely to make. Evolutionary psychologists, in contrast, think about the "environment of evolutionary adaptedness" (Barkow et al. 1992) in the Pleistocene, which leads to consideration of what behaviors might be currently maladaptive (see Sherman and Reeve 1997, Buss 1999).

37. R. Smuts 1992.

38. See Irons 1998, Cronk 1994, 1995, Boyd and Richerson 1992, Flinn 1997.

39. Social competition fostering intelligence: Humphrey 1976, 1983, Jolly 1966, and Alexander 1979, 1987; costs of large brains: e.g., Byrne 1995: 213.

40. Machiavellian intelligence: Byrne and Whitten 1988.

41. Social complexity and neocortical enlargements: Dunbar 1992; tactical deception: Byrne and Whitten 1992, Byrne 1993; communication as manipulation: Dawkins and Krebs 1978, Cronk 1994, Alexander 1979, 1987.

42. Cross and Guyer 1980, Cross 1983, Costanza 1987.

43. Quote: Dawes 1988: 103; group biases: Plouse 1993, ch. 17, 18. Status has a self-reinforcing pattern: we use interactions to make previously irrelevant traits into status traits, and dominants individuals use status in interactions to pressure subordinates into accepting beliefs that may be disadvantageous to them: Webster and Hyson 1998, Ridgeway et al. 1998.

44. See, e.g., Kahneman and Tversky 1982, Nisbett 1980 and Dawes 1988, who give examples.

45. Cf. Cheng et al. 1986, Sperber et al. 1995, Cosmides and Tooby 1992, 1996.

46. Larrick et al. 1993, Cosmides and Tooby 1992, 1996, Cheng et al. 1986, Nisbett et al. 1987.

47. For a readable overview, see Blum 1997.

48. Don't forget, however, the cultural influences of how adults treat and train boys compared to girls (chapter 4).

CHAPTER 11. SEX AND COMPLEX COALITIONS

1. There is evidence that males do in fact coerce females and physically keep them from escaping by biting them or doing a full-body slam, and that it is the pressures of controlling access to females that leads to temporary supercoalitions among males (Connor et al. 1992).

2. De Waal 1989 (especially diagram 4), 1996b. Later, coalitions reformed, and in a bloody conflict, the male Luit was killed. Matt Ridley (1996: 159) draws a wonderful parallel with the War of the Roses: Margaret of Anjou as Luit, Edward VI as Nikkie, and Warwick the Kingmaker as Yeroen. The roles do fit.

3. Tiger and Fox 1971.

4. E.g. Low 1989b, 1990a,b, 1992, 1993b, 1994a,b. Men, cooperating in the community sphere, often with nonrelatives, can in many systems gain substantial direct reproductive benefits—resources to buy another wife, for example. Because women, like other female mammals, make the most substantial direct

reproductive gains not by maximizing mates but by maximizing the survival of their children, they may historically have had more to gain by seeking parental investment from males and by cooperating with other women in the familial, rather than the community, sphere. This will obviously be influenced by the distribution and predictability of resources, and by the reproductive consequences of cooperation versus defection (chapter 3, Emlen and Oring 1977, Daly and Wilson 1983, Krebs and Davies 1991, Low 1990a, 1992).

5. Chapter 4; also see B. Smuts 1996: 236, Hooks and Green 1993.

6. In the game as given, in a single evening, there are two Nash equilibria: baseball-baseball, and Stravinsky-Stravinsky, and payoffs are symmetrical.

7. See Maynard Smith 1982. Chicken: $T > R > S > P$; BoS: $S > T > R > P$.

8. Common examples are polygynous species with internal fertilization, resulting in female-only care; see Clutton-Brock 1991 for additional influences determining which parent does the care. Examples of monitoring and desertion: Trivers 1972. Beissinger 1987 gives one of the most interesting examples, that of the Snail Kite (in the U.S., residing in the Florida Everglades and endangered). The kites hunt large snails, and both parents can feed the young. The breeding season is long, and a parent who deserts may be able to have another clutch with a new mate. For a few weeks before fledging, both sexes play "chicken," staying away longer before returning with food. If one parent consistently stays with the nestlings when the other parent is absent, the absent parent is extremely likely to desert. What is unusual about mate desertion in this species is that either parent may desert.

9. E.g., Pusey and Packer 1987, Greenwood 1980.

10. This very allocation of effort to harassing subordinates and maintaining dominance can itself have a cost, as Packer et al. 1995 have shown. Resisting male sexual coercion: rhesus, Bernstein and Ehardt 1985; general patterns: B. Smuts 1987b, 1996, B. Smuts and R. Smuts 1993.

11. Redtails and patas: Cords 1986; gorillas: Fossey 1983; gelada baboons: Dunbar 1983. In multimale groups, both canines and testes are likely to be comparatively larger (chapter 4).

12. Nishida and Hiraiwa-Hasegawa 1986. See also McGrew et al. 1996, Wrangham and Peterson 1996, van Schaik 1996.

13. Struhsaker and Leland 1987, B. Smuts 1987a,b.

14. Importance of reproductive resources: de Waal 1982, 1984, 1986, 1996a,b. De Waal 1996b contrasts the more egalitarian nature of chimpanzee coalitions with the more hierarchical coalitions of, e.g., macaques.

15. Quote: de Waal 1989: 53; lethal coalitions: de Waal 1989: 61–69; chimpanzees and humans: Wrangham and Peterson 1996: 24. De Waal 1996a,b notes that chimpanzees show a more "egalitarian" dominance style than many other primates, with interventions on behalf of the subordinate more often than, for example, in rhesus macaques (which he defines as having despotic dominance style) or stumptail macaques (tolerant).

16. E.g., Parish 1996. As in chimpanzees, females are smaller than males (fe-

male bonobos are four-fifths the size of males); their control comes from coalitions, not size. G-G rubbing: Parish 1996; low level of male-male aggression over copulations: Kano 1996.

17. Prevention of others' success: B. Smuts 1985, 1987b; harassment: Wasser 1983. See also Hrdy 1976, 1978, Packer et al. 1995.

18. E.g., Hrdy 1978, Wasser 1983, Gouzoules et al. 1982, Pusey et al. 1997. Pattern of female variance: If the most successful female in a year has one offspring and the least successful has none, statistically the variance in reproduction is less than among a group of polygynous males, in which the (rare or unique) top male(s) may have ten or more offspring, while more than half the males have none. Further, in at least some primates, dominance striving itself carries reproductive costs (Packer et al. 1995).

19. Wasser 1983.

20. Bonnet macaques: Silk and Boyd 1983; reproductive costs for dominants: Packer et al. 1995, de Waal 1996b.

21. De Waal 1982, 1984.

22. Bygott 1974, Bygott et al. 1979, Goodall 1986, Clutton-Brock and Harvey 1976, 1978.

23. Wrangham 1979, Halperin 1979, Nishida 1979, Goodall 1986.

24. Wrangham 1979, Nishida 1979.

25. E.g., Watts 1996. Coalitions against nonkin: Stewart and Harcourt 1986; coalitions for maternal kin: Silk 1987; sex of participants in aggressive encounters: B. Smuts 1985, 1987b.

26. Wasser 1983, Chapais 1983.

27. This situation is reminiscent of the competition problem: if two species compete, there are four outcomes: A wins, B wins, unstable coexistence, and stable coexistence. The conditions under which stable coexistence is the outcome are restricted, yet we can look around and see that coexistence abounds. So, we must ask, how do the costs and benefits play out?

28. Fallacy of the Twins: Binmore 1992: 310 et seq. And remember (chapter 2): even siblings are likely to kill one another if the stakes are high enough.

29. Interestingly, this works out to be a group-selection argument of the Wynne-Edwards sort (chapter 9). It shows up also in popular antiwar and proenvironment movements.

30. Ostrom et al. 1994, discussed below; see especially their table 9.1. Perhaps communication is a proximate trigger that signals a repeated interactor; here is a problem for evolutionary psychologists. This phenomenon, that it pays more to cooperate when interactions are repeated and one doesn't know when interactions will end, underlies many observations: that cooperation and self-sacrifice are more common in small, related, stable-membership groups, that successful management of "commons" resources is easier in just such groups—for these are the group conditions that make the payoffs work for cooperation.

31. Ostrom et al. 1994. They tested both relatively low-stakes (low-endow-

ment) and high-endowment commons situations; tested no-communication versus communication options; fine (sanction) versus no-sanction environments. They then calculated the average "yield" (production minus fees and fines).

32. The biologist Garrett Hardin in 1968 described the classic English grazing commons: a village green on which the sheep of all villagers grazed. Obviously, any pastoralist who could add a sheep or two to his own herd got all the profit of an extra sheep. The whole village shared the cost of an extra sheep—so there was great incentive for each individual to graze more sheep on the commons than was good for the village as a whole—exactly the sort of "levels of selection" problem introduced in chapters 9 and 10. Modern commons problems include oceanic whaling, ocean fisheries, and atmospheric problems like acid rain, global warming, and ozone layer depletion. In each of these cases, individuals make more profit if they behave in ways that cost the group as a whole (e.g., taking more than the sustainable amount of fish; cheap-but-polluting manufacturing). Not surprisingly, such cases are notoriously hard to solve for the group's long-term good. See Ostrom 1990, 1998a,b, Ostrom et al. 1999, Ledyard 1995.

33. This, of course, is precisely what one would expect after reviewing the "levels of selection" issues of chapter 9.

34. Interestingly, experimental economists are finding a "taste for punishment" in enforcing reciprocity (perhaps related to the issues of chapters 9 and 10: not only punish cheaters, but also those who fail to punish). See, e.g., Fehr et al. 1997, Fehr and Gächter 1998.

35. See especially Ostrom 1998a,b. Ostrom et al. 1999.

36. Ostrom et al. 1994, Ostrom seminar to the Complex Systems Program, University of Michigan, February 1997.

37. Rewarded competition: Axelrod and Hamilton 1981, Nowak and Sigmund 1994. See also Pusey and Packer 1997, de Waal 1996b. Nowak et al. 1995 have a readable review.

38. Thus game theory converges with behavioral ecological theory. Behavioral ecological theory predicts that cooperation will occur under specific conditions: first, among relatives who live together; second, among nonrelatives in long-lived social species, capable of recognizing individuals and likely to have repeated interactions.

39. Hill and Hurtado 1996.

40. The Ache again provide an interesting example. "Tolerated theft" is common (e.g., see Winterhalder 1996a,b,c for a recent treatment). D. S. Wilson (1998b) has argued a group-level selective pressure for this phenomenon, but Hawkes, Bleige Bird, et al. (1998) and E. A. Smith (1998b) clarify the confusion.

41. Low 1989b, 1996; see also chapters 4, 7.

42. Nomadic societies, however, tend to teach children to be generous more than expected, a phenomenon perhaps related to the uncertainly of resource acquisition in nomadic life (Low 1989b). Information sharing and turn-taking with

respect to food gathering may lower the variance (lower risk of failure), and for poorer hunters, increase the average return (e.g., Kaplan and Hill 1985). It may well be that many societies treat unpredictable variables not as uncertainties about which little is known, but as risks (i.e., assigning heuristically some probability of failure).

43. Pathogen stress shows a number of strong relationships with social factors (fig. 4.1); it has even been proposed as a source of kin assistance (Lewis 1998). It appears to change the costs and benefits of a variety of social relationships. Importantly, as the risk of infection by serious pathogens increases, fewer men are suitable, healthy husbands, and the degree of polygyny increases (and, thus, the proportion of unmarried men and the degree of male-male competition).

44. It is probably important that animal husbandry and frequency of internal warfare (which also is associated with low trust) are positively related—that is, wars are frequent over resources such as cattle, used for brideprice. Like polygyny, these suggest heightened male-male competition. See Ember and Ember (1992) for cross-cultural studies, and Richard Nisbett's (1996) book on the associations between pastoralism and male violence in the United States.

45. Today we face growing problems related to these issues, since reciprocity and cooperation within a group are easier to promote when groups are small, have many kin, and have stable membership—all more typical of small traditional societies than most communities in modern nations (see chapter 15 and Ostrom 1990, 1998a,b).

46. E.g., see Keohane and Ostrom 1995.

47. Flinn and Low 1986, Alexander 1977, 1979.

48. Or brother-sister in societies with low male confidence of paternity; Alexander 1979.

49. E.g., Chagnon 1982, 1988, Whyte 1978.

50. Flinn and Low 1986, M. Ross 1983. Note, also, that patrilocality probably increases the strength of male-male coalitions and makes strong female-female coalitions less probable.

51. Divorce and co-wife conflict: Betzig 1989, White 1988; female-female coalitions: e.g., Irons 1983b.

52. E.g., Beals 1961, Blood 1960.

53. See, e.g., Daly and Wilson 1984, 1985, 1987 on child abuse patterns.

54. An interesting exception to this general pattern exists among the South African Herrero (Harpending, pers. comm.).

Men and women act, and form groups, to influence events in different spheres. In most societies, there is little ritual and cultural conceptualization of female roles. Collier and Rosaldo (1981) suggested that "marriage organizes obligations, and . . . such obligations shape political life." Their discussion of the centrality of marriage to men's lives highlights the importance of male-male coalitions, in the community sphere, as mating effort. Male-male and female-

female coalition differences, related to reproductive gain, are broader than the various asymmetries created by different marriage arrangements.

55. Freedman 1980, Draper 1976, Whiting and Edwards 1973, 1988.

56. Draper 1980.

57. Szal 1972.

58. Piaget 1932: 32.

59. Freedman 1980, Lever 1976, 1978.

60. Lever 1976: 482.

CHAPTER 12. POLITICS AND REPRODUCTIVE COMPETITION

1. De Waal 1982: 212; quote: de Waal 1982: 19.

2. The "glass ceiling" effect; for example, only 8 percent of law partners in the United States today are women, according to a survey by the American Bar Association. See also Browne 1995.

3. See Hart and Pilling 1960, Goodale 1971; review by Irons 1983b.

4. Goodale 1971: 65.

5. The Oxford English Dictionary gives as a primary definition: "the science and art of government" and defines government as "the action of ruling; continuous exercise of authority over the actions of subjects or inferiors." Political anthropologists (e.g., Swartz et al. 1966, from which this definition comes; Lewellen 1983) focus on group processes: a typical definition is "the study of the processes involved in determining and implementing public goals and in the differential achievement and power by the members of the groups concerned with these goals."

6. Acknowledging that politics is about resource acquisition and power sometimes leads to expressions of moral censure. It may not surprise anyone that Ambrose Bierce defined a politician as "an eel in the slime upon which the superstructure of organized society is reared," but even Abraham Lincoln was curt about politicians: "a set of men who have interests aside from the interests of the people and who . . . are, taken as a mass, at least one step removed from honest men." Such sly witticisms reflect that there is individual gain to be had from politics. Yet, aside from Machiavelli, few authors treating politics formally, including political anthropologists, have considered the possibility that there may be individual gains in political activity.

7. Whyte (1978, 1979), who used the odd-numbered societies ($n = 93$, around the world) of the Standard Cross-Cultural Sample.

8. Information is missing in published studies for a fifth of the societies, which I suspect means that women's power is slightly overrepresented here, since a major reason for not reporting relative status is the nonexistence of women as leaders, thus I suspect a number of "no information" societies are actually male power societies.

9. Whyte 1978, 1979. Women's greater say in kin leadership—Africa: Ibo, Ashanti; Pacific Islands: Truk; Manchuria: Manchu; North America: Haida; South America: Cubeo. Equal influence—Africa: Nubians; Asia: Semang; Pacific Islands: Iban, Marquesans.

10. Convergence of reproductive interest: Flinn and Low 1986, Low 1988b; women's political office: Low 1992.

11. Low 1992.

12. Nama, Montagnai, Mbundu.

13. Whyte 1978.

14. In patrilocal societies, the couple lives in villages with the man's male relatives; in patrilineal societies, goods are inherited from father to son.

15. Schapera 1930.

16. Schapera 1930: 150.

17. Schapera 1930: 332.

18. Childs 1949, McCullough 1952.

19. Childs 1949: 178.

20. Rattray 1923.

21. Rattray 1923: 81.

22. Rattray 1923: 83.

23. Rattray 1923: 82.

24. E.g., Charlton 1984.

25. Richards 1951.

26. This is not common, but occurs elsewhere in central Africa as well.

27. Richards 1951: 174. Although descent was matrilineal, rank cut across descent lines. Descent was counted back on the mother's side, but the father's name was added as a sort of surname to both sons' and daughters' names, and the father's as well as the mother's ancestral spirit was honored. Not surprisingly, men were likely to recount their father's lineage if it outranked their mother's.

28. Women's literacy, etc.: Carcopino 1941: 84; married women's emancipation: Pellison 1897: 45.

29. Friedlander 1907: 239.

30. Friedlander 1907: 240.

31. E.g., Gardner 1986, Hallett 1984.

32. Linton 1939.

33. Linton 1939: 155.

34. Hershkovits and Hershkovits 1934.

35. Hershkovits and Hershkovits 1934: 192.

36. Speck 1931, Leacock 1955, Lips 1947.

37. Leacock 1955, Strong 1929.

38. Lips 1947.

39. Lips 1947. Note that the importance of male kin is similar to that in the Yanomamö (chapter 7).

40. Lips 1947: 400.

41. Swanton 1928a: 337.

42. Chief's power: Swanton 1928b: 696; women as chiefs: ibid., 696, 700; elections: Swanton 1928a: 331.

43. Bossu, cited in Swanton 1928a: 312.

44. Mead 1935; Gewertz 1981, 1983; Gewertz and Errington 1991.

45. O'Brien 1977.

46. Shilluk: Farran 1963; Nyoro: Roscoe 1923, Beattie 1960, 1963.

47. E.g., Trivers 1985, Clutton-Brock et al. 1982, 1986.

48. In no sense do any of these results suggest genetic determinism; there is no assumption or implication that particular alleles render males or females more fit for particular kinds of coalitions, or likely to respond in particular, rigid ways. Rather, the question is: For different kinds of individuals (males, females, high/low status, etc.) in particular environments (social as well as physical), what are the costs and benefits of different strategies? This argument does not hinge on the fact that men are generally bigger and stronger than women. In primates, and in human societies, the social complexities so far outweigh the impact of physical size that size alone is a poor predictor of success in getting power or influence. Similarly, this does not reduce to an assertion that women are bound by the constraints of pregnancy, nursing, and child care (e.g., because women who become politically active tend to do so later in life than men). If that were true, sterile women as well as postmenopausal women might broadly be expected to have and wield more public political power. In fact, most men do not achieve significant power when they are young, and even though women may spend more hours per day working in child care while men are "idle" (perhaps doing politics?), there are hours remaining. If there were net reproductive profit to political activities, women's political networks, or bisexual politics, should be more broadly seen.

49. Betzig 1986, Betzig and Weber 1993.

CHAPTER 13. SEX, RESOURCES, AND EARLY WARFARE

1. E.g., Ferrill 1985, Keeley 1996. Quote: Keeley 1996: 175; proportion of men: Keeley 1996: 34; death rates: Keeley 1996: 64.

2. Huntingford and Turner 1987.

3. In humans, too, within-population lethal individual encounters are a male phenomenon—and driven largely by reproductive conflicts, though we don't usually think of homicide that way (Daly and Wilson 1988).

4. Reproductive roots of human homicide: Daly and Wilson 1988; assessment of combatant's strengths: e.g., Parker 1974, 1984, Maynard Smith and Parker 1976; quote: Parker 1974; adult male red deer mortality from fights: Clutton-Brock et al. 1982

5. See review by Huntingford and Turner 1987.

6. E.g., Manson and Wrangham 1991, Low 1992, 1993b, 1994a.

7. Perhaps the closest I can think of is Barbara Fritchie, who was a childhood icon for me, for hoisting the new American flag under occupation in the American Revolutionary War, and saying, supposedly, "Shoot if you will, this old gray head, but spare my country's flag."

8. Darwin 1871. See Andersson 1994.

9. Aggressiveness cross-culturally: Ember 1981, Barry et al. 1976, Low 1989b; homicides: Daly and Wilson 1988; scope of male versus female politics: Ross 1983, Low 1990a, 1992; women's relative gain: Low 1990a, 1992; gender differences in aggression: Hyde 1996.

10. During the seventeenth, eighteenth, and nineteenth centuries women occasionally passed themselves off as men and fought with their men (Holmes 1985: 102). In fact, this pattern appears to have existed from at least the time of Alexander; in Alexander's time, at the end of the campaign, soldiers' children were legitimized (Keegan 1987).

11. Even in ungulate species like red deer, in which status and resource control are mediated through physical combat and there is no evidence of reciprocal "political" alliances, size is not the only determinant of status (e.g., Clutton-Brock et al. 1982).

12. Note that under these conditions, women face a conflict of interest with their husbands, for their husbands may be making war upon their fathers and brothers; Adams 1983.

13. The "Handicap Principle": Zahavi 1975, Pomiankowski 1987a,b, Zahavi and Zahavi 1997.

14. This is the "Sexy Son" hypothesis. See review by Andersson 1994: 44–45.

15. Runaway sexual selection: Fisher 1958, quote p. 152.

16. Chagnon 1988. See also Chagnon 1997 for the way ecological factors affect possible payoff in warfare.

17. Lorenz 1966; variations of this argument are found in Eibl-Eibesfeldt 1979 and Ardrey 1966. The term "instinctive" has typically been used for traits that seem to appear without learning or gradual development, though often as we discover more about development, we uncover the process. Thus "instinctive" behaviors are not qualitatively different from "learned" behaviors, simply more likely to occur when relatively fixed and simple responses are appropriate. Alexander (1979) uses the probably more correct but infelicitous term "cryptontogenetic" (hidden ontogeny).

18. Refer back to chapter 1 for the reasons a behavioral ecologist would reject this notion.

19. E.g., Harris et al. 1998. Once again evidence from a pathological condition gives some information. In the early 1900s, a Dutch family was identified in which several men had a point mutation resulting in a selective deficiency of enzymatic activity of monoamine oxidase A; these men showed borderline men-

tal retardation and considerable impulsive aggression, arson, attempted rape, and exhibitionism (Brunner et al. 1993). A "gene for aggression"? Genetic influence, certainly, but the data suggest that, here too, genes, environmental conditions, and developmental circumstances interact (Brunner 1996; see also Miles and Carey 1997, Cairns 1996, Cadoret et al. 1997).

20. Tooby and Cosmides 1988.

21. Further, although Cosmides and Tooby discuss allocation of rewards to participants (that they may be unequal) they fail to address the fact that these payoffs must be compared to payoffs from other strategies (the dilemma of disenfranchised males).

22. Ember and Ember 1992.

23. Chagnon 1997. This is an excellent study of the interplay of evolutionary trends influenced by ecological conditions.

24. Sex ratio and aggression: Mesquida and Wiener 1996, manuscript. Aggression in young males: Daly and Wilson 1988. The *U.N. World Population Prospects,* 1996 revision, suggests that in developing nations, young males will increase as a proportion of population, increasing by 25 million; in the developed world, young males are expected to decrease by 4 million. As the proportion of young males increases in resource-limited environments, the potential for violent solutions in both local and larger conflicts seems likely to increase.

25. Wolves: Mech 1977; hyenas: Kruuk 1972; lions: Packer 1986.

26. See Byrne 1995, especially ch. 9, 14.

27. Gorillas: Harcourt 1978; intergroup aggression in primates: Cheney 1987; chimpanzee raids: Manson and Wrangham 1991, Goodall 1986, Goodall et al. 1979, Nishida et al. 1985; chimpanzee female distribution: Goodall 1986, Nishida 1979. See also chapters 4, 10 and references therein.

28. E.g., Pusey and Packer 1987, Goodall 1986, Manson and Wrangham 1991. Chimpanzee fision-fusion groupings vary in the strength of male-male bonds (Boesch 1996), and we might profit from more inquiry into variation in chimpanzees.

29. E.g., see Chagnon 1988, 1997.

30. Cheney and Seyfarth 1987, Packer et al. 1995.

31. E.g., Hrdy 1976, Wasser 1983, Silk and Boyd 1983. These situations typically involve harassment of subordinate females and infanticide, with little risk to the aggressors.

32. Several anthropologists (Durham 1976, Harris 1979, Divale and Harris 1976) have suggested that wars were fought to secure scarce animal protein from the hunting grounds accruing to the winning side. Yet there is a strong argument that it was not the means of production but the means of *re*production that led to such serious escalation of competition (Chagnon and Hames 1979, Chagnon 1979; see also Ember and Ember 1992). Frequency of war: Ross 1983, Ember 1978, Ember and Ember 1992.

33. Manson and Wrangham 1991.

34. Jivaro: e.g., Karsten 1923.

35. Blackfoot: e.g., Denig 1961, Ewers 1958. It seems to be currently fashionable among some anthropologists to argue that cultural disruption (e.g., the introduction of horses) upset some balance and generated warfare. Yet the Blackfoot case is functionally identical to the Meru case (and many other pastoral societies) in which no such "disruption" could be postulated.

36. Fadiman 1982; cf. Flinn and Low 1986.

37. Manson and Wrangham 1991.

38. Land: Ember and Ember 1992; adultery and wife-stealing: Divale 1973; capture of women: White 1988; Maori: e.g., Vayda 1960.

39. Hartung 1992.

40. Chagnon 1997: 191.

CHAPTER 14. SOCIETAL COMPLEXITY AND THE ECOLOGY OF WAR

1. Keegan 1987. Relatedness among the Companions: Keegan 1987: 34.

2. Stiller 1973.

3. Warfare and state formation: Carniero 1970; political complexity and war: Strate 1982, Otterbein 1970.

4. For example, an early debate (e.g. Carniero 1970 vs. Wright 1977, Webster 1975) centered on whether population growth is demonstrated to precede warfare. However, if competition can be driven to lethal levels by reproductive conflicts, this is no longer a relevant question. Similarly, the military historian's distinction between "pre-heroic" and "heroic" leadership (e.g. Keegan 1987) may be a matter of scale rather than function.

5. Axelrod and Dion 1988 have shown formally that increasing the number of actors makes cooperation more difficult; and "noise" (misperceptions, incorrect information) can invite exploitation. In fact, no strategy is evolutionarily stable if the "shadow of the future" is too short. See also chapter 9 on the tensions between individual and group interests.

6. Portuguese nobles: Boone 1986, 1988; disenfranchised males generally: Dickemann 1979; Cheyenne: Moore 1990. In a spectacular misunderstanding of the principles of behavioral ecology (chapter 1), Moore uses the lack of reproduction in war chiefs to argue "no gene for warfare" and posits it as a response to Chagnon's work (which did not propose such a gene).

7. *U.N. World Population Prospects,* 1996 revision.

8. Sex ratio and aggression: Mesquida and Wiener 1996, manuscript; aggression in young males: Daly and Wilson 1988; increase/decrease in young males: *U.N. World Population Prospects,* 1996 revision.

9. Kin selection: Shaw and Wong 1989; local low-intensity warfare: Brogan 1990, Keegan and Wheatcroft 1986, Dunnigan and Bay 1986.

10. E.g., Hanson 1989 argued that classical Greek warfare has "left us with

what is now a burdensome legacy in the West: a presumption that battle under any guise other than a no-nonsense, head-to-head confrontation between sober enemies is or should be unpalatable." Political consequences: McNeill 1963; seasonal constraints: Hanson 1989.

11. General's importance to troop morale: Hanson 1989; generals moving to the rear: Keegan 1987.

12. Violent housekeeping: Hale 1985: 13; knights fighting individually: Hackett 1983: 28; birth order and risk: Boone 1988; socially acceptable property acquisition: Hale 1985: 22; insignias representing social status: Vale 1981: 148.

13. E.g., see McNeill 1982.

14. Hale 1985: 75, Miller 1975, Elton 1975.

15. Savorgnan quote: Hale 1985: 109; Thomas Wilson quote: Hale 1985: 125.

16. Chamberlin 1965: 151.

17. These patterns are parallel to the training of boys in nonstratified polygynous societies in which male kin groups are important (chapter 6). See also Johnson 1986, Johnson et al. 1987.

18. E.g., Holmes 1985, Dixon 1976. In this context it is interesting to return to the sex differences in child training noted in earlier chapters: highly polygynous groups with no stratification train sons to be obedient—and these are the societies in which small-scale warfare is common: men fighting alongside male kin over women.

19. Hackworth 1989: 633–35.

20. Cited in Gies 1984: 157.

21. *Detroit News and Free Press*, February 10, 1991.

22. E.g., Watson 1978, Hackworth 1989, Schein 1957.

23. E.g., Wilkinson 1986; see also Jacobson and Zimmerman 1993.

24. Low-intensity warfare reviews: Gutteridge 1986, Klare and Kornbluh 1988; genetic interests in warfare: Shaw and Wong 1989; paternalistic terrorist leaders: e.g., Aston 1986; ethnic components in large-scale war: Aston 1986, Brogan 1990, Keegan and Wheatcroft 1986, Dunnigan and Bay 1986, Goose 1988. See also Maechling 1988, Barnet 1988.

25. Deception in other primates: Byrne 1995; faulty information: Parker 1974, Alexander 1987: 239.

26. E.g., Putnam 1993b.

27. E.g., Alexander 1979.

28. David Hackworth (1989: 634) called the failure to mimic the familial structure of preindustrial warfare (which resulted in high casualties and an increasing need for replacements, followed by a lowering of induction requirements) "the most blatant example of the use and misuse of the poor and disadvantaged in America's wars." In the Vietnam War, these men, the Project 100,000 soldiers, were those who had (or would have) previously failed the armed services' physical or mental requirements; they proceeded through the Army "as they proceed through life, walking wounded in the center of a monstrous joke, forced to

struggle with basic training as they are forced to struggle with everything else" (Just 1970: 62). The Vietnam example may simply be the most recent and extreme case (Watson 1978: 34).

29. Cf. Holmes 1985: 93.

30. E.g., the Comanche: Cabello y Robles 1961: 178; the Natchez: Swanton 1911: 104. See also Betzig 1986.

31. *New York Times,* October 3, 1992.

32. Overviews: e.g., Singer 1980, 1989, Huth 1988; conflict causes: Richardson 1960, Stoessinger 1982; pacifying influence of shared ethnicity: Shaw and Wong 1989, Richardson 1960; common government and recent alliance: Richardson 1960; military strength: Huth 1988. Leader's personality: Stoessinger 1982, Winter 1989.

33. Brogan 1990; other atlases: Dunnigan and Bay 1986, Keegan and Wheatcroft 1986.

34. Cf. Stoessinger 1982.

35. Shaw and Wong 1989: 208.

36. E.g., Keohane 1984, Oye 1986.

37. Alexander 1987: 240.

38. E.g., Goldstein 1989, Keohane 1984, Oye 1986, Groebel and Hinde 1989.

CHAPTER 15. WEALTH, FERTILITY, AND THE ENVIRONMENT IN FUTURE TENSE

1. Of course, caveats are important here; see Hawkes, O'Connell, and Rogers 1997.

2. This is a particular problem in some common-pool resources—those in which only a few people get benefits, and everyone shares the costs (or the reverse). A number of current environmental problems are common-pool (really open-access) resources: ocean fish and whale populations, global temperature, the ozone layer. The demographer Ronald Lee (1990) argued that, in a resource-limited world, fertility itself is a common-pool resource issue: the children you have are children I can no longer afford to have.

Neither the welfare of strangers nor the far-distant future has been of much concern in human evolution. Common-pool resources and collective action have been problems for traditional societies as well as modern ones (Hames 1988b), and getting individuals to cooperate for the group good has typically required some individual benefit for a stable solution. Hawkes (1993), in examining why men hunt and share their kills, suggested that these individuals supplied meat as "public goods" because of individual incentives, including direct reproductive benefits for men who demonstrated hunting skill. In light of these issues, it is of interest that even in extremely simple models (e.g., the origins of cellular cooperation) we find that cooperation for the group good (e.g., energy spent for cellular regulation) is more frequent when r (degree of relatedness) is high among cellular components, and/or the costs are paid from the group resources rather than individual components' resources (S. A. Frank 1998).

3. Trade-offs: e.g., Blurton Jones 1989, 1997, Rogers 1995. Most papers (e.g., Vining 1986 and others) that see a negative or flat relationship between wealth and fertility use aggregate data, inappropriate for answering the questions I raise here (see below, note 9). Fertility decisions when parental investment counters mortality: e.g., Blurton Jones 1997, Cigno 1998.

4. Biological version of r-K selection theory: MacArthur and Wilson 1967; human explorations: Barkow 1977, Beauchamp 1994, Irons 1983a, Low 1993a, 1999, Low et al. 1992, Low and Clarke 1991, Voland 1984, 1989, 1990, Voland and Dunbar 1995. See also Wood 1998, Cronk 1991a, Borgerhoff Mulder 1991, 1992, 1998b. Becker (1981) considered the costs of children explicitly; Freedman and Thornton (1982) and Johnson and Lean (1985) discussed heuristic rules like those proposed in chapter 9 (e.g., how am I doing compared to my peers?) as they affect fertility decisions. Thai schooling: Knodel and Wongsith 1991; market training: Kaplan et al. 1995. Note: we typically do not have enough data to tell whether human demographic transitions are really an example of the biological "r- and K-selection"; we have tantalizing clues like Knodel's Thai data and Kaplan's market analysis, and Mace's 1998 study of the coevolution of fertility and wealth inheritance strategies. Mace, blending empirical data and dynamic simulation, found that increasing costs of children decreased the optimum fertility and increased optimum inheritance—classic r-K selection.

5. "Externalization" is the economic term for keeping the benefits and sending away the costs of one's actions (e.g., shipping one's garbage or toxic wastes to poor communities). A classic example exists of a memo, for example, in which an official argued that the United States should export toxic wastes to poor African countries: after all, the population density was low (so few would be harmed), and besides they were poor and powerless if any ill effects followed.

6. I write as if we all agreed that there *is* a dilemma. I realize that we don't all agree, but my own area of research convinces me. Despite the variety of opinion, daily, in the popular press, politics, and academe, we read and hear about environmental consequences of the combined effects of population growth and per capita consumption. The *Limits to Growth* study of 1972 argued that current growth trends would reach serious limits, but that it was possible to alter patterns and "establish a condition of ecological and economic stability that is sustainable into the future." Since then, a number of environmentalists and political leaders have adopted sustainable development as a goal (e.g., see edited reader of Mazur 1994). The Brundtland Report in 1987 defined sustainable development as "development that seeks to meet needs and aspirations of the present without compromising the ability to meet those of the future." For additional examples, see Ehrlich 1997, MacKellar 1996, 1997.

7. Birdsall 1980. Total Fertility Rate is a projection that asks: If all women in the population lived all the way through their reproductive lives, and at every

age had the average number of babies (and all lived), what would the result be? Though widely used, it ignores both infant and maternal mortality (both higher in poorer than wealthier countries).

8. The possible implications of these patterns are interesting in terms of population policies around the world. The required investment may limit the number of children a couple can afford. Within subgroups, however, those with more than sufficient resources may be able to support additional children and still have all be adequately invested (e.g., Hughes 1986, Low et al. 1992).

9. Three difficulties are most common with aggregate data (e.g., Wrong 1980, 1985): (1) fertility itself is not measured; data are "household" and one cannot tell, e.g., step- and foster children from genetic children; (2) sometimes fertility is measured, but those who never marry, or those hard to census, are excluded from the analysis; (3) only averages are calculated, with no idea of variability, so it is hard to tell if differences are meaningful. So, we have a dilemma: it is likely that fertility sometimes declines with income, but we seldom have the measurements to tell.

Aggregate data tell us nothing about variation. For example, if fertility is 1.8 children per couple in much of western Europe, and 2.1 in the United States, is that a "real," significant difference? There is no way to tell. Aggregate data for different occupational groups within countries have varied patterns—but without knowing the variance within groups, and whether wealth in each group is normally distributed or skewed, we cannot make any inference.

Further, aggregate data typically undersample a very important group in any population: the poor and the homeless. These are probably the least likely to have children, and statistically the biggest predictor of variance is the size of the zero-success class. For example, U.S. census data are taken from households. Suppose poorer households have more children in them than wealthier households. Does that mean wealth and fertility are inversely related in the U.S. today? By no means. Relationships are not given; income is not apportioned to individuals: so there is no way to know father's earnings versus mother's work effort, with its complex effects.

Finally, the poorest of the poor have both very low fertility and very low income—and seldom appear in census data; if the homeless comprise a sufficient percentage of the unsampled population, the statistical analysis is compromised. Unless we ask about them as well, we have no reliable information about the overall pattern of wealth and fertility in the United States. These issues are currently being debated about the last U.S. census, in which it is almost certain that serious undercounts of exactly the sort I am describing have occurred.

Evolutionary anthropologists: e.g., Kaplan 1994, Kaplan et al. 1995, Kaplan and Lancaster 1999; r- and K- (high-investment, low numbers/low-investment, high numbers) strategy: MacArthur and Wilson 1967; see also Betzig 1996, Low 1993a, Low et al. 1992. The investment required to produce successful offspring varies with environment, and specifically with the threshold level of investment required for a child's success, and fertility varies across natural-fertility societies (Bentley et al. 1992). How harsh is the competition offspring face? In this sense, environmentally influenced differences in successful investment in children is precisely analogous to the proper use of MacArthur and Wilson's (1967) concept of r- and K-selection (see reviews by Low et al. 1992, Low 1993a, 1999, also Watkins 1989, Mason 1997 and Borgerhoff Mulder 1998b. For empirical descriptions of developing world transitions and the failure of demographic transition theory, see Schofield and Coleman 1986, Robey et al. 1993). Do relatively wealthy women who have late and low fertility nonetheless have greater net lineage increase? I suspect not, but we have few data. In postdemographic transition societies, in which it appears that successful children cost more per capita in parental investment and fertility is relatively low, women might actually profit reproductively at the lineage level by spending more effort bringing in resources rather than filling traditional roles of child production.

10. Positive empirical correlations: e.g., Irons 1983a, Chagnon 1988, Mueller 1991, Mueller 1993, Mueller and Mazur 1996, Low and Clarke 1992, Low 1990e, 1991, Essock-Vitale 1984, Hill 1984, Essock-Vitale and Maguire 1988. Disappearing correlations: Kaplan et al 1995, Kaplan and Lancaster 1999, Low 1989a. See also Rogers 1990 and Harpending and Rogers 1990.

11. Wealth accumulation and success: Rogers 1995, unpublished (reported in Borgerhoff Mulder 1998b); competition as factor: Woods 1998. Small family size and ability to accumulate wealth: Havanon et al. 1992. We are still unsure whether constraint of fertility in competitive environments is a general picture in modern societies. It seems likely to be true for a family with some wealth; in contrast, a family with no resources may do best to maximize fertility and let children's own initiative be the deciding factor—yet really poor women in the U.S. today have low total fertility.

12. Models on the value of children: Becker 1981; see also Merrigan and St.-Pierre 1998, Easterlin, 1978; children's economic value: Caldwell 1976, 1982, 1983. Note that, while the only good empirical data we have suggest that children in today's hunter-gatherer societies never produce true net upward wealth flows (e.g., Hawkes, O'Connell et al. 1998, 1999, Turke 1992), children's value as genetic currencies means that even when they have a net economic cost, they have a total net benefit, and this answers the quote in the paragraph (from Schoen et al. 1997).

13. Support for Easterlin's hypothesis: review by Macunovich 1998; effects of parity and marital status: e.g., Schoen et al. 1997; effects of government policies: e.g., Gauthier and Hatzius 1997, Zhang et al. 1994.

14. Women's workforce participation: e.g., Kasarda et al. 1986. I suspect (but know of no data) that women's work and education also correlate positively with market consumption. Delayed fertility is the strongest correlate of slowed population growth, even when it does not mean lower total lifetime fertility.

15. Consider the question of welfare payments and fertility. For non-white

women in the United States today, it appears that the AFDC program increases the likelihood of having a first child (Taşiran 1995, his table 30; see also Luker 1996). However, it has *no effect* on probability of second and subsequent births; thus the AFDC program is unlikely to lead to high total fertility among poor women. In fact, contrary to widespread beliefs about the fertility of welfare mothers, today in the United States poor women have early but *low* total fertility. Women on welfare in Wisconsin have 29.4 percent fewer children than women across the United States, and 32.8 percent fewer children than women in Wisconsin; such women tended to be unmarried, black, and less educated than others (Rank 1989).

16. Poor women's fertility in the United States tends to be low but early, which is adaptive for several reasons. Consider poor urban black women in the United States today: as they age, they suffer "excess mortality" and a variety of sublethal health problems that make it not only more difficult to conceive, but to raise children successfully. High mortality among black men means that the probability of having a male partner declines with a woman's age. Even female assistance (mothers, aunts, etc.) declines. In the poor black populations of Harlem, central Detroit, and Watts today, a teenage mother has about a 75 percent chance that her mother will be alive and able to help when her child is five years old; for women who postpone childbearing until age twenty, that figure is only 40 percent. Not surprisingly, some scholars suggest that early childbearing may mitigate the threat to family economies and care-taking systems (and reproductive success) imposed by the heavy mortality and disease burdens of these women (e.g., Geronimus et al. 1996, 1997, Geronimus 1996a,b. Lancaster 1986 made similar suggestions for more traditional societies). Further, there is evidence that these disadvantages affect the life chances of such women's children, exacerbating disparities over generations (Duncan et al. 1998).

These findings fit the predictions from life history theory remarkably well: the most important variable in setting the timing of fertility is adult mortality (chapter 6; see also Coale and Trussell 1974, Ellison 1991). Early fertility is common among poorer women who face uncertain futures, and teenage fertility is strongly associated with low educational achievement for women—in contrast to the age-specific fertility of American women in general, which is shifting to later in life. From 1976 to 1988, overall fertility peaked at ages 20–24, but compared to 1976, pregnancy rates in 1988 were substantially higher for women aged 30–39 (Rank 1989, Luker 1996).

The most puzzling fertility pattern is that of relatively wealthy women who earn a high proportion of their household income, and who routinely delay fertility past the mid-twenties. My own suspicion is that two things interact. First, when very high offspring-specific true parental investment is required to produce competitive and successful offspring—that is, when environments are highly competitive, women's mate value shifts from largely-to-entirely reproductive value, to relatively more resource value. Women, like their husbands, may concentrate on getting resources for offspring rather than on producing more offspring; such production of fewer-but-better-invested offspring seems likely to be more profitable under these circumstances. Obviously, any family that can produce both more and better-invested children would profit, but such families would be a very small percentage of the population, and a linear pattern of "high wealth–low fertility" is not likely. Under these circumstances, we predict declining family sizes, and greater participation of women not only in the job market but in the higher-paying professional market. But I suspect a second factor operates as well: the proximate cues (chapter 2) of "doing well" are wealth, health, possessions; when both partners are highly educated, a likely result is that it will prove enticing to continue to strive and accumulate—and this will decrease fertility. So we may see a very mixed bag: some small but highly successful families who concentrate resources in few successful children, some families that attempt to pattern and fail, and some highly consumptive ("yuppie") couples who may have no children.

17. Taşiran 1995. Taşiran compares empirical studies and models for Sweden (pre-1935 to 1965 cohorts), and white and non-white American women (pre-1945 to 1966 cohorts), using macro-, micro-, and integrated data. Women's education, another "lost time" trade-off, has a negative effect on fertility: the longer women stay in school after high school, the fewer children they are likely to have. This effect is also complex. For Swedish women, more education means they are less likely to have children, and if they have children they are likely to stop at two (Taşirann 1995, table 30). More-educated white American women are also less likely to begin having families. For non-white American women, 5–9 years of education makes having a first child less likely; ten or more years of education makes having a first child more likely, though later in life.

18. Some of these differences appear to be simple cultural preferences; others seem to reflect cultural influences on the costs and benefits of having a child. In Sweden, whether a woman is married or cohabiting (versus living alone) has no effect on her likelihood of having one or more children; significant social services exist for parental benefits. A man's wages in Sweden have a positive effect on the likelihood of having a first and second child. Interestingly, in the United States, a man's income has a negative relationship to a woman's probability of having a first, second, or third child. We have no information (e.g., lineage persistence) to help us understand whether this is a strategy shift (higher per capita investment per child, fewer children) or simply a reproductively costly cultural preference. Further, this ignores the effects of delayed fertility. If highly educated and wealthy men tend to marry highly educated women in the United States (because social services do not affect fertility), then we might see a spillover effect . . . what an interesting question!

19. Taşiran 1995, Rank 1989, Kasarda et al. 1986, Low 1993a, Low and Clarke 1992.

20. Cohen (1995) and Lutz (1994) are excellent nonpolemic, analytic treatments. See also MacKellar (1997) and Cohen (1998) for explicit treatments of the role of equity in sustainability.

21. Dryden's was the first recorded use of the term, in *The Conquest of Granada* (1672):

> I am as free as nature first made man,
> Ere the base laws of servitude began,
> When wild in woods the noble savage ran.

Rousseau, of course, used the concept effectively to anathematize civilization—and in fact, even today, there is a strong streak of romantic distaste for human company and human machinations in various environmental movements. A peculiar conjunction of historical events may have exacerbated possible misinterpretations. Considerable data were collected during the nineteenth-century explorations, at the height of the Romantic approach to ethnography and the growth of "manifest destiny" in the United States; there is much imputation, but few data, about the precontact practices of traditional peoples. A mythology developed, probably fueled by the Romantic view, of a pristine America barely peopled, and populated by societies virtually without impact (Denevan 1992). American primitivist writers like Cooper and Thoreau probably fueled this image. Threatened by constraint, even extinction, it may have paid indigenous peoples to acquiesce in an overstatement of the globally conserving long-term nature of their policies. Today we have remnants of this history. For example, the words of Chief Seattle regarding an impending ecological crisis, reminding us that "the earth does not belong to man, man belongs to the earth" are widely quoted—yet those words were written by a scriptwriter in the early 1970s, and there is no empirical evidence suggesting that Chief Seattle had any such attitude (e.g., see Budiansky 1995: 32–34). And, as anyone who remembers George Bush's desire to be remembered as the "Environmental President" can attest, stated attitudes and actual practice may remain far apart (e.g., Tuan 1968, 1970).

Noble ecological savage: Redford 1991. Oelschlanger (1991) reviews romantic history views in which reverence is required. At its extreme, the Romantic view surfaces as the Gaia concept, which argues that the earth is a homeostatic, self-regulating unitary entity especially suited for human life (Lovelock and Margulis 1974, Lovelock 1988). See Williams (1992b) for a cogent debunking.

Such romantic misconceptions might not matter, except that they generate normative prescriptions rather than understanding. Several current strategies of environmental and conservation education reflect our faith in this wisdom (see review by Budiansky 1995), but without any logical or empirical evidence. Some muddle the pragmatic and the normative: if only we could recapture the reverence and cooperativeness of traditional societies, and expand it, we could solve our problems. And this "wisdom" is simply not true. Empirical tests of the Noble Savage hypothesis suggest that individuals strive to get sufficient resources efficiently, in competition with other (also striving) individuals (e.g., Hames 1979, 1988a,b, 1989, 1991, Alvard 1994, 1998). "Excess" resources, over and above what's needed for health, are sought if they are reproductively useful. The cross-cultural data are about what an evolutionary cynic would predict, but not what a Noble Savage advocate would hope (Low 1996).

Traditional peoples often do express concerns about the long-term "health" of the ecosystems on which they depend. They often have unique knowledge of sometimes subtle ecological relationships because such knowledge has had survival value (e.g., Atran 1993). And they often, particularly when the interests of kin or local community are vested in sustainable use, can manage resources sustainably, sometimes even when new, highly efficient technologies make overexploitation easy (White 1988). But these are possible outcomes, not driving forces. People take resources from the environment to feed themselves and their families, to garner favorable attention (especially for men in sexual selection; chapter 7), and to use reciprocity as insurance in the face of uncertainty. Such reciprocity is typically biased by kinship relations, with closer kin receiving more aid (e.g., Hames 1989). Since empirical studies suggest that people take as much as they need (or sometimes as much as they can), novel technological changes do not give them any reason to consider potential impacts of maximum use— it has never mattered before. This, I suspect, is the source of the pattern in the cross-cultural data that degradation frequently follows technological change.

For traditional peoples as well as modern, impact is simply a function of the population density, the relative impact of the technology used, and the consumption level (the biggest influence here is the profit from taking the resource, if any, above subsistence). Several combinations of population, profit, and technology will yield low environmental impact but do not imply conservation: low population density plus inefficient technology (e.g., stone axes) plus no market, for example. To infer that low impact must imply conservation ethic is simply not logical; we can infer deliberate conservation only when impact is low even though there is high profit to be made, the technology is sufficient, and/or the population density is moderate to high.

22. I am indebted to Professor Emilio Moran, Indiana University, for this quotation.

23. See, for example, Lancaster 1991, Hrdy 1997, and Gowaty 1997, an edited volume of broad scope.

24. E.g., Simon 1996. Note, however, that the lesson we draw from this correlation is very different if we think, as did Simon, that more people lead to more comfort, versus the interpretation that inventions increasing the standard of living tend to increase fertility, and thus population growth (see, e.g., Ahlburg 1998). The long-standing argument between Julian Simon and Paul and Ann

Ehrlich really involved different scales of analysis. Simon noted that as world population increased, so did technology and standard of living; the Ehrlichs noted that no species ever has been able to grow without limits.

25. The role of environmental uncertainty (e.g., Low 1990b, 1996, Alvard 1998) meant that saving for the distant future never paid. Would one expect conservative management for one's great-grandchildren? Probably not, under most conditions. Remember that, in an outbred population, the relatedness of an individual to her/his direct descendants decreases by 0.5^k over k generations. So ego would share only about 12 percent of genes with a great-grandchild; by five generations, the shared genes are 3 percent! So the far-distant future of one's descendants was probably of little concern, even if our ancestors had been able to control their environments. This emphasis in our evolutionary past on short-term optimization may be a serious stumbling block in our ability to ask ourselves for any significant sacrifices now for the sake of future protection.

26. Bechara et al. 1997.

27. Getting information does not confront the issue of interests—that we may not all want the same things, or that we may want the same ends in theory but are unable to agree on who should pay, and how. The problems most easily countered by simple information are usually small-scale and with relatively great confluence of interests (kin groups, small isolated populations).

Communication, coalitions, and social incentives: these social approaches are surprisingly underrated, especially at the local level. Participation in local recycling is a common success story: the costs and benefits are local, and we see results quickly. Successful programs of course have information, but importantly, they usually also have both economic (if we don't recycle, garbage will cost $1 per bag) and social incentives (the Big Blue Box out front advertises that you are a good recycler and a desirable neighbor). People cooperate, not surprisingly, *if* they can establish communication and get to know the others in the group (a mimic of the reciprocity so important in our past), and the personal cost is not seen as too high (e.g., Ostrom's lab examples in chapters 9, 10).

A subset of social incentives includes exhortation and inculcation. Can it help, to exhort ourselves (others) to "do the right thing"? Perhaps; partly because we have evolved in complex social groups, we are very social and the opinion of others matters. Remember (chapter 9) that social norms can be powerful if the costs are low enough. We can, and do, use this to our advantage: children raised in an environment strong in any particular ethic are more likely to be active in promoting that ethic. Once again, this tactic works best in small, stable communities; conformity is more easily obtained, and the opinions of others can be important.

Conservationists are fond of quoting recycling education and exhortation as examples of success stories and "small wins," but perhaps a note of caution is important: while today's elementary school children in the United States are far more enthusiastic recyclers, on average, than their parents, it will be some years before these children confront the costs associated with recycling goals. Further, until infrastructure makes lost-cost manufacturing of postconsumer materials reasonable, stuff will accumulate in warehouses.

Economic incentives seem likely to be effective, for all the obvious reasons, in complex or costly problems. We all strive to avoid costs, including monetary costs, to ourselves. In fact, here lies a danger: many locally successful solutions achieve success by exporting the problem ("externalizing" costs), like paying poor communities to take hazardous wastes, or not cleaning emissions from smoke stacks because it takes your neighbors (who pay the cost) a long time to figure out that you're the source of their problems—and even then they may not be able to make you pay. Businesses count success by profit, at least in part, so making it possible to "do well by doing good" is an effective strategy in "green" business (e.g., Hart 1997). Here is a topic on which there are examples, but little analysis yet, and a fruitful focus for investigation.

Regulations are a formalized version of the coerced cooperation of chapter 9. Economists prefer strategies that work in the marketplace but keep people from externalizing costs. For example, set no pollution limits, but impose a cost-per-unit to pollution production; then companies can trade their pollution permits. These are extraordinarily complex issues, and it's hard even to mention them without being drawn into more detail than is appropriate here. Not all problems are easily solved this way. Regulations, usually established by external agencies, can do much to make costs and benefits explicit, and thus predictable. However, no regulatory agency is a singular body: individual conflicts can complicate things here, too. Getting agreement can be difficult precisely because of the sorts of interest conflicts of earlier chapters. Even when we get agreement, governmental regulations can become outstripped by technological advances, yet be hard to change. This can have perverse effects such as making it illegal to adopt newer, cheaper, more effective technology.

All of the difficulties above are exacerbated, I suspect, when short-horizon, First World specialists are called in to solve problems in cultures of developing nations (e.g., see Kottak 1990, Kottak and Costa 1993, Atran 1993).

28. Of course, natural selection has never been "pleasant"—extinction is the most probable outcome for any species. It is certainly true that extinction rates have increased greatly in the last hundred years, due to human activity; it is also true that 95 percent of all species ever in existence are extinct today. In the case of our own species, many of us have more than an academic interest in whether we can avoid, or delay, our extinction.

Glossary

adaptation—a structure, or physiological or behavioral phenomenon shaped by natural selection, that renders the possessor more likely to survive and reproduce than its competitors. Also, the evolutionary process leading to such a trait. The concept is onerous; no phenomenon should be labeled an adaptation without effort to determine that the benefit seen is in fact a true function (q.v.) of the trait (Williams 1966).

aggregation—a temporary grouping of animals, usually around some limited and localized resource. Animals in aggregations, in contrast to animals in longer-term social groups, generally show no, or limited, social interactions.

allele—a particular form of a gene, distinguishable from other forms or alleles of the same gene; thus at a locus for eye color, there may be competition between alleles for different eye color to occupy the locus.

allopatric—referring to populations or species occupying different geographical areas. (*Cf.* sympatric).

alpha—referring to the highest-ranked individual in a dominance hierarchy.

altricial—pertaining to young animals which are helpless for a significant period after birth or hatching (*cf.* precocial).

altruism—in common parlance, any helpful behavior. Here I restrict its use to behavior that genetically costs the performer and benefits some other unrelated and nonreciprocating individual. Most behavior labeled as altruism should more properly be specified as phenotypic altruism: behavior that appears altruistic but may well be genetically selfish (as parental behavior, reciprocity), to distinguish it from the usage I employ here, genetic altruism.

anisogamy—condition in which the gametes of two reproducing individuals are of unequal size. In general, we call the large gamete an egg and its producer a female; and the small gamete a sperm and its producer a male.

anthropomorphism—interpretation of what is not human or personal in terms of human or personal characteristics.

aposematic—equipped with conspicuous "warning" coloration, usually associated with poisonous or distasteful characteristics (e.g., Monarch butterflies).

avunculocal residence—a pattern of residence in which a married couple lives with or near the husband's mother's brother. (*See* matrilocal, patrilocal, neolocal.)

Barr body—a condensed, inactive X chromosome found in interphase nuclei of XX individuals in mammalian species.

bridewealth—a substantial gift of goods, money or service (bride service) given by the groom or his family to the bride's family at or before marriage. (*See* dowry.)

brood—n. litter or clutch; v. to incubate eggs by sitting on them.

carrying capacity—the largest number of individuals of a particular species the environment can support, theoretically for an indefinite period of time; usually symbolized by K. Although treated as a constant, it is clear that the carrying capacity of all but the most stable environments must change, because environmental conditions change.

character displacement—the process by which two species (usually newly in contact) interact in ways that favor divergence of characteristics (e.g., males with mating calls of different, rather than similar, frequencies in the two species; different foraging patterns in individuals of the two species).

character release—the process by which a species, on entering an area free of its previous competitors, encounters a relaxed selection on important characteristics; increased variance in those characteristics results.

clone—a population of individuals all derived asexually from a single parent.

clutch—the number of eggs laid by a female at one time.

coefficient of relatedness—the fraction of genes identical by descent in two individuals; symbolized by *r*. (*See* inclusive fitness and chapter 2.)

commons—a term loosely used to mean either open-access resources (for which there are no access rules), which are typically soon depleted as people pursue their individual interests at the cost of group resources; and also the more restricted common-pool resources, which are controlled and used by a well-defined group in common, while outsiders are excluded.

conjugation—joining of cells via a cytoplasmic bridge, followed by transfer or exchange of DNA to accomplish reproduction.

constancy—the extent to which an environment is predictable because conditions remain the same with regard to the parameter measured; symbolized by C. (*Cf.* contingency.)

contingency—the extent to which an environment is predictable because some parameter is correlated with (contingent on) some other parameter; e.g., the seasonality of rainfall; symbolized by M. (*Cf.* constancy.)

cross cousins—children of siblings of the opposite sex. One's cross-cousins are father's sister's children and mother's brother's children. (*See* parallel cousins.)

culture—socially transmitted information, including behaviors, beliefs, attitudes, customs, and ideas.

deme—a small, reproductively isolated population, within which reproduction is generally assumed to be random. (*See* chapter 9.)

dimorphism—having characteristically two (di-) "morphs," or physical types.

For example, sexually dimorphic species are those in which the two sexes look different.

diploid—having a chromosome complement consisting of two copies of each chromosome. In mammals, for example, most body cells are diploid. (*Cf.* haploid.)

directional selection—selection that favors individuals with one extreme of a characteristic, and disfavors individuals with the other extreme; thus the presence of coyotes as predators for rabbits is likely to favor fast-running rabbits, and disfavor slow rabbits. (*Cf.* disruptive and stabilizing selection.)

disruptive selection—selection that operates against the middle of the range of variation in any characteristic, favoring the two extremes, thus tending to split the population. It is often proposed as the mechanism for sympatric speciation, for example in allochronic speciation, and for the evolution of two sexes with divergent characteristics. (*Cf.* stabilizing and directional selection.)

dominance hierarchy—the behavioral domination of some members of a group by other members, in relatively long-lasting patterns.

dowry—goods paid by the bride's family to the bride. (*See* bridewealth and chapter 6.)

effect—any result of a characteristic not produced by natural selection, as opposed to functional adaptations produced by natural selection. The function of an apple is reproduction; an effect was the derivation of Newtonian physics (Williams, 1966: 9).

epideictic display—displays in which members of a population show themselves and "allow others to assess the population density" (Wynne-Edwards 1962). Wynne-Edwards proposed that the function (q.v.) of such displays was the regulation of population numbers. (*See* chapter 9.)

epistasis—masking effect or interference of one gene on the phenotypic expression of a non-allelic gene or mutation in the same genome.

equilibrium frequency—symbolized by q; the frequency of any specified gene after the population has reached genetic equilibrium.

estrus—a cyclic condition in female mammals, in which the female has ovulated (and thus can conceive) and is sexually receptive.

evolution—any change in relative frequencies of genes over time, as a result of natural selection (q.v.), genetic drift (q.v.), mutation (q.v.), and genetic recombination (q.v.). Sometimes the terms *organic evolution* and *biotic evolution* are used to specify changes over time of the kind of individuals seen, and kind of groups seen, respectively.

evolutionarily stable strategy (ESS)—a strategy that cannot be beaten, or invaded by, another.

exogamy—cultural rule specifying marriage to a person outside one's own kin or community group.

fecundity—denotes the theoretical reproductive potential.

fertility—a term reflecting the actual, or realized, reproduction.

fitness—as originally used by Darwin, the term "fitness" meant roughly the ability to survive and reproduce. There are at least five definitions including this one (Dawkins 1982). Two of the most commonly used are *W*, the fitness of the genotype (the contribution to the next generation of one genotype relative to the contributions of other genotypes); and "classical fitness," an individual measure (an individual's relative lifetime reproductive success). The remaining two definitions are Hamilton's inclusive fitness (q.v.) and what Hamilton called "neighbor-modulated fitness," thought to be too unwieldy to calculate (an individual's relatives' effect on its reproduction).

function—any characteristic produced by natural selection as a result of adaptation; the purpose of the characteristic. (*Cf.* effect.)

gamete—haploid cell specialized to fuse with another haploid cell (fertilization) in a sexual life cycle (e.g., egg or sperm).

game theory—mathematical analysis of optimal choice of strategies when one's payoffs are affected by other actors.

gene—a segment of DNA encoding a polypeptide or one of the RNAs involved in translation.

genotype—the genetic constitution of an individual organism; may be used in reference to a single trait, a set of traits, or the entire genetic constitution (genome) of an organism. (*Cf.* phenotype.)

group selection—several quite distinct arguments are all called by this term: (1) selection favoring or disfavoring whole groups rather than genes or individuals (Wynne-Edwards); (2) the effects of group structure on gene frequency (Sewall-Wright); (3) the effects of coalitions on the interests of others within a group; (4) the effects of cultural transmission on gene frequency. (*See* chapter 9.)

Hamilton's rule—predicts that helping behavior among relatives will be favored by natural selection whenever $rb - c > 0$, where r = degree of relatedness, c = cost to the helper, and b = benefit to the recipient.

haplodiploidy—the condition seen in many social insects, in which the males are derived from unfertilized eggs (and are thus haploid), while females are derived from fertilized eggs (and are thus diploid).

haploid—having a chromosome complement consisting of just one copy of each chromosome. (*Cf.* diploid.)

harem—a social group, usually consisting of one mature male, several females, and sometimes immature males and females. Typically, harem-holding males attempt to keep females from leaving the group and other mature males from joining.

herbivore—an animal that eats principally vegetable material. Organisms may be more specifically categorized as frugivores (fruit eaters), granivores (grain eaters), folivores (leaf eaters), etc.

heterogametic—having different sex chromosomes, as XY in male mammals. (*See* homogametic.)

heterozygous—having, at any specified genetic locus, two dissimilar alleles.

homogametic—having two sex chromosomes of the same type, as XX in female mammals. (*See* heterogametic.)

homozygous—having, at any specified genetic locus, two identical alleles.

inbreeding—the mating of closely related individuals.

inclusive fitness—the sum of an individual's contribution to the next generation through the effects of the individual's actions on genes identical by descent (IBD) in other individuals. (*See* Grafen 1991, ch. 1.)

interdemic selection—when groups of reproducing individuals are small, isolated, and breed largely within their group, this structuring of the population can affect gene frequency. The conditions for effectiveness of interdemic selection are stringent and seldom met. (*See also* group selection, chapter 9.)

intrinsic rate of increase—the maximum growth rate of a population under specified (close to optimal) conditions; symbolized by r, this parameter is usually wrongly assumed for analytic purposes to be a constant for any species.

iteroparous—reproducing repeatedly. (*Cf.* semelparous.)

K-selection—selection imposed by consistently high density of conspecific competitors. MacArthur and Wilson (1967) noted that a high density of competitors favored parents who were efficient in converting resources into offspring. In non-human species high density of competitors is the strongest pressure favoring production of fewer, but better-invested offspring. (*See* r-selection.)

kin selection—(= inclusive fitness) selection on genes resulting from those genes' presence in relatives other than descendant relatives; e.g., individuals may increase their inclusive fitness (q.v.) not only through offspring, but also through helping nondescendant relatives. What counts and what doesn't is a bit tricky, and has often been miscalculated (Grafen 1991). (*See* inclusive fitness.)

lek—an area traditionally used for communal sexual or courtship displays. Typically, males display only their ability to get and hold a territory; no resources useful to females, such as food or nest sites, are involved. Males on particular (often central, hard to defend) territories attract the most mates.

levirate—cultural custom in which a man marries his brother's widow.

life history—an organism's entire ontogeny, including time to maturity, degree of iteroparity, degree of sexual dimorphism, breeding system, clutch or litter size, degree and kind of parental care. *Life-history strategy analysis* is the term used to describe investigation into the costs and benefits of each life-history parameter.

lineage—a kinship group. In anthropology, the group is further defined by tracing the relationship either through males (a patrilineage) or females (matrilineage).

litter—all offspring born to a female at the same time.

locus—the location of a gene on a chromosome.

marsupial—mammals (e.g., kangaroos, koalas, wombats) that lack a placenta and give birth to young after a short period (e.g., 30 days). The young typically complete their development in the marsupium, a pouch on the female's abdomen.

mating effort—that portion of reproductive effort (q.v.) devoted to gaining a mate. Typically assumed to include both caloric and risk expenditure. Includes territorial behavior, mating displays, dominance fights, etc. (*Cf.* parental effort.)

matri—of or through the mother, as matrilocal, living with mother's kin; or matrilineal, tracing descent through the female line.

matrilocal residence—a pattern of residence in which a married couple lives with or near the wife's parents. (*See* patrilocal, avunculocal, neolocal.)

monogamy—the condition in which equal numbers of males and females contribute to the next generation. Typically, a single male and female will cooperate to raise at least one brood. (*Cf.* polygyny, polyandry.)

mutation—a chance alteration in the code of a gene, often as the result of environmental insult.

natural selection—Darwin observed that (1) in any environment, not all individuals will survive and reproduce equally well, and (2) variation that is heritable means that, over time, individuals who survive and reproduce better will come to comprise a larger and larger proportion of the population. We call this process natural selection: the selective, or filtering influence of environmental conditions. Natural selection is the only one of the contributing forces leading to evolution for which there is any predictive ability.

neolocal residence—a pattern of residence in which a married couple lives separately, and usually at some distance, from the kin of both spouses. (*See* matrilocal, patrilocal, avunculocal.)

neoteny—attaining sexual maturity while still in larval form, as in *Necturus*, the mud puppy.

omnivore—an animal that eats both animal and vegetable material.

paradigm—in general, an outstandingly clear or typical pattern. In research, a logical structure or pattern used to solve problems.

parallel cousins—children of siblings of the same sex. One's parallel cousins are one's father's brother's children and mother's sister's children. (*See* cross cousins.)

parental effort (*PE*)—that portion of reproductive effort (q.v.) devoted to the production of offspring; typically, egg and sperm production, lactation, etc., are included (Low 1978). Some activities (e.g., nest construction) may function as parental effort if specifically done for the use of raising offspring, or as mating effort (q.v.) if used to attract a mate.

parental investment—that portion of parental effort (q.v.) received by any offspring (Trivers 1972). ΣPI = PE for any period.

parental manipulation—adjustment or manipulation of parental investment by parents to reduce the reproduction of some offspring in the interests of maximizing the parent's inclusive fitness. Examples might be sex ratio adjustment in social insects, or the enforced sterility of mature wolf offspring in a pack.

parsimony—the requirement that no theory be more complicated than necessary to explain phenomena observed.

parturition—the act or process of giving birth.

patri—of or through the father, as in patrilocal (living with husband's parents) or patrilineal (tracing descent through father).

patrilocal residence—a pattern of residence in which a married couple lives with or near the husband's parents. (*See* matrilocal, avunculocal, neolocal.)

phenotype—the appearance of a trait.

placental—a mammal (e.g., deer, horses, humans) which has a vascular organ (the placenta) uniting the fetus and its mother. Nutrients are carried through the placenta. (*Cf.* marsupials.)

pleiotropy—the condition in which a single gene has multiple effects (e.g., as in senescence; see chapter 6).

polyandry—a biological mating system in which fewer females than males contribute to the next generation, e.g., jacanas, Arctic sanderlings. Sexual competition between females is intensified, and female variance in reproductive success is greater than male variance. Socially, polyandry is a marriage system in which a woman may have more that one husband at a time. (*Cf.* monogamy, polygyny.)

polygamy—a general term describing the possession during a lifetime of more than one mate. (*See* polygyny and polyandry.)

polygynandry—a biological mating system in which both males and females may have multiple mates, as in Dunnocks or the Ache; note that the Ache marriage system is serial monogamy. Male variance in reproductive success probably exceeds female variance.

polygyny—a biological mating system in which fewer males than females contribute to the next generation (i.e., a few males do most of the breeding, most fail). Male sexual competition is intensified. In anthropology, polygyny is a marriage system in which at least some men are permitted more than one wife simultaneously; it is further distinguished by whether the wives are sisters (sororal polygyny) or not (nonsororal polygyny). (*Cf.* polyandry, monogamy.)

precocial—referring to young animals that can move about and forage very quickly after birth or hatching. (*Cf.* altricial.)

predator—an organism that kills other animals for food.

predictability—the extent to which forecasts can be made about the condition of an environment with regard to important parameters. Symbolized by P, predictability is mathematically defined (Colwell 1974), and is the sum of constancy, C (q.v.), and contingency, M (q.v.).

proximate mechanism—environmental (including social) conditions triggering any response. (*See* ultimate cause.)

r-selection—selection imposed by environment in which density of conspecific competitors is usually low. Typically low density of competitors favors high production of offspring (as opposed to fewer, better-invested offspring).

reciprocity—mutual exchange of benefits, often at different times. Sometimes called *reciprocal altruism.*

recombination—also called *crossing over,* this is the process of exchange of genetic material between homologous chromosomes during meiosis.

Red Queen—in biology, the concept that all competitive progress is relative, because individuals exert reciprocal selective pressure on one another. Taken from the statement by the Red Queen (whom Alice meets in *Through the Looking-Glass*) that here you must run as fast as you can just to keep up, because everything else is moving also.

reproductive effort—any energy or risk devoted to achieving genetic representation in the next generation; consists of mating effort (q.v.) and parental effort (q.v.). (*See* somatic effort.)

reproductive success—the number of surviving (and presumably reproductive) offspring of an individual.

reproductive value—the number of female offspring remaining to be born to a female at any point in life given the prevailing age-specific fertility and mortality schedules; symbolized by v_x, where x is the age of the female under consideration.

rut—term applied to the breeding season of ungulates (hoofed mammals).

selective pressure—any feature of the environment resulting in natural selection. Usually reducible to Darwin's "hostile forces"—e.g., shortage of resources (including mates), risk of disease, and parasites.

semelparous—reproducing only once in the organism's lifetime. (*Cf.* iteroparous.)

sexual dimorphism—any consistent difference between males and females beyond the basic functional anatomy of the reproductive organs; may include appearance and behavioral differences.

social group—a group of individuals of the same species that interacts frequently in competitive and cooperative ways, and often remains together much of the time; for example, a pack of wolves, a school of fish.

somatic effort—that portion of an individual's lifetime risk and energy budget not directly devoted to reproduction. (*Cf.* reproductive effort.)

stabilizing selection—selection that favors individuals having characteristics near the mean of some array, and disfavoring individuals with extremes. Typical of moderate unchanging environments.

sustainability—combinations of fertility and resource consumption that result in persistence of the population through time.

sympatric—occupying the same geographical area. (*Cf.* allopatric.)

sympatric speciation—the formation of two or more reproductively isolated species from a single parental species without a period of geographic separation (e.g., *Gryllus pennsylvanicus* and *G. veletus*).

trophic—pertaining to food.

Turner's syndrome—a human female condition characterized by XO genotype, resulting in sterility and disturbances of secondary sex characteristics.

ultimate cause—the cost-benefit ratio, or selective conditions, leading to any response. The ultimate cause of migratory behavior in a species, for example, is likely to be the consistent shift in relative goodness of two areas at different times of the year, while the proximate mechanism (q.v.) or proximate cause of migration may be a change in day length, temperature, or hormonal level.

ungulate—a hoofed mammal (e.g., deer, goats, cattle, horses). There are two subgroups: artiodactylids (even-toed ugulates such as deer), and perisodactylids (odd-toed ungulates such as horses).

viscosity—in population genetics, a low rate of gene flow due to dispersal by individuals; a viscous population has low dispersal and low gene flow.

zygote—diploid cell resulting from the union (fertilization) of two haploid gametes.

References

Abernethey, V., and R. Yip. 1990. Parent characteristics and infant mortality: The case in Tennessee. *Human Biology* 62(2): 279–290.

Adams, D. B. 1983. Why there are so few women warriors. *Behav. Sci. Res.* 18(3): 196–212.

Adams, G. D. 1997. Abortion: Evidence of an issue evolution. *Amer. J. Pol. Sci.* 41(3): 718–737.

Ahlburg, D. 1998. Julian Simon and the population growth debate. *Pop. and Devel. Rev.* 24: 317–327.

Åkerman, S. 1981. The importance of remarriage in the seventeenth and eighteenth centuries. In J. Dupâquier, E. Hélin, P. Laslett, M. Livi-Bacci, and S. Sogner (eds.), *Marriage and Remarriage in Populations of the Past,* 163–175. New York: Academic Press.

Alcock, J. 1998a. Unpunctuated equilibrium in the *Natural History* essays of Stephen Jay Gould. *Evo. and Human Behav.* 19: 321–336.

Alcock, J. 1998b. *Animal Behavior: An Evolutionary Approach.* 5th ed. Sunderland, Mass.: Sinauer.

Alexander, R. D. 1971. The search for an evolutionary philosophy of man. *Proc. Roy. Soc. Victoria* (Melbourne) 84: 99–120.

Alexander, R. D. 1974. The evolution of social behavior. *Ann. Rev. Ecol. and Syst.* 5: 325–383.

Alexander, R. D. 1977. Natural selection and the analysis of human sociality. In C. E. Goulden (ed.), *Changing Scenes in the Natural Sciences: 1776–1976.* Bicentennial Symposium Monograph, Philadelphia Academy of Natural Science, Special Publication 12.

Alexander, R. D. 1979. *Darwinism and Human Affairs.* Seattle: University of Washington Press.

Alexander, R. D. 1987. *The Biology of Moral Systems.* New York: Aldine de Gruyter.

Alexander, R. D. 1988. Evolutionary approaches to human behavior: What does the future hold? In L. Betzig, M. Borgerhoff Mulder, and P. Turke (eds.), *Human Reproductive Behaviour: A Darwinian Perspective,* 317–341. New York: Cambridge University Press.

Alexander, R. D. 1990. How did humans evolve? Reflections on the uniquely unique species. Museum of Zoology, University of Michigan, Special Publication 1.

Alexander, R. D., and G. Borgia. 1978. On the origin and basis of the male-female phenomenon. In M. F. Blum and N. Blum (eds.), *Sexual Selection and Reproductive Competition in Insects*, 417–440. New York: Academic Press.

Alexander, R. D., and K. Noonan. 1979. Concealment of ovulation, parental care, and human social evolution. In N. A. Chagnon and W. Irons (eds.), *Evolutionary Biology and Human Social Behavior: An Anthropological Perspective*, 436–453. North Scituate, Mass.: Duxbury Press.

Alexander, R. D., John L. Hoogland, R. D. Howard, K. M. Noonan, and P. W. Sherman. 1979. "Sexual dimorphism and breeding systems in pinnipeds, ungulates, primates, and humans." In N. A. Chagnon and W. Irons (eds.), *Evolutionary Biology and Human Social Behavior: An Anthropological Perspective*, 402–435. North Scituate, Mass.: Duxbury Press.

Altonji, J. G., F. Hayashi, and L. J. Kotlikoff. 1997. Parental altruism and inter vivo transfers: Theory and evidence. *J. Pol. Econ.* 105: 1121–1166.

Alvard, M. 1994. Conservation by native peoples. *Human Nature* 5: 127–154.

Alvard, M. 1998. Evolutionary ecology and resource conservation. *Evolutionary Anthropology* 7: 62–74.

Anderies, J. M. 1996. An adaptive model for predicting !Kung reproductive performance: A stochastic dynamic programming approach. *Ethol. and Sociobiol.* 17: 221–245.

Andersson, M. 1994. *Sexual Selection*. Princeton, N.J.: Princeton University Press.

Aoki, K., and M. W. Feldman. 1994. Cultural transmission of a sign language when deafness is caused by recessive alleles at two independent loci. *Theor. Pop. Biol.* 45: 101–120.

Ardrey, R. 1966. *The Territorial Imperative*. New York: Atheneum.

Arthur, W. B. 1991. Designing economic agents that act like human agents: A behavioral approach to bounded rationality. *AEA Papers and Proceedings* 81(2): 353–359.

Arthur, W. B. 1994. Inductive reasoning and bounded rationality. *AEA Papers and Proceedings* 84(2): 406–411.

Aston, C. C. 1986. Political hostage-taking in western Europe. In William Gutteridge (ed), *Contemporary Terrorism*, 57–83. New York: Facts on File.

Aswad, B. 1971. *Property Control and Social Strategies in Settlers in a Middle Eastern Plain*. University of Michigan Museum of Anthropology, Anthropological Papers no. 44.

Atran, S. 1993. Itza Maya tropical agro-forestry. *Current Anthropology* 34: 633–689.

Axelrod, R. 1984. *The Evolution of Cooperation*. New York: Basic Books.

Axelrod, R. 1986. An evolutionary approach to norms. *Amer. Pol. Sci. Rev.* 80(4): 1095–1111.

Axelrod, R., and D. Dion. 1988. The further evolution of cooperation. *Science* 242: 1385–1390.

Axelrod, R., and W. D. Hamilton. 1981. The evolution of cooperation. *Science* 211: 1390–1396.

Bailey, R. C., M. R. Jenike, P. T. Ellison, G. R. Bentley, A. M. Harrigan, and N. R. Peacock. 1992. The ecology of birth seasonality among agriculturalists in central Africa. *J. Biosoc. Sci.* 24: 393–412.

Bailey, R. E., and M. J. Chambers. 1998. The impact of real wage and mortality fluctuations on fertility and nuptiality in precensus England. *J. Pop. Econ.* 11: 413–434.

Ball, S., and C. C. Eckel. 1998. The economic value of status. *J. Socio-Economics* 27: 495–514.

Barber, P. 1988. *Vampires, Burial, and Death: Folklore and Reality.* New Haven: Yale University Press.

Barkow, J. H. 1977. Conformity to ethos and reproductive success in two Hausa communities: An empirical evaluation. *Ethos* 5: 409–425.

Barkow, J. H., L. Cosmides, and J. Tooby. 1992. *The Adapted Mind.* New York: Oxford University Press.

Barnet, R. J. 1988. The costs and perils of intervention. In Michale T. Klare and Peter Kornbluh (eds.), *Low-Intensity Warfare: Counterinsurgency, Proinsurgency, and Anti-terrorism in the Eighties,* 207–221. New York: Pantheon.

Barry, H. III, M. K. Bacon, and I. L. Child. 1957. A cross-cultural study of some sex differences in socialization. *J. Abnormal and Soc. Psych.* 55: 327–332.

Barry, H. III, L. Josephson, E. Lauer, and C. Marshall. 1976. Traits inculcated in childhood. 5. Cross-cultural codes. *Ethnology* 15: 83–114.

Barth, F. 1956. *Models of Social Organization.* Occasional Paper No. 23, Royal Anthropological Institute, London.

Barton, N., and B. Charlesworth. 1998. Why sex and recombination? *Science* 281: 1986–1990.

Barton, R. A., R. W. Byrne, and A. Whiten. 1996. Ecology, feeding competition, and social structure in baboons. *Behav. Ecol. and Sociobiol.* 38: 3211–329.

Bateman, A. J. 1948. Intrasexual selection in Drosophila. *Heredity* 2: 349–368.

Beall, C. M., and M. C. Goldstein. 1981. Tibetan fraternal polyandry: A test of sociobiological theory. *American Anthropologist* 83: 5–12.

Beals, A. R. 1961. Cleavage and internal conflict: An example from India. *Conflict Resolution* 5(1): 27–34.

Bean, L. L., and G. P. Minneau. 1986. The polygyny-fertility hypothesis: A reevaluation. *Population Studies* 40: 67–81.

Beattie, J.H.M. 1960. *Bunyoro: An African Kingdom.* New York: Holt, Rinehart and Winston.

Beattie, J.H.M. 1963. Aspects of Nyoro symbolism. *Africa* 38: 413–442.

Beauchamp, G. 1994. The functional analysis of human fertility decisions. *Ethol. and Sociobiol.* 15: 31–53.

Bechara, A., H. Damasio, D. Tranel, and A. R. Damasio. 1997. Deciding advantageously before knowing the advantageous strategy. *Science* 275: 1293–1294.

Becker, G. 1981. *A Treatise on the Family.* Cambridge, Mass.: Harvard University Press.

Becker, G., and R. J. Barro. 1988. Reformulating the economic theory of fertility. *Quar. J. Econ.* 103: 1–25.

Becker, G., and H. G. Lewis. 1974. Interaction between quantity and quality of children. In T. W. Schultz (ed.), *Economics of the Family: Marriage, Children and Human Capital,* 81–90. Chicago: University of Chicago Press.

Beissinger, S. R. 1987. Mate desertion and reproductive effort in the Snail Kite. *Animal Behaviour* 35: 1504–1519.

Bell, G. 1978. The evolution of anisogamy. *J. Theor. Biol.* 73: 247–270.

Bell, G. 1982. *The Masterpiece of Nature: The Evolution and Genetics of Sexuality.* Berkeley: University of California Press.

Bell, G. 1997a. *Selection: The Mechanism of Evolution.* New York: Chapman and Hall.

Bell, G. 1997b. *The Basics of Selection.* New York: Chapman and Hall.

Belovsky, G. 1987. Hunter-gatherer foraging: A linear programming approach. *J. Anthropol. Archaeol.* 6: 29–76.

Belovsky, G. E., J. B. Slade, and J. M. Chase. 1996. Mating strategies based on foraging ability: An experiment with grasshoppers. *Behavioral Ecology* 7: 438–444.

Bentley, G. R., T. Goldberg, and G. Jasienska. 1992. The fertility of agricultural and non-agricultural traditional societies. Working Paper No. 1992–02, Population Issues Research Center, Pennsylvania State University, 22 Burrowes Bldg., University Park, PA.

Berard, J. D., P. Nurnberg, J. T. Epplen, and J. Schmidtke. 1993. Male rank, reproductive behavior, and reproductive success in free-ranging rhesus macaques. *Primates* 34: 481–489.

Bernstein, I., and C. Ehardt. 1985. Agonistic aiding: Kinship, rank, age, and sex influences. *Amer. J. Primatol.* 8: 37–52.

Bertoni, M. 1941. *Los Guayakies.* Asunción: Revista de la Sociedad Cientifica del Paraguay.

Betzig, L. 1986. *Despotism and Differential Reproduction: A Darwinian View of History.* New York: Aldine de Gruyter.

Betzig, L. 1989. Causes of conjugal dissolution: A cross-cultural sudy. *Current Anthropology* 30(5): 654–676.

Betzig, L. 1996. Not whether to count babies but which. In C. Crawford and D. Krebs (eds.), *Evolution and Human Behavior: Issues, Ideas, and Applications.* Hillsdale, N.J.: Erlbaum.

Betzig, L., and L. H. Lombardo. 1991. Who's pro-choice and why. *Ethol. and Sociobiol.* 13: 49–71.

Betzig, L., and S. Weber. 1993. Polygyny in American politics. *Pol. and Life Sci.* 12: 45–52.

Bideau A. 1980. A demographic and social analysis of widowhood and remarriage: The example of the Castellany of Thoissey-en-Dombes, 1670–1840. *J. Family Hist.* 5(1): 28–43.

Bideau, A., and A. Perrenoud. 1981. Remariage et fécondité. Contribution à l'étude des mécanismes de récuperation des populations anciennes. In J. Dupâquier, E. Hélin, P. Laslett, M. Livi-Bacci, and S. Sogner (eds.), *Marriage and Remarriage in Populations of the Past,* 547–449. New York: Academic Press.

Binmore, K. 1992. *Fun and Games: A Text on Game Theory.* Lexington, Mass.: D. C. Heath.

Birdsall, N. 1980. Population growth and poverty in the developing world. *Population Bulletin* 35(5): 3–46.

Birkhead, T. R., and A. P. Møller. 1992. *Sperm Competition in Birds: Evolutionary Causes and Consequences.* London: Academic Press.

Bliege Bird, R. L., and D. W. Bird. 1997. Delayed reciprocity and tolerated theft: The behavioral ecology of food-sharing strategies. *Current Anthropology* 38: 49–78.

Blood, R. O. 1960. Resolving family conflicts. *Conflict Resolution* 4(2): 209–219.

Blum, D. 1997. *Sex on the Brain: The Biological Differences between Men and Women.* New York: Viking.

Blurton Jones, N. B. 1984. A selfish origin for human food sharing: Tolerated theft. *Ethol. and Sociobiol.* 5: 1–3.

Blurton Jones, N. 1986. Bushman birth spacing: A test for optimal interbirth intervals. *Ethol. and Sociobiol.* 7: 91–105.

Blurton Jones, N. 1987a. Bushman birth spacing: Direct tests of some simple predictions. *Ethol. and Sociobiol.* 8:183–203.

Blurton Jones, N. 1987b. Tolerated theft, suggestions about the ecology and evolution of sharing, hoarding, and scrounging. *Soc. Sci. Infor.* 26: 31–54.

Blurton Jones, N. 1989. The costs of children and the adaptive scheduling of births: Towards a sociobiological perspective on demography. In A. Rasa, C. Vogel, and E. Voland (eds.), *Sexual and Reproductive Strategies,* 265–282. Kent, U.K.: Croom Helm.

Blurton Jones, N. 1997. Too good to be true? Is there really a tradeoff between number and care of offspring in human reproduction? In L. L. Betzig (ed.), *Human Nature: A Critical Reader,* 83–86. Oxford: Oxford University Press.

Blurton Jones, N., and M. J. Konner. 1973. Sex differences in the behavior of Bushman and London two- to five-year-olds. In J. Crook and R. Michael (eds.), *Comparative Ecology and Behavior of Primates.* New York: Academic Press.

Blurton Jones, N., and R. M. Sibley. 1978. Testing adaptiveness of culturally determined behavior: Do Bushman women maximize their reproductive success by spacing births widely and foraging seldom? In N. Blurton Jones and V. Reynolds (eds.), *Human Behavior and Adaptation.* Symposium No. 18 of the Society for the Study of Human Biology. London: Taylor and Francis.

Boesch, C. 1996. Social grupings in Taï chimpanzees. In W. C. McGrew, L. F. Marchant, and T. Nishida (eds.), *Great Ape Societies*, 101–113. Cambridge, U.K.: Cambridge University Press.

Bongaarts, J. 1978. A framework for analyzing the proximate determinants of fertility. *Pop. and Devel. Rev.* 4(1): 105–132.

Bongaarts, J. 1982. The fertility inhibiting effects of the intermediate fertility variables. *Stud. in Family Planning* 13(6/7): 179–189.

Bonner, J. T. 1980. *The Evolution of Culture in Animals.* Princeton, N.J.: Princeton University Press.

Boone, J. L. III. 1986. Parental investment and elite family structure in preindustrial states: A case study of late medieval–early modern Portuguese genealogies. *American Anthropologist* 88: 859.

Boone, J. L. III. 1988. Parental investment, social subordination and population processes among the 15th and 16th century Portuguese nobility. In Laura Betzig, Monique Borgerhoff Mulder, and Paul Turke (eds.), *Human Reproductive Behaviour: A Darwinian Perspective*, 201–219. Cambridge, U.K.: Cambridge University Press.

Borgerhoff Mulder, M. 1987. On cultural and biological success: Kipsigis evidence. *American Anthropologist* 89: 617–634.

Borgerhoff Mulder, M. 1988a. Reproductive success in three Kipsigis cohorts. In T. H. Clutton-Brock (ed.), *Reproductive Success*, 419–435. Chicago: University of Chicago Press.

Borgerhoff Mulder, M. 1988b. Kipsigis bridewealth payments. In Laura Betzig, Monique Borgerhoff Mulder, and Paul Turke (eds.), *Human Reproductive Behaviour: A Darwinian Perspective*, 65–82. Cambridge, U.K.: Cambridge University Press.

Borgerhoff Mulder, M. 1990. Kipsigis women's preferences for wealthy men: Evidence for female choice in mammals? *Behav. Ecol. and Sociobiol.* 27: 255–264.

Borgerhoff Mulder, M. 1991. Human behavioral ecology. In J. R. Krebs and N. B. Davies (eds.), *Behavioural Ecology*, 69–98. 3d ed. London: Blackwell.

Borgerhoff Mulder, M. 1992. Reproductive decisions. In E. A. Smith and B. Winterhalder (eds.), *Evolutionary Ecology and Human Behavior*, 339–374. New York: Aldine de Gruyter.

Borgerhoff Mulder, M. 1995. Bridewealth and its correlates. *Current Anthropology* 36(3): 573–603, including commentary.

Borgerhoff Mulder, M. 1997. Marrying a married man: A postscript. In L. Betzig (ed.), *Human Nature: A Critical Reader*, 115–117. Oxford: Oxford University Press.

Borgerhoff Mulder, M. 1998a. Brothers and sisters: How sibling interactions affect optimal parental allocations. *Human Nature* 9: 119–162.

Borgerhoff Mulder, M. 1998b. The demographic transition: Are we any closer to an evolutionary explanation? *Trends in Ecol. and Evol.* 13: 266–270.

Borgia, G. 1979. Sexual selection and the evolution of mating systems. In M. F.

Blum and N. Blum (eds.), *Sexual Selection and Reproductive Competition in Insects*. New York: Academic Press.

Borgia, G. 1985. Bower quality, number of decorations and mating success of male satin bowerbirds (*Ptilinorynchus violaceous*): An experimental analysis. *Animal Behaviour* 33: 266–271.

Boswell, J. 1990. *The Kindness of Strangers: Abandonment of children in Western Europe from Late Antiquity to the Renaissance*. New York: Vintage Books. (First published by Pantheon, 1988.)

Boyd, R., and P. J. Richerson. 1985. *Culture and the Evolutionary Process*. Chicago: University of Chicago Press.

Boyd, R., and P. J. Richerson. 1988. The evolution of reciprocity in sizeable groups. *J. Theor. Biol.* 132: 337–356.

Boyd, R., and P. J. Richerson. 1990. Culture and cooperation. In J. J. Mansbridge (ed.), *Beyond Self-Interest*, 111–132. Chicago: University of Chicago Press.

Boyd, R., and P. J. Richerson. 1992. Punishment allows the evolution of cooperation (or anything else) in sizeable groups. *Ethol. and Sociobiol.* 13: 171–195.

Boyd, R., and P. J. Richerson. 1996. Why culture is common, but cultural evolution is rare. *Proc. Brit. Acad.* 88: 77–93.

Boyd, R., and J. B. Silk. 1997. *How Humans Evolved*. New York: Norton.

Brändström, A. 1984. De Kärlekslösa mödrarna: Spädbarnsdödligheten I Sverige under 1800-talet med särskild hänsyn till Nedertorneå. *Acta Universitatis Umensis* 62: 1–271.

Brogan, P. 1990. *The Fighting Never Stopped: A Comprehensive Guide to World Conflict Since 1945*. New York: Vintage Books.

Brown, J. L. 1983. Cooperation—a biologist's dilemma. In J. S. Rosenblatt (ed.), *Advances in Behavior*, 1–37. New York: Academic Press.

Brown, J. L. 1997. A theory of mate choice based on heterozygosity. *Behav. Ecol.* 8: 60–65.

Browne, K. R. 1995. Sex and temperament in modern society: A Darwinian view of the Glass Ceiling and the Gender Gap. *Arizona Law Review* 37(4): 971–1106.

Brunner, H. G. 1996. MAOA deficiency and abnormal behaviour: Perspectives on an association. *Ciba Found. Symp.* 194: 155–164 (discussion 164–167).

Brunner, H. G., M. Nelen, X. O. Breakfield, H. H. Rogers, and B. A. Oost. 1993. Abnormal behavior associated with a point mutation in the structural gene for monoamine oxidase A. *Science* 262: 578–580.

Budiansky, S. 1995. *Nature's Keepers: The New Science of Nature Management*. New York: Free Press.

Bueno de Mesquita, B. 1981. *The War Trap*. New Haven: Yale University Press.

Bugos, P. E., and L. M. McCarthy. 1984. Ayoreo infanticide: A case study. In G. Hausfater and S. B. Hrdy (eds.), *Infanticide: Comparative and Evolutionary Perspectives*, 503–520. New York: Aldine de Gruyter.

Burley, N. 1979. The evolution of concealed ovulation. *American Naturalist* 114: 835–858.

Buss, D. M. 1985. Human mate selection. *American Scientist* 73: 47–51.

Buss, D. M. 1989. Sex differences in human mate preferences: Evolutionary hypotheses tested in 37 cultures. *Behav. and Brain Sci.* 12: 1–49.

Buss, D. M. 1994. *The Evolution of Desire.* New York: Basic Books.

Buss, D. M. 1999. *Evolutionary Psychology: The New Science of the Mind.* Boston: Allyn and Bacon.

Bygott, D. 1974. Agonistic behavior and dominance in wild chimpanzees. Ph.D. dissertation, University of Cambridge, Cambridge, U.K.

Bygott, J. D., B.C.R. Bertram, and J. P. Hanby. 1979. Male lions in large coalitions gain reproductive advantages. *Nature* 282: 839–841.

Byrne, R. 1993. Do larger brains mean greater intelligence? *Behav. and Brain Sci.* 16: 696–697.

Byrne, R. 1995. *The Thinking Ape: Evolutionary Origins of Intelligence.* Oxford: Oxford University Press.

Byrne, R., and A. Whiten. 1988. *Machiavellian Intelligence: Social Expertise and the Evolution of Intellect in Monkeys, Apes, and Humans.* Oxford: Clarendon Press.

Byrne, R., and A. Whiten. 1992. Cognitive evolution in primates: Evidence from tactical deception. *Man* 27: 609–627.

Cabello y Robles, D. 1961. A description of the Comanche Indians in 1786 by the governor of Texas. West Texas Historical Association *Yearbook* 37: 177–182.

Cabourdin, G. 1981. Le remariage en France sous l'ancien régime (seizième–dix-huitième siècles). In J. Dupâquier, E. Hélin, P. Laslett, M. Livi-Bacci, and S. Sogner (eds.), *Marriage and Remarriage in Populations of the Past,* 273–285. New York: Academic Press.

Cadoret, R. J., L. D. Leve, and E. Devor. 1997. Gentics of aggressive and violent behavior. *Psychiatr. Clin. North. Amer.* 20: 301–322.

Cain, M. 1985. On the relationship between landholding and fertility. *Population Studies* 39: 5–15.

Cairns, R. B. 1996. Aggression from a developmental perspective: genes, environments and interactions. *Ciba Found. Symp.* 194: 45–56 (discussion 57–60).

Caldwell, J. C. 1976. Toward a restatement of demographic transition theory *Pop. and Devel. Rev.* 2: 321–366.

Caldwell, J. C. 1982. *Theory of Fertility Decline.* New York: Academic Press.

Caldwell, J. C. 1983. Direct economic costs and benefits of children. In R. A. Bulatao and R. D. Lee (eds.), *Determinants of Fertility in Developing Countries,* vol. 1, 458–493. New York: Academic Press.

Campbell, A. 1995. A few good men: Evolutionary psychology and female adolescent aggression. *Ethol. and Sociobiol.* 16: 99–123.

Campbell, D. T. 1975. Conflicts between biological and social evolution and between psychology and moral tradition. *American Psychologist* 30: 21–37.

Campbell, D. T. 1979. Comments on the sociobiology of ethics and moralizing. *Behavioral Science,* 24: 37–45.

Campbell, D. T. 1983. Legal and primary group controls. In M. Gruter and P. Bohanon (eds.), *Law, Biology, and Culture: The Evolution of Law,* 159–171. Santa Barbara, Calif.: Ross-Ericson.

Campbell, K. L., and J. W. Wood. 1988. Fertility in traditional societies. In P. Diggory, M. Potts, and S. Teper (eds.), *Natural Human Fertility: Social and Biological Determinants,* pp. 39–69. London: Macmillan, in cooperation with the Eugenics Society.

Cancian, F. A. 1973. *What Are Norms? A Study of Beliefs and Action in a Maya Community.* Cambridge, U.K.: Cambridge University Press.

Carcopino, J. 1941. *Daily Life in Ancient Rome.* London: Routledge & Sons.

Carniero, R. L. 1970. A theory of the origin of the state. *Science* 169: 733–738.

Case, T. J. 1978. Endothermy and parental care in the terrestrial vertebrate. *American Naturalist* 112: 861–874.

Cashdan, E. 1985. Coping with risk: Reciprocity among the Basarwa of northern Botswana. *Man* 20: 454–474.

Cavalli-Sforza, L. L., and M. W. Feldman. 1981. *Cultural Transmission and Evolution.* Princeton, N.J.: Princeton University Press.

Cavalli-Sforza, L. L., M. W. Feldman, K. H. Chen, and S. Dornbusch. 1982. Theory and observation in cultural transmission. *Science* 218: 19–27.

Chagnon, N. 1979. Is reproductive success equal in egalitarian societies? In N. A. Chagnon and W. Irons (eds.). *Evolutionary Biology and Human Social Behavior: An Anthropological Perspective.* North Scituate, Mass.: Duxbury Press.

Chagnon, N. 1982. Sociodemographic attributes of nepotism in tribal populations: Man the rule-breaker. In Kings' College Sociobiology Group (eds.), *Current Problems in Sociobiology.* Cambridge, U.K.: Cambridge University Press.

Chagnon, N. 1988. Life histories, blood revenge, and warfare in a tribal population. *Science* 239: 985–992.

Chagnon, N. 1997. *Yanomamö.* 5th ed. Fort Worth, Tex.: Harcourt Brace.

Chagnon, N. A., and R. Hames. 1979. Protein deficiency and tribal warfare in Amazonia: New data. *Science* 203: 910–913.

Chamberlin, E. R. 1965. *Everyday Life in Renaissance Times.* London: B. T. Batsford.

Chapais, B. 1983. Dominance, relatedness, and the structure of female relationships in rhesus monkeys. In R. A. Hinde (ed.), *Primate Social Relationships: An Integrated Approach.* Oxford: Blackwell.

Charlesworth, B. 1978. The population genetics of anisogamy. *J. Theor. Biol.* 73: 347–357.

Charlesworth, B. 1980. *Evolution in Age-structured Populations.* Cambridge, U.K.: Cambridge University Press

Charlton, S.E.M. 1984. *Women in Third World Development.* Boulder: Westview Press.

Charnov, E. L. 1982. *The Theory of Sex Allocation.* Princeton, N.J.: Princeton University Press.

Charnov, E. L. 1991. Evolution of life history variation among female mammals. *Proc. Nat. Acad. Sci. USA* 88: 1134–1137.

Charnov, E. L. 1993. *Life History Invariants: Some Explorations of Symmetry in Evolutionary Ecology.* Oxford: Oxford University Press.

Cheney, D. 1987. Interactions and relationships between groups. In Barbara B. Smuts, Dorothy L. Cheney, Robert M. Seyfarth, Richard W. Wrangham, and Thomas T. Struhsaker (eds.), *Primate Societies,* 267–281. Chicago: University of Chicago Press.

Cheney, D., and R. Seyfarth. 1987. The influence of inter-group competition on the survival and reproduction of female vervet monkeys. *Behav. Ecol. and Sociobiol.* 21: 375–386.

Cheney, D., and R. Seyfarth. 1990. *How Monkeys See the World.* Chicago: University of Chicago Press.

Cheney, D., and R. W. Wrangham. 1987. Predation. In Barbara B. Smuts, Dorothy L. Cheney, Robert M. Seyfarth, Richard W. Wrangham, and Thomas T. Struhsaker (eds.), *Primate Societies,* 227–239. Chicago: University of Chicago Press.

Cheng, P. W., K. J. Holyoak, R. E. Nisbett, and L. M. Oliver. 1986. Pragmatic versus syntactic approaches to training deductive reasoning. *Cognitive Psychology* 18: 293–328.

Childs, G. M. 1949. *Umbundu Kinship and Character.* Published for the International African Institute by Oxford University Press, Oxford.

Chisholm, J., and V. Burbank. 1991. Monogamy and polygyny in southeast Arnhem Land: Male coercion and female choice. *Ethol. and Sociobiol.* 12: 291–313.

Chitty, D. 1996. *Do Lemmings Commit Suicide? Beautiful Hypotheses and Ugly Facts.* Oxford: Oxford University Press.

Cigno, A. 1998. Fertility decisions when infant survival is endogenous. *J. Pop. Econ.* 11: 21–28.

Clark, D. C., S. J. DeBano, and A. J. Moore. 1997. The influence of environmental quality on sexual selection in *Nauphoeta cineria* (Dictyoptera: Blaberidae). *Behavioral Ecology* 8: 46–53.

Clarke, A. L., and B. Low. 1992. Ecological correlates of human dispersal in 19th century Sweden. *Animal Behaviour* 44: 677–693.

Clastres, P. 1972a. *Chronique des Indiens Guayaki. Ce que Sevant les Aché, Chasseurs.* Nomades du Paraguay. Paris: Pion.

Clastres, P. 1972b. The Guayaki. In M. Bicchieri (ed.), *Hunters and Gatherers Today.* New York: Holt, Reinhart, and Winston.

Clutton-Brock, T. H. 1983. Selection in relation to sex. In D. S. Bendall (ed.), *From Molecules to Men,* 457–481. Cambridge, U.K.: Cambridge University Press.

Clutton-Brock, T. H. (ed.). 1988. *Reproductive Success: Studies of Individual Variation in Contrasting Breeding Systems.* Chicago: University of Chicago Press.

Clutton-Brock, T. H. 1991. *The Evolution of Parental Care.* Princeton, N.J.: Princeton University Press.

Clutton-Brock, T. H., and P. Harvey. 1976. Evolutionary rules and primate societies. In P.P.G. Bateson and R. A. Hinde (eds.), *Growing Points in Ethology*, 195–237. Cambridge, U.K.: Cambridge University Press.

Clutton-Brock, T. H., and P. Harvey. 1978. Mammals, resources, and reproductive strategies. *Nature* 273: 191–195.

Clutton-Brock, T. H., and G. A. Parker. 1995a. Sexual coercion in animal societies. *Animal Behaviour* 49: 1345–1365.

Clutton-Brock, T. H., and G. A. Parker. 1995b. Punishment in animal societies. *Nature* 373: 209–216.

Clutton-Brock, T. H., S. Albon, and F. Guinness. 1986. Great expectations: Dominance, breeding success and offspring sex ratio in red deer. *Animal Behaviour* 34: 460–471.

Clutton-Brock, T. H., F. E. Guinness, and S. D. Albon. 1982. *Red Deer: The Ecology of Two Sexes*. Wildlife Behavior and Ecology Series. Chicago: University of Chicago Press.

Coale, A. J., and S. C. Watkins. 1986. *The Decline of Fertility in Europe*. Princeton, N.J.: Princeton University Press.

Coale, A. J., and T. J. Trussell. 1974. Model fertility schedules: Variations in the age structure of childbearing in human populations. *Population Index* 40: 185–258. (See also Erratum, *Population Index* 41: 572.)

Coale, A. J., and T. J. Trussell. 1978. Technical note: Finding the two parameters that specify a model schedule of fertility. *Population Index* 44: 203–213.

Cohen, J. E. 1995. *How Many People Can the Earth Support?* New York: Norton.

Cohen, J. E. 1998. Can a more equal world support more or fewer people than a less equal one? Development Discussion Paper No. 628, Harvard Institute for International Development.

Cohen, R. 1967. *The Kanuri*. New York: Rinehart and Winston.

Coleman D., and R. Schofield (eds.). 1986. *The State of Population Theory: Forward from Malthus*. London: Blackwell.

Collier, J. F., and M. Rosaldo. 1981. Politics and gender in simple societies. In S. B. Ortner and H. Whitehead (eds.), *Sexual Meanings: The Cultural Construction of Gender and Sexuality*. Cambridge, U.K.: Cambridge University Press.

Comfort, A. 1956. *Ageing: The Biology of Senescence*. New York: Holt, Rinehart, and Winston.

Connor, R. 1986. Pseudo-reciprocity: Investment in mutualism. *Animal Behaviour* 34: 1562–1566.

Connor, R. 1995a. The benefits of mutualism: A conceptual framework. *Biology Review* 70: 427–457.

Connor, R. 1995b. Altruism among non-relatives: Alternatives to the "prisoner's dilemma." *TREE* 10: 84–86.

Connor, R., R. A. Smolker, and A. F. Richards. 1992. Two levels of alliance formation among male bottlenose dolphins (*tursiops* sp.). *Proc. Nat. Acad. Sci. USA* 89: 987–990.

Cooney, T. M., and P. Uhlenburg. 1992. Support from parents over the life course: The adult child's perspective. *Social Forces* 71: 63–84.

Cords, Marina. 1986. Forest guenons and patas monkeys: Male-male competition in one-male groups. In B. B. Smuts, D. L. Cheney, R. M. Seyfarth, R. W. Wrangham, and T. T. Struhsaker (eds.), *Primate Societies*, 98–11. Chicago: University of Chicago Press.

Corsini, C. A. 1981. Why is remarriage a male affair? Some evidence from Tuscan villages during the eighteenth century. In J. Dupâquier, E. Hélin, P. Laslett, M. Livi-Bacci, and S. Sogner (eds.), *Marriage and Remarriage in Populations of the Past*, 385–395. New York: Academic Press.

Cosmides, L., and J. Tooby. 1992. Cognitive adaptations for social exchange. In J. Barkow, L. Cosmides, and J. Tooby (eds.), *The Adapted Mind*, 163–228. New York: Oxford University Press.

Cosmides, L., and J. Tooby. 1996. Are humans good intuitive statisticians after all? Rethinking some conclusions from the literature on judgement under uncertainty. *Cognition* 58: 1–73.

Costanza, Robert. 1987. Social traps and environmental policy. *Bioscience* 37: 407–12.

Cowlishaw, G., and R. I. Dunbar. 1991. Dominance rank and mating success in male primates. *Animal Behaviour* 41: 1045–1056.

Cowlishaw, G., and R. Mace. 1996. Cross-cultural patterns of marriage and inheritance: A phylogenetic approach. *Ethol. and Sociobiol.* 17: 87–97.

Crimmins, E. M., and R. A. Easterlin. 1984. The estimation of natural fertility: A micro approach. *Social Biology* 31: 160–70.

Cronin, H. 1991. *The Ant and the Peacock.* Cambridge, U.K.: Cambridge University Press.

Cronk, L. 1989. Low socioeconomic status and female-biased parental investment: The Mukogodo example. *American Anthropologist* 91: 414–429.

Cronk, L. 1991a. Human behavioral ecology. *Ann. Rev. Anthropol.* 20: 25–53.

Cronk, L. 1991b. Low socioeconomic status and female-biased parental investment: The Mukogodo example. *American Anthropologist* 91: 414–429.

Cronk, L. 1991c. Wealth, status, and reproductive success among the Mukogodo of Kenya. *American Anthropologist* 93(2): 345–360.

Cronk, L. 1991d. Intention versus behavior in parental sex preferences among the Mukogodo of Kenya. *J. Biosoc. Sci.* 23: 229–240.

Cronk, L. 1991e. Preferential parental investment in daughters over sons. *Human Nature* 2: 387–417.

Cronk, L. 1993. Parental favoritism toward daughters. *American Scientist* 81: 272–279.

Cronk, L. 1994. Evolutionary theories of morality and the manipulative use of signals. *Zygon* 29: 81–101.

Cronk, L. 1995. Is there a role for culture in human behavioral ecology? *Ethol. and Sociobiol.* 16: 181–205.

Cronk, L. 1999. Female-biased parental investment and growth performance among Mukogodo children. In L. Cronk, N. Chagnon, and W. Irons (eds.), *Adaptation and Human Behavior: An Anthropological Perspective*. Hawthorne, N.Y.: Aldine de Gruyter.

Crook, J. H., and S. J. Crook. 1988. Tibetan polyandry: Problems of adaptation and Fitness. In L. Betzig, M. Borgerhoff Mulder, and P. Turke (eds.), *Human Reproductive Behavior: A Darwinian Approach*, 97–114. Cambridge, U.K.: Cambridge University Press.

Cross, J. 1983. *A Theory of Adaptive Economic Behavior*. Cambridge, U.K.: Cambridge University Press.

Cross, J. , and M. Guyer. 1980. *Social Traps*. Ann Arbor: University of Michigan Press.

Crow, J. F. 1958. Some possibilities for measuring selection intensities in man. *Human Biology* 30: 1–13.

Croze, H., A. K. Hillman, and E. M. Lang. 1981. Elephants and their habits: How do they tolerate each other? In C. W. Fowler and T. Smith (eds.), *Dynamics of Large Mammal Populations*, 297–316. New York: Wiley.

Curtis, H. J. 1963. Biological mechanisms underlying the aging process. *Science* 141: 686–694.

Daly, M., and M. Wilson. 1983. *Sex, Evolution, and Behavior.* 2d ed. Boston: Willard Grant.

Daly, M., and M. Wilson. 1984. A sociobiological analysis of human infanticide. In G. Hausfater and S. B. Hrdy (eds.), *Infanticide: Comparative and Evolutionary Perspectives*, 487–502. New York: Aldine de Gruyter.

Daly, M., and M. Wilson. 1985. Child abuse and other risks of not living with both parents. *Ethol. and Sociobiol.* 6: 197–210.

Daly, M., and M. Wilson. 1987. Children as homicide victims. In R. J. Gelles and Jane B. Lancaster (eds.), *Child Abuse and Neglect: Biosocial Dimensions*, 201–214. New York: Aldine de Gruyter.

Daly, M., and M. Wilson. 1988. *Homicide.* Hawthorne, N.Y.: Aldine de Gruyter.

Darwin, C. 1859. *On the Origin of Species by Means of Natural Selection.* Facsimile of the first edition, with an introduction by Ernst Mayr, published 1987. Cambridge, Mass.: Harvard University Press.

Darwin, C. 1871. *The Descent of Man and Selection in Relation to Sex.* 2 vols. London: John Murray.

Das Gupta, M., and P. N. Mari Bhat. 1997. Fertility decline and increased manifestation of sex bias in India. *Population Studies* 51: 307–315.

Davies, N. B. 1991. Mating systems. In J. R. Krebs and N. B. Davies. (eds.), *Behavioral Ecology: An Evolutionary Approach*, chap. 9. 3d ed. London: Blackwell Scientific.

Davies, N. B. 1992. *Dunnock Behaviour and Social Evolution.* Oxford: Oxford University Press.

Dawes, R. 1988. *Rational Choice in an Uncertain World.* Fort Worth, Tex.: Harcourt Brace Jovanovich.

Dawkins, R. 1979. Twelve misunderstandings of kin selection. *Z. Tierpsych.* 47: 61–76.

Dawkins, R. 1982. *The Extended Phenotype: The Gene as the Unit of Selection.* Oxford: Oxford University Press.

Dawkins, R. 1986. *The Blind Watchmaker.* New York and London: Norton.

Dawkins, R. 1989. *The Selfish Gene.* New ed. Oxford: Oxford University Press.

Dawkins, R., and J. R. Krebs. 1978. Animal signals: Information or deception? In J. R. Krebs and N. B. Davies (eds.), *Behavioural Ecology: An Evolutionary Approach,* 282–309. Oxford: Blackwell.

DeFries, J. C., R. P. Corley, R. C. Johnson, S. G. Vandenberg, and J. R. Wilson. 1982. Sex-by-generation and ethnic group-by-generation interactions in the Hawaii Family Study of Cognition. *Behavior Genetics* 12: 223–230.

DeNault, L. K., and D. A. McFarland. 1995. Reciprocal altruism between male vampire bats, *Desmodus rotundus. Animal Behaviour* 49: 855–856.

Denevan, W. M. 1992. The pristine myth: The landscape of the Americas in 1492. *Ann. Assoc. of Amer. Geographers* 82: 369–385.

Denig, E. T. 1961. *Five Indian Tribes of the Upper Missouri: Sioux, Arickaras, Assiniboines, Cree, Crows.* Edited and with an introduction by John C. Ewers. Norman: University of Oklahoma Press.

Dennett, D. C. 1991. *Consciousness Explained.* Boston: Little, Brown.

deRuiter, J., J. van Hoof, and W. Scheffrahn. 1994. Social and genetic aspects of paternity in wild long-tailed macaques (*Macaca fascicularis*). *Behaviour* 129: 203–224.

Deutscher, I. 1973. *What We Say/What We Do: Sentiments and Acts.* Glenview: Scott, Foresman.

de Waal, F. 1982. *Chimpanzee Politics.* New York: Harper & Row.

de Waal, F. 1984. Sex differences in the formation of coalitions among chimpanzees. *Ethol. and Sociobiol.* 5: 239–255.

de Waal, F. 1986. The integration of dominance and social bonding in primates. *Quart. Rev. Biol.* 61: 459–479.

de Waal, F. 1989. *Peacemaking among Primates.* Cambridge, Mass.: Harvard University Press.

de Waal, F. 1996a. *Good Natured: The Origins of Right and Wrong in Humans and Other Animals.* Cambridge, Mass.: Harvard University Press.

de Waal, F. 1996b. Conflict as negotiation. In W. C. McGrew, L. Marchant, and T. Nishida (eds.), *Great Ape Societies,* 159–172. Cambridge, U.K.: Cambridge University Press.

de Waal, F. 1997. The chimpanzee's service economy: Food for grooming. *Evol. and Human Behav.* 18: 375–386.

Dewsbury, D. A. 1982. Dominance rank, copulatory behavior, and differential reproduction. *Quar. Rev. Biol.* 57: 135–159.

Diamond, J. M. 1992. *The Third Chimpanzee: The Evolution and Future of the Human Animal.* New York: HarperCollins.

Diamond, J. 1992. *The Third Chimpanzee.* New York: HarperCollins.

Diamond, J. 1997. *Why Sex is Fun.* New York: Basic Books.

Dickemann, M. 1979. The reproductive structure of stratified societies: A preliminary model. In N. A. Chagnon and W. Irons (eds.), *Evolutionary Biology and Human Social Organization: An Anthropological Perspective,* 331–367. North Scituate, Mass.: Duxbury Press.

Dickemann, M. 1981. Paternal confidence and dowry competition: A biocultural analysis of purdah. In R. D. Alexander, D. W. Tinkle (eds.), *Natural Selection and Social Behavior.* New York: Chiron Press.

Dienske, H. 1986. A comparative approach to the question of why human infants develop so slowly. In J. G. Else and P. C. Lee (eds.), *Primate Ontogeny, Cognition, and Social Behavior,* 147–154. London: Cambridge University Press.

Divale, W. 1973. *Warfare in Primitive Societies: A Bibliography.* Santa Barbara, Calif.: American Bibliographic Center Clio.

Divale, W., and M. Harris. 1976. Population, warfare, and the male supremacist complex. *American Anthropologist* 80(1): 21–41.

Dixon, N. F. 1976. *On the Psychology of Military Incompetence.* London: Cape.

Dobzhansky, T. 1961. Discussion. In J. S. Kennedy (ed.), *Insect Polymorphism,* 111. London: Royal Entomological Society.

Dominey, W. J. 1980. Female mimicry in male bluegill sunfish—a genetic polymorphism? *Nature* 284: 546–548.

Dorjahn, V. R. 1958. Fertility, polygyny, and their interrelations in Temne society. *American Anthropologist* 60: 838–860.

Downhower, J. F., and E. L. Charnov. 1998. A resource range invariance rule for optimal offspring size predicts patterns of variability in parental phenotypes. *Proc. Nat. Acad. Sci. USA* 95: 6208–6211.

Drake, M. 1969. *Population and Society in Norway: 1735–1865.* Cambridge, U.K.: Cambridge University Press.

Draper, P. 1976. Social and economic constraints on child life among the !Kung. In R. B. Lee and I. DeVore (eds.), *Kalihari Hunger-Gatherers.* Cambridge, Mass.: Harvard University Press.

Draper, P. 1980. The interaction of behavior variables in the development of dominance relations. In D. R. Omark, F. F. Strayer, and D. G. Freedman (eds.), *Dominance Relations: An Ethological View of Human Conflicts and Social Interaction.* New York: Garland STPM Press.

Drury, W. H. 1973. Population changes in New England seabirds. *Bird Banding* 44: 267–313.

Dugatkin, L. A. 1996. Interface between culturally based preferences and genetical preferences: Female mate choice in *Poecilia reticulata. Proc. Nat. Acad. Sci. USA* 93(7): 2770–2773.

Dugatkin, L. A. 1997. *Cooperation among Animals: An Evolutionary Approach.* Oxford: Oxford University Press.

Dugatkin, L. A. 1998. Game theory and cooperation. In L. A. Dugatkin and H. K. Reeve (eds.), *Game Theory and Animal Behavior,* chap. 3. Oxford: Oxford University Press.

Dugatkin, L. A., L. Farrand, R. Wilkins, and D. S. Wilson. 1994. Altruism, Tit for Tat, and "outlaw" genes. *Evol. Ecology* 8: 431–437.

Dugatkin, L. A., and H. K. Reeve. 1994. Behavioral ecology and the levels-of-selection debate: Dissolving the group selection controversy. *Adv. in Study of Behav.* 23: 101–133.

Dugatkin, L. A., and H. K. Reeve (eds.). 1998. *Game Theory and Animal Behavior.* Oxford: Oxford University Press.

Dunbar, R.I.M. 1983. Structure of gelada baboon reproductive units, 4. Integration at group level. *Z. Tierpsych.* 63: 265–283.

Dunbar, R.I.M. 1987. Demography and reproduction. In Barbara B. Smuts, Dorothy L. Cheney, Robert M. Seyfarth, Richard W. Wrangham, and Thomas T. Struhsaker (eds.), *Primate Societies,* 240–249. Chicago: University of Chicago Press.

Dunbar, R.I.M. 1988. *Primate Social Systems.* London: Croom Helm.

Dunbar, R.I.M. 1992. Neocortex size as a constraint on group size in primates. *J. Human Evol.* 20: 469–493.

Dunbar, R.I.M., and E. P. Dunbar. 1977. Dominance and reproductive success among female gelada baboons. *Nature* 266: 351–352.

Dunbar, R.I.M., A. Clark, and N. L. Hurst. 1995. Conflict and cooperation among the Vikings: Contingent behavioral decisions. *Ethol. and Sociobiol.* 16: 233–246.

Duncan, G., W. J. Yeung, J. Brooks-Gunn, and J. R. Smith. 1998. How much does childhood poverty affect the life chances of children? *Amer. Sociol. Rev.* 63: 406–423.

Dunnigan, J. F., and A. Bay. 1986. *A Quick and Dirty Guide to War: Briefings on Present and Potential Wars.* Updated Edition. New York: Quill/William Morrow.

Dupâquier, J. 1972. De l'animal à l'homme: Le mechanisme autorégulateur des populations traditionelles. *Rev. de l'Inst. Sociol. de l'Univ. Libre de Bruxelles* 2: 177–211.

Durham, W. H. 1976. Resource competition and human aggression, part 1: A review of primitive war. *Quar. Rev. Biol.* 51: 385–415.

Durham, W. H. 1991. *Coevolution: Genes, Culture, and Human Diversity.* Stanford, Calif.: Stanford University Press.

Dweck, C. S. 1975. The role of expectations and attributions in the alleviation of learned helplessness. *J. Personality and Soc. Psych.* 31: 674–685.

Dweck, C. S., and T. E. Goetz. 1978. Attributions and learned helplessness. In J. H. Harvey, W. Ickes, and R. F. Kidd (eds.), *New Directions in Attribution Theory.* Hillsdales, N.J.: Erlbaum.

Dweck, C. S., and C. B. Wortman. 1982. Learned helplessness, anxiety, and achievement motivation. In W. H. Krohne and L. Lauz (eds.), *Achievement, Stress, and Anxiety*. New York: Hemisphere.

Dweck, C. S., W. Davidson, S. Nelson, and B. Enna. 1978. Sex differences in learned helplessness: II. The contingencies of evaluative feedback in the classroom, and III. An experimental analysis. *Developmental Psychology* 14: 268–276.

Dyson-Hudson, R., and E. A. Smith. 1978. Human territoriality. *American Anthropologist* 80: 21–42.

Eals, M., and I. Silverman. 1994. The hunter-gatherer theory of spatial sex differences: Proximate factors mediating the female advantage in recall of object arrays. *Ethol. and Sociobiol.* 15: 95–105.

Early, J. D., and J. F. Peters. 1990. *The Population Dynamics of the Mucajai Yanomamö*. New York: Academic Press.

Easterlin, R. 1978. The economics and sociology of fertility: A synthesis. In C. Tilly (ed.), *Historical Studies of Changing Fertility*, 57–134. Princeton, N.J.: Princeton University Press.

Easterlin, R., and E. Crimmins. 1985. *The Fertility Revolution: A Supply-Demand Analysis*. Chicago: University of Chicago Press.

Ebert, D., and W. D. Hamilton. 1996. Sex against virulence: The coevolution of parasitic diseases. *TREE* 11: 79–82.

Eckel, C. C., and P. J. Grossman. 1988. Are women less selfish than men? Evidence from dictator experiments. *Economic Journal* 108: 726–735.

Economos, A. C. 1980. Brain-life span conjecture: A re-evaluation of the evidence. *Gerontology* 26: 82–89.

Edgerton, R. B. 1992. *Sick Societies: Challenging the Myth of Primitive Harmony*. New York: Free Press.

Ehrlich, P. R. 1997. *A World of Wounds: Ecologists and the Human Dilemma*. Oldendorf/Luhe, Germany: Ecology Institute.

Ehrlich, P. R., and A. H. Ehrlich. 1981. *Extinction: The Causes and Consequences of the Disappearance of Species*. New York: Random House.

Eibl-Eibesfeldt, I. 1979. *The Biology of Peace and War*. London: Viking Press.

Elgar, M. A., and N. E. Pierce. 1988. Mating success and fecundity in an ant-tended Lycaenid butterfly. In T. H. Clutton-Brock (ed.), *Reproductive Success: Studies of Individual Variation in Contrasting Breeding Systems*, 59–75. Chicago: University of Chicago Press.

Ellis, L. 1993. Dominance and reproductive sucess among non-human animals: A cross-species comparison. *Ethol. and Sociobiol.* 16: 257–333.

Ellison, P. 1991. Reproductive ecology and human fertility. In C.G.N. Mascie-Taylor and G. W. Lasker (eds.), *Applications of Biological Anthropology to Human Affairs*, 14–54. Cambridge, U.K.: Cambridge University Press.

Elton, G. R. 1975. Taxation for war and peace in early-Tudor England. In J. M. Winter, (ed.), *War and Economics in Development*. Cambridge, U.K.: Cambridge University Press.

Ember, C. R. 1978. Myths about hunter-gatherers. *Ethnology* 17: 439–448.

Ember, C. R. 1981. A cross-cultural perspective on sex differences. In R. H. Monroe, R. L. Monroe, and B. B. Whiting (eds.), *Handbook of Cross-Cultural Human Development,* 531–580. New York: Garland Press.

Ember, C. R., and M. Ember. 1992. Resource unpredictability, mistrust, and war: A cross-cultural study. *J. Conflict Resol.* 36(2): 242–262.

Emlen, S. T. 1982a. The evolution of helping: I. An ecological constraints model. *American Naturalist* 119: 29–39.

Emlen, S. T. 1982b. The evolution of helping: II. The role of behavioral conflict. *American Naturalist* 119: 40–53.

Emlen, S. T. 1995. An evolutionary theory of the family. *Proc Nat. Acad. Sci. USA* 92: 8092–8099.

Emlen, S. T. 1997. Predicting family dynamics in social vertebrates. In J. R. Krebs and N. B. Davies (eds.), chap. 10. *Behavioural Ecology.* 4th ed. Oxford: Blackwell.

Emlen, S. T., and L. W. Oring. 1977. Ecology, sexual selection, and the evolution of mating systems. *Science* 197: 215–223.

Endler, J. A. 1986. *Natural Selection in the Wild.* Princeton, N.J.: Princeton University Press.

Essock-Vitale, S. 1984. The reproductive success of wealthy Americans. *Ethol. and Sociobiol.* 5: 45–49.

Essock-Vitale, S., and M. T. McGuire. 1988. What 70 million years hath wrought: Sexual histories and reproductive success of a random sample of American women. In in L. L. Betzig, M. Borgerhoff Mulder, and P. Turke, eds., *Human Reproductive Behavior: A Darwinian Perspective,* 221–236. Cambridge, U.K.: Cambridge University Press.

Ewers, J. C. 1958. *The Blackfeet.* Norman: University of Oklahoma Press.

Fadiman, J. A. 1982. *An Oral History of Tribal Warfare: The Meru of Mt. Kenya.* Athens: Ohio University Press.

Fairbanks, L. A., and M. T. McGuire. 1986. Determinants of fecundity and reproductive success in captive vervet monkeys. *Amer. J. Primatol.* 7: 27–38.

Falconer, D. S. 1981. *Introduction to Quantitative Genetics.* 2d ed. London and New York: Longman.

Farran, C. 1963. *Matrimonial Laws of the Sudan.* London: Buttersworth.

Fathauer, G. 1961. Trobrianders. In D. Schneider and K. Gough (eds.), *Matrilineal Kinship,* 234–269. Berkeley: University of California Press.

Fedigan, L. M. 1983. Dominance and reproductive success in primates. *Yearbook of Phy. Anthropol.* 26: 91–129.

Fehr, E., S. Gächter, and G. Kirchsteiger. 1997. Reciprocity as a contract enforcement device: Experimental evidence. *Econometrica* 65: 833–860.

Fehr, E., and S. Gächter. 1998. Reciprocity and economics: The economic implications of *Homo reciprocans. Eur. Econ. Rev.* 42: 845–859.

Feldman, M. W., L. L. Cavalli-Sforza, and L. A. Zhivotovsky. 1994. On the complexity of cultural transmission and evolution. In G. Cowan, D. Pines, and D. Melzer (eds.), *Complexity: Metaphors, Models, and Reality,* 47–62. SFI Studies in the Science of Complexity, Proceedings, vol. 19. Reading, Mass.: Addison Wesley.

Ferrill, A. 1985. *The Origins of War.* London: Thames and Hudson.

Festing, M.F.W., and D. K. Blackmore. 1971. Life span of specified pathogen-free (MRC Category 4) mice and rats. *Lab. Animal Bull.* 5: 179–192.

Fisher, R. A. 1958. *The Genetical Theory of Natural Selection.* New York: Dover Books.

Fiske, P., P. T. Rintamäki, and E. Karvonen. 1998. Mating success in lekking males: A meta-analysis. *Behavioral Ecology* 9: 328–338.

Flandrin, J. L. 1979. *Families in Former Times: Kinship, Household, and Sexuality.* Transl. Richard Southern. Cambridge, U.K.: Cambridge University Press.

Flinn, M. V. 1981. Uterine versus agnatic kinship variability and associated cousin marriage preferences. In R. D. Alexander and D. W. Tinkle (eds.), *Natural Selection and Social Behavior,* 439–475. New York: Chiron Press.

Flinn, M. V. 1997. Culture and the evolution of social learning. *Evol. and Human Behav.* 18: 23–67.

Flinn, M. V., and B. S. Low. 1986. Resource distribution, social competition, and mating patterns in human societies. In D. Rubenstein and R. Wrangham (eds.), *Ecological Aspects of Social Evolution.* Princeton, N.J.: Princeton University Press.

Fossey, D. 1983. *Gorillas in the Mist.* Boston: Houghton Mifflin.

Frank, R. 1985. *Choosing the Right Pond.* Oxford: Oxford University Press.

Frank, R. 1988. *Passions within Reason.* New York: Norton.

Frank, S. A. 1986. Hierarchical selection theory and sex ratios. I. General solutions for structured populations. *Theor. Pop. Biol.* 29: 312–342.

Frank, S. A. 1990. When to copy or avoid an opponent's strategy. *J. Theor. Biol.* 170: 41–46.

Frank, S. A. 1995. Mutual policing and repression of competition in the evolution of cooperative groups. *Nature* 377: 520–522.

Frank, S. A. 1996. The design of natural and artificial adaptive systems. In Michael R. Rose and George V. Lauder (eds.), *Adaptation,* 451–505. San Diego: Academic Press.

Frank, S. A. 1998. *Foundations of Social Evolution.* Princeton, N.J.: Princeton University Press.

Freedman, D. G. 1974. *Human Infancy: An Evolutionary Perspective.* Hillsdale, N.J.: Erlbaum.

Freedman, D. G. 1980. Sexual dimorphism and the status hierarchy. In D. R. Omark, F. F. Strayer, and D. G. Freedman, *Dominance Relations,* 261–271. New York: Garland Press.

Freedman, D. S., and A. Thornton. 1982. Income and fertility: The elusive relationship. *Demography* 19(1): 65–78.

Friedlander, L. 1907. *Roman Life and Manners under the Early Empire,* vol 1. 2d ed. Reprinted 1979. New York: Arno Press.

Fuchs, R. 1984. *Abandoned Children: Foundlings and Child Welfare in Nineteenth-Century France.* Albany: State University of New York Press.

Gagneaux, P., D. Woodruff, and C. Boesch. 1997. Furtive mating in female chimpanzees. *Nature* 387: 358–359.

Galloway, P. R. 1986. Differentials in demographic responses to annual price variations in pre-revolutionary France: A comparison of rich and poor areas in Rouen, 1681–1787. *European J. Pop.* 2(1986): 269–305.

Gallup, G. G. 1982. Permanent breast enlargement in human females: A sociobiological analysis. *J. Human Evol.* 11: 597–601.

Gallup, G. G. 1986. Unique features of human sexuality in the context of human evolution. In D. Byrne and K. Kelley (eds.), *Alternative Approaches to the Study of Sexual Behavior,* 13–42. Hillsdale, N.J.: Erlbaum.

Gangestad, S. W., R. Thornhill, and R. A. Yeo. 1994. Facial attractiveness, developmental stability, and fluctuating asymmetry. *Ethol. and Sociobiol.* 15: 73–85.

Gardner, J. F. 1986. *Women in Roman Law and Society.* Bloomington: Indiana University Press.

Garenne, M., and E. van de Walle. 1989. Polygyny and fertility among the Sereer of Senegal. *Population Studies* 43: 267–283.

Gaulin, S.J.C. 1980. Sexual dimorphism in the post-reproductive lifespan: Possible causes. *Human Evolution* 9: 227–232.

Gaulin, S.J.C., and J. S. Boster. 1990. Dowry as female competition. *American Anthropologist* 92: 994–1005.

Gaulin, S.J.C., and J. Boster. 1997. When are husbands worth fighting for? In L. Betzig (ed.), *Human Nature: A Critical Reader,* 372–374. Oxford: Oxford University Press.

Gaulin, S.J.C., and R. W. Fitzgerald. 1986. Sex differences in spatial ability: An evolutionary hypothesis and test. *American Naturalist* 127: 74–88.

Gaulin, S.J.C., and H. A. Hoffman. 1988. Evolution and development of sex differences in spatial ability. In Laura Betzig, Monique Borgerhoff Mulder, and Paul Turke, (eds.), *Human Reproductive Behaviour: A Darwinian Perspective,* 129–152. Cambridge, U.K.: Cambridge University Press.

Gaulin, S.J.C., and C. J. Robbins. 1991. Trivers-Willard effect in contemporary North American society. *Amer. J. Phys. Anthropol.* 85: 61–68.

Gaulin, S.J.C., D. H. McBurney, and S. L. Brakeman-Wartell. 1997. Matrilateral biases in the investment of aunts and uncles: A consequence and measure of paternity uncertainty. *Human Nature* 8: 139–151.

Gaunt, D. 1983. The property and kin relations of retired farmers. In R. Wall, J. Robin, and P. Laslett (eds.), *Family Forms in Historic Europe.* Cambridge, U.K.: Cambridge University Press.

Gaunt, D. 1987. Rural household organization and inheritance in Northern Europe. *J. Family Hist.* 12: 121–141.

Gauthier, A. H., and J. Hatzius. 1997. Family benefits and fertility: An econometric analysis. *Population Studies* 51: 295–306.

Georgellis, Y., and H. J. Wall. 1992. The fertility effect of dependent tax exemptions: Estimates for the United States. *Applied Economics* 24: 1139–1145.

Gerger, T., and G. Hoppe. 1980. Education and society: The geographer's view. *Acta Universitatis Stockholmensis* 1: 1–124.

Geronimus, A. T. 1996a. Black/white differences in the relationship of maternal age to birthweight: A population-based test of the Weathering Hypothesis. *Soc. Sci. and Med.* 42(4): 589–597.

Geronimus, A. T. 1996b. What teen mothers know. *Human Nature* 7(4): 323–352.

Geronimus, A. T., John Bound, and T. A. Waidmann. 1997. Health inequality, family caretaking systems, and population variation in fertility-timing. Paper presented at 1997 Annual meeting of the Population Association of America, Washington, D. C.

Geronimus, A. T., John Bound, T. A. Waidmann, M. M. Hillemeier, and P. B. Burns. 1996. Excess mortality among Blacks and Whites in the United States. *New England J. Med.* 335: 1552–1558.

Gewertz, D. 1981. A historical reconstruction of female dominance among the Chambri of Papua New Guinea. *American Ethnologist* 8: 94–106.

Gewertz, D. 1983. *Sepik River Societies: A Historical Ethnography of the Chambri and Their Neighbors.* New Haven: Yale University Press.

Gewertz, D., and F. K. Errington. 1991. *Twisted Histories, Altered Contexts: Representing the Chambri in a World System.* Cambridge, U.K.: Cambridge University Press.

Ghiselin, M. T. 1969. The evolution of hermaphroditism among animals. *Quart. Rev. Biol.* 44: 189–208.

Gibbard, A. 1992. *Apt Feelings and Wise Choices.* Cambridge, Mass.: Harvard University Press.

Gies, Frances. 1984. *The Knight in History.* New York: Harper and Row.

Glick, P. C., and S.-L. Lin. 1986. Recent changes in divorce and remarriage. *J. Marriage and Family* 48: 737–747.

Godin, J.-G., and L. A. Dugatkin. 1996. Female mating preference for bold males in the guppy, *Poecilia reticulata. Proc. Nat. Acad. Sci. USA* 93(19): 10262–10267.

Goldizen, A. W. 1987. Tamarins and marmosets: Communal care of offspring. In B. B. Smuts, D. L. Cheney, R. M. Seyfarth, R. W. Wrangham, and T. T. Strusaker (eds.), *Primate Societies,* 34–43. Chicago: University of Chicago Press.

Goldstein, M. 1976. Fraternal polyandry and fertility in a high Himalayan valley in Northwest Nepal. *Human Ecology* 4: 223–233.

Goldstein, A. P. 1989. Aggression reduction: Some vital steps. In J. Groebel and R. A. Hinde (eds.), *Aggression and War: Their biological and social bases,* 112–131. Cambridge, Mass.: Cambridge University Press.

Goodale, J. C. 1971. *Tiwi Wives.* Seattle: University of Washington Press.

Goodall, J. 1986. *The Chimpanzees of Gombe: Patterns of Behavior.* Cambridge, Mass.: Harvard University Press.

Goodall, J., A. Bandora, E. Bergmann, C. Busse, H. Matama, E. Mpongo, A. Pierce, and D. Riss. 1979. Intercommunity interactions in the chimpanzee population of the Gombe National Park. In D. A. Hamburg and E. R. McCown (eds.), *The Great Apes.* Menlo Park, Calif.: Benjamin Cummings.

Goose, S. D. 1988. Low-intensity warfare: The warriors and their weapons. In M. T. Klare and Peter Kornbluh (eds.), *Low-Intensity Warfare: Counterinsurgency, Proinsurgency, and Anti-Terrorism in the Eighties,* 80–111. New York: Pantheon.

Gorer, G. 1967. *Himalayan Village.* 2d ed. New York: Basic Books.

Gould, S. J., and R. C. Lewontin. 1979. The spandrels of San Marcos and the Panglossian paradigm: A critique of the adaptationist programme. *Proc. Roy. Soc. London* B 205: 581–598.

Gouzoules, H., S. Gouzoules, and L. Fedigan. 1982. Behavioral dominance and reproductive success in female Japanese monkeys (*Macaca fuscata*). *Animal Behaviour* 30: 1138–1150.

Gowaty, P. A. (ed.). 1997. *Feminism and Evolutionary Biology.* New York: Chapman and Hall.

Grafen, A. 1984. Natural selection, kin selection and group selection. In J. R. Krebs and N. B. Davies (eds.), *Behavioural Ecology: An Evolutionary Approach,* 62–84. 2d ed. Oxford: Blackwell Scientific.

Grafen, A. 1991. Modelling in behavioural ecology. In J. R. Krebs and N. B. Davies (eds.), *Behavioural Ecology: An Evolutionary Approach,* 5-31. Oxford: Blackwell.

Gray, R. 1983. The impact of health and nutrition on natural fertility. In R. A. Bulatao and R. D. Lee (eds., with P. E. Hollerbach and John Bongaarts), *Determinants of Fertility in Developing Countries,* 139–162. New York: Academic Press.

Greenwood, P. J. 1980. Mating systems, philopatry, and dispersal in birds and mammals. *Animal Behavior* 28: 1140–1162.

Griffith, J. D. 1980. Economy, family and remarriage: Theory of remarriage and application to preindustrial England. *J. Family Issues* 1(4): 479–496.

Groebel, J., and R. A. Hinde. 1989. A multi-level approach to the problems of aggression and war. In Jo Groebel and Robert A. Hinde (eds.), *Aggression and War: Their Biological and Social Bases,* 223–229. Cambridge, U.K.: Cambridge University Press.

Guttentag, M., and P. Secord. 1983. *Too Many Women? The Sex Ratio Question.* New York: Sage.

Gutteridge, W. (ed.) 1986. *Contemporary Terrorism.* New York: Facts on File.

Hackett, J. 1983. *The Profession of Arms.* New York: Macmillan.

Hackworth, D. H. 1989. *About Face: The Odyssey of an American Warrior.* With Julie Sherman. New York: Simon and Schuster.

Haig, D. 1992. Genomic imprinting and the theory of parent-offspring conflict. *Sem. Devel. Biol.* 3: 153–160.

Haig, D. 1993. Genetic conflicts in human pregnancy. *Quart. Rev. Biol.* 68(4): 495–532.

Hajnal, J. 1965. European marriage patterns in perspective. In D. V. Glass and D.E.C. Eversley (eds.), *Population in History: Essays in Historical Demography.* London: Edward Arnold.

Haldane, J.B.S. 1932. *The Causes of Evolution.* New York: Harper and Brothers.

Haldane, J.B.S. 1955. Population genetics. *New Biology* 18: 34–51.

Hale, J. R. 1985. *War and Society in Renaissance Europe, 1450–1620.* New York: St. Martin's Press.

Hallett, J. P. 1984. *Fathers and Daughters in Roman Society.* Princeton, N.J.: Princeton University Press.

Halperin, S. 1979. Temporary association patterns in free-ranging chimpanzees. In D. Hamburg and E. McCown (eds.), *The Great Apes,* 491–499. Menlo Park, Calif.: Benjamin Cummings.

Hamer, Dean, and Peter Copeland. 1998. *Living with Our Genes.* New York: Doubleday.

Hames, R. B. 1979. A comparison of the shotgun and the bow in Neotropical forest hunting. *Human Ecology* 7: 219–252.

Hames, R. B. 1988a. The allocation of parental care among the Ye'kwana. In L. Betzig, M. Borgerhoff Mulde, and P. Turke (eds.), *Human Reproductive Behaviour: A Darwinian Perspective.* Cambridge, U.K.: Cambridge University Press.

Hames, R. B. 1988b. Game conservation or efficient hunting? In B. J. McCay and J. M. Acheson (eds.), *The Question of the Commons: The Culture and Ecology of Communal Resources.* Tucson: University of Arizona Press.

Hames, R. B. 1989. Time, efficiency, and fitness in the Amazonian protein quest. *Res. Econ. Anthropol.* 11: 43–85.

Hames, R. B. 1991. Wildlife conservation in tribal societies. In M. L. Oldfield and J. B. Alcorn (eds.), *Biodiversity: Culture, Conservation, and Ecodevelopment.* Boulder, Colo.: Westview Press.

Hames, R. B. 1996. Costs and benefits of monogamy and polygyny for Yanomam" women. *Ethol. and Sociobiol.* 17: 181–199.

Hamilton, W. D. 1964. The genetical evolution of social behaviour I, II. *J. Theor. Biol.* 7: 1–52.

Hamilton, W. D. 1966. The moulding of senescence by natural selection. *J. Theore. Biol.* 12: 12–45.

Hamilton, W. D. 1970. Selfish and spiteful behaviour in an evolutionary model. *Nature* 228: 1218–1220.

Hamilton, W. D. 1975. Innate social aptitudes of man: An approach from evolutionary genetics. In R. Fox (ed.), *ASA Studies 4: Biosocial Anthropology,* 133–153. London: Malaby Press.

Hamilton, W. D. 1979. Wingless and fighting males in fig wasps and other insects. In M. S. and N. A. Blum (eds.). *Sexual Selection and Reproductive Competition in Insects*. New York: Academic Press.

Hamilton, W. D. 1980. Sex versus non-sex versus parasite. *Oikos* 35: 282–290.

Hamilton, W. D. 1996. *Narrow Roads of Gene Land: The Collected Papers of W. D. Hamilton,* vol. 1, *Evolution of Social Behaviour.* Oxford: W. H. Freeman.

Hamilton, W. D., and M. Zuk. 1982. Heritable true fitness and bright birds: A role for parasites? *Science* 218: 384–387.

Hamilton, W. D., P. A. Henderson, and N. Moran. 1981. Fluctuation of environment and coevolved antagonist polymorphism as factors in the maintenance of sex. In R. D. Alexander and D. W. Tinkle (eds.), *Natural Selection and Social Behavior: Recent Research and Theory.* New York: Chiron Press.

Hammel, E. A., S. Johansson, and C. Gunsberg. 1983. The value of children during industrialization: Sex ratios in childhood in nineteenth-century America. *J. Family Hist.* Winter 1983, 400–417.

Hammerstein, P. 1996. Darwinian adaptation, population genetics and the streetcar theory of evolution. *J. Math. Biol.* 34: 511–532.

Hampson, E., and D. Kimura. 1988. Reciprocal effects of hormonal fluctuations on human motor and perceptual-spatial skills. *Behav. Neurosci.* 102: 456–459.

Hanson, V. D. 1989. *The Western Way of War: Infantry Battle in Ancient Greece.* Oxford: Oxford University Press.

Harcourt, A. H. 1978. Strategies of emigration and transfer by primates, with particular reference to gorillas. *Z. Tierpsychol.* 48: 401–420.

Hardin, G. 1968. The tragedy of the commons. *Science* 162: 1243–1248.

Harpending, H. 1998. Commentary on D. S. Wilson's "Hunting, sharing and multilevel selection." *Current Anthropology* 39: 88–89.

Harpending, H., P. Draper, and A. Rogers. 1987. Human sociobiology. *Yearbook Phys. Anthropol.* 30: 127–150.

Harpending, H., P. Draper, and A. Rogers. 1990. Fitness in stratified societies. *Ethol. and Sociobiol.* 11: 497–509.

Harris, J. A., P. A. Vernon, and D. I. Boomsma. 1998. The heritability of testosterone: A study of Dutch adolescent twins and their parents. *Behavioral Genetics* 28: 165–171.

Harris, M. 1979. *Cultural Materialism.* New York: Random House.

Hart, C. W., and A. R. Pilling. 1960. *The Tiwi of North Australia.* New York: Holt.

Hart, S. L. 1997. Beyond greening: Strategies for a sustainable world. *Harvard Bus. Rev.* 75: 66–76.

Hartung, J. 1982. Polygyny and the inheritance of wealth. *Current Anthropology* 23: 1–12.

Hartung, J. 1983. In defense of Murdock: A reply to Dickemann. *Current Anthropology* 24(1): 125–126.

Hartung, J. 1992. Getting real about rape. *Behav. and Brain Sci.* 15(2): 390–392.

Hartung, J. 1997. If I had it to do over. In L. Betzing (ed.), *Human Nature: A Critical Reader,* 344–348. Oxford: Oxford University Press.

Harvey, P., and A. H. Harcourt. 1984. Sperm competition, testis size and breeding systems in primates. In R. L. Smith (ed.), *Sperm Competition and the Evolution of Animal Mating Systems,* 589–600. New York: Academic Press.

Harvey, P., R. D. Martin, and T. H. Clutton-Brock. 1986. Life histories in comparative perspective. In B. B. Smuts, D. L. Cheney, R. M. Seyfarth, R. W. Wrangham, and T. Struhsaker (eds.), *Primate Societies,* 181–196. Chicago: University of Chicago Press.

Hasegawa, T., and M. Hiraiwa-Hasegawa. 1983. Opportunistic and restrictive matings among wild chimpanzees in the Mahale Mountains. *J. of Ethology* 1: 75–85.

Hausfater, G., and S. B. Hrdy. 1984. *Infanticide: Comparative and Evolutionary Perspectives.* New York: Aldine de Gruyter.

Havanon, N., J. Knodel, and W. Sittitrai. 1992. The impact of family on wealth accumulation in rural Thailand. *Population Studies* 46: 37–51.

Hawkes, K. 1990. Why do men hunt? Some benefits for risky strategies. In E. Cashdan (ed.), *Risk and Uncertainty in Tribal and Peasant Economies,* 15–166. Boulder, Colo.: Westview Press.

Hawkes, K. 1991. Showing off: Tests of an hypothesis about men's foraging goals. *Ethol. and Sociobiol.* 12: 29–54.

Hawkes, K. 1992. Sharing and collective action. In E. A. Smith and B. Winterhalder (eds.), *Evolutionary Ecology and Human Behavior,* 269–300. New York: Aldine de Gruyter.

Hawkes, K. 1993. Why hunter-gatherers work: An ancient version of the problem of public goods. *Current Anthropology* 34: 341–361.

Hawkes, K., and E. L. Charnov. 1988. Human fertility: Individual or group benefit? *Current Anthropology* 20: 469–471.

Hawkes, K., R. L. Bleige Bird, and D. W. Bird. 1998. Commentary on D. S. Wilson's "Hunting, sharing and multilevel selection." *Current Anthropology* 39: 89–90.

Hawkes, K., J. F. O'Connell, and N. G. Blurton Jones. 1997. Hadza women's time allocation, offspring provisioning, and the evolution of long postmenopausal life spans. *Current Anthropology* 18(4): 551–577.

Hawkes, K., J. F. O'Connell, N. G. Blurton Jones, H. Alvarez, and E. L. Charnov. 1998. Grandmothering, menopause, and the evolution of human life histories. *Proc. Natl. Acad. Sci. USA* 95: 1336–1339.

Hawkes, K., J. F. O'Connell, N. G. Blurton Jones, H. Alvarez, and E. L. Charnov. 1999. The grandmother hypothesis and human evolution. In L. Cronk, N. Chagnon, and W. Irons (eds.), *Adaptation and Human Behavior: An Anthropological Perspective,* Hawthorne, N.Y.: Aldine de Gruyter.

Hawkes, K., J. F. O'Connell, and L. Rogers. 1997. The behavioral ecology of modern hunter-gatherers, and human evolution. *TREE* 12: 29–31.

Hawkes, K., A. R. Rogers, and E. L. Charnov. 1995. The male's dilemma: Increased offspring production is more paternity to steal. *Evolutionary Ecology* 9: 662–677.

Hayami, A. 1980. Class differences in marriage and fertility among Tokugawa villagers in Mino Province. *Keio Econ. Stud.* 17(1): 1–16.

Heinemann, U. R. 1990. *Eunuchs for the Kingdom of Heaven.* New York: A. Deutsch, Ltd.

Heinrich, B. 1989. *Ravens in Winter.* New York: Summit.

Heinsohn, R., and C. Packer. 1995. Complex cooperative strategies in groups-territorial African lions. *Science* 269: 1260–1262.

Hemelrijk, C. K. 1994. Support for being groomed in long-tailed macaques, *Macaca fascicularis. Animal Behaviour* 48: 479–481.

Henn, W., K. D. Zang, and D. Skuse. 1997. Mosiacism in Turner's syndrome. *Naure* 390: 569.

Henry, L. 1961. Some data on natural fertility. *Social Biology* 8: 81–91.

Hershkovitz, M. J., and F. S. Hershkovitz. 1934. *Rebel Destiny.* Freeport, N.Y.: Books for Libraries Press.

Hewlett, B. S., and L. L. Cavalli-Sforza. 1986. Cultural transmission among Aka Pygmies. *American Anthropologist* 88: 922–934.

Hiatt, L. R. 1981. Polyandry in Sri Lanka: A test case for parental investment theory. *Man* 15: 583–602.

Hill, C. M., and H. L. Ball. 1996. Abnormal births and other "ill omens": The adaptive case for infanticide. *Human Nature* 7: 381–401.

Hill, D. 1987. Social relationships between adult male and female rhesus macaques: 1. Sexual consortships. *Primates* 28: 439–456.

Hill, E. and B. Low. 1991. Contemporary abortion patterns: A life-history approach. *Ethol. and Sociobiol.* 13: 35–48.

Hill, J. 1984. Prestige and reproductive success in man. *Ethol. and Sociobiol.* 5: 77–95.

Hill, K., and A. M. Hurtado. 1991. The evolution of premature reproductive senescence and menopause in human females: An evaluation of the "Grandmother Hypothesis." *Human Nature* 2: 313–350.

Hill, K., and A. M. Hurtado. 1996. *Ache Life History: The Ecology and Demography of a Foraging People.* New York: Aldine de Gruyter.

Hill, K., and H. Kaplan. 1988a. Tradeoffs in male and female reproductive strategies among the Ache: Part 1. In L. Betzig, M. Borgerhoff Mulder, and P. Turke (eds.), *Human Reproductive behaviour: A Darwinian Perspective,* 277–289. Cambridge, U.K.: Cambridge University Press.

Hill, K., and H. Kaplan. 1988b. Tradeoffs in male and female reproductive strategies among the Ache: Part 2. In L. Betzig, M. Borgerhoff Mulder, and P. Turke (eds.), *Human Reproductive Behaviour: A Darwinian Perspective,* 291–305. Cambridge, U.K.: Cambridge University Press.

Hirschleifer, J. 1977. Economics from a biological viewpoint. *J. Law and Econ.* 20: 1–52.

Hoekstra, R. F. 1987. The evolution of sexes. In S. Stearns (ed.), *The Evolution of Sex and Its Consequences*, 59–92. Basel: Birkhäuser.

Hofbauer, J., and K. Sigmund. 1998. *Evolutionary Games and Population Dynamics*, Cambridge, U.K.: Cambridge University Press.

Hofman, M. A. 1983. Energy metabolism, brain size, and longevity in mammals. *Quart. Rev. Biol.* 58: 495–512.

Höglund, J., and R. V. Alatalo. 1995. *Leks.* Princeton, N.J.: University Press.

Holekamp, K. E., and P. W. Sherman. 1989. Why male ground squirrels disperse. *American Scientist* 77: 232–239.

Holmes, R. 1985. *Acts of War: The Behavior of Men in Battle.* New York: Free Press.

Hooks, B. L., and P. A. Green. 1993. Cultivating male allies: A focus on primate females, including *Homo sapiens. Human Nature* 4: 81–107.

Howard, R. D. 1979. Estimating reproductive success in natural populations. *American Naturalist* 114: 221–231.

Howell, N. 1979. *Demography of the Dobe !Kung.* New York: Academic Press.

Hrdy, S. B. 1974. Male-male competition and infanticide among the lemurs (*Presbytis entellus*) of Abu Rajasthan. *Folia Primatologia* 22: 19–58.

Hrdy, S. B. 1976. The care and exploitation of nonhuman primate infants by conspecifics other than the mother. *Adv. Study of Behav.* 6: 101–158.

Hrdy, S. B. 1978. Allomaternal care and the abuse of infants among Hanuman langurs. In D. J. Chivers and P. Herbert (eds.), *Recent Advances in Primatology*, vol. 1. New York: Academic Press.

Hrdy, S. B. 1979. Infanticide among animals: A review classification and implications for the reproductive strategies of females. *Ethol. and Sociobiol.* 1: 13–40.

Hrdy, S. B. 1981. *The Woman That Never Evolved.* Cambridge, Mass.: Harvard University Press.

Hrdy, S. B. 1992. Fitness tradeoffs in the history and evolution of delegated mothering with special reference to wet-nursing, abandonment, and infanticide. *Ethol. and Sociobiol.* 13: 409–442.

Hrdy, S. B. 1997a. Raising Darwin's consciousness: Female sexuality and the prehominid origins of patriarchy. *Human Nature* 8: 1–49.

Hrdy, S. B. 1997b. Mainstreaming Medea. In L. L. Betzig (ed.), *Human Nature: A Critical Reader*, 423–426. Oxford: Oxford University Press.

Hrdy, S. B., and D. S. Judge. 1993. Darwin and the puzzle of primogeniture. *Human Nature* 4: 1–45.

Hughes, A. 1986. Reproductive success and occupational class in eighteenth-century Lancashire, England. *Social Biology* 33: 109–115.

Hughes, A. L. 1988. *Evolution and Human Kinship.* Oxford: Oxford University Press.

Hull, T. H. 1990. Recent trends in sex ratios at birth in China. *Pop. and Devel. Rev.* 16(1): 63–83.

Humphrey, N. K. 1976. The social function of intellect. In P.P.G. Bateson and R. A. Hinde (eds.), *Growing Points in Ethology.* London: Cambridge University Press.

Humphrey, N. K. 1983. *Consciousness Regained: Chapters in the Development of Mind.* Oxford: Oxford University Press.

Huntingford, F., and A. Turner. 1987. *Animal Conflict.* London: Chapman and Hall.

Hurst, L. 1991. Sex, slime, and selfish genes. *Nature* 354: 23–4.

Hurst, L. 1992. Intragenomic conflict as an evolutionary force. *Proc. Roy. Soc. London* B 244: 91–99.

Hurst, L. D. 1995. Selfish genetic elements and their role in evolution: The evolution of sex and some of what that entails. *Phil. Trans. R. Soc. London* B 349: 321–332.

Hurst, L. D. 1996. Why are there only two sexes? *Proc. R. Soc. London* B 263: 415–422.

Hurst, L., and W. D. Hamilton. 1992. Cytoplasmic fusion and the nature of sexes. *Proc. Roy. Soc. London* B 247: 189–207.

Hurst, L., and G. T. McVean. 1997. Growth effects of uniparental disomies and the conflict theory of genomic imprinting. *Trends in Genetics* 13: 436–443.

Hurtado, M., K. Hawkes, and K. Hill. 1985. Female subsistence strategies among Ache hunter-gatherers of eastern Paraguay. *Human Ecology* 13: 1–28.

Hurtado, M., K. Hill, H. Kaplan, and I. Hurtado. 1992. Tradeoffs between food acquisition and child care among Hiwi and Ache women. *Human Nature* 3: 185–216.

Hurtado, M., and K. Hill. 1990. Seasonality in a foraging society: Variation in diet, work effort, fertility, and the sexual division of labor among the Hiwi of Venezuela. *J. Anthropol. Res.* 46: 293–346.

Huth, P. K. 1988. *Extended Deterrence and the Prevention of War.* New Haven: Yale University Press.

Hutt, C. 1972. *Males and Females.* London: Penguin.

Hyde, J. S. 1996. Where are the gender differences? Where are the gender similarities? In D. Buss and N. Malamuth (eds.), *Sex, Power, Conflict: Evolutionary and Feminist Perspectives,* 107–118. New York: Oxford University Press.

Imhof, A. E. 1981. Remarriage in rural populations and in urban middle and upper strata in Germany from the sixteenth to the twentieth century. In J. Dupâquier, J. E. Hélin, P. Laslett, M. Livi-Bacci, and S. Sogner (eds.), *Marriage and Remarriage in Populations of the Past,* pp. 335–346. New York: Academic Press.

Inger, G. 1980. *Svensk Rattshistoria.* Lund, Sweden: Liber Laromedel.

Inoue, M., F. Mitsunaga, H. Ohsawa, A. Takenaka, Y. Sugiyama, S. Gaspard, and O. Takenaka. 1991. Male mating behaviour and paternity discrimination by DNA fingerprinting in a Japanese macaque group. *Folia Primatologia* 56: 202–210.

Irons, W. 1975. *The Yomut Turkmen: A Study of Social Organization among a Central Asian Turkic-Speaking Population.* Museum of Anthropology, University of Michigan, Anthropological Paper 58.

Irons, W. 1979a. Natural selection, adaptation, and human social behavior. In N. A. Chagnon and W. Irons (eds.), *Evolutionary Biology and Human Social Behavior: An Anthropological Perspective.* North Scituate, Mass.: Duxbury Press.

Irons, W. 1979b. Emic and reproductive success. In N. A. Chagnon and W. Irons (eds.), *Evolutionary Biology and Human Social Behavior: An Anthropological Perspective.* North Scituate, Mass.: Duxbury Press.

Irons, W. 1980. Is Yomut social behavior adaptive? In G. Barlow and J. Silverman (eds.), *Sociobiology: Beyond Nature/Nurture?* Boulder, Colo.: Westview Press.

Irons, W. 1983a. The cultural and reproductive success hypothesis: A synthesis of empirical tests. Paper presented at annual meeting of American Anthropological Association, 1993.

Irons, W. 1983b. Human female reproductive strategies. In Samuel K. Wasser (ed.), *Social Behavior of Female Vertebrates,* 169–213. New York: Academic Press.

Irons, W. 1991. How did morality evolve? *Zygon* 26: 49–85.

Irons, W. 1996a. Morality, religion, and human nature. In W. M. Richardson and W. Wildman (eds.), *Religion and Science: History, Method, Dialogue,* 375–399. New York: Routledge.

Irons, W. 1996b. In our own self image: The evolution of morality, deception, and religion. *Skeptic* 4: 50–61.

Irons, W. 1998. Adaptively relevant environments versus the environment of evolutionary adaptedness. *Evolutionary Anthropology* 6: 194–204.

Iwasa, Y. 1998. The conflict theory of genomic imprinting: How much can be explained? *Curr. Top. Dev. Biol.* 40: 255–293.

Jacobson, H., and W. Zimmerman. 1993. *Behavior, Culture, and Conflict in World Politics.* Ann Arbor: University of Michigan Press.

James, T. W., and D. Kimura. 1997. Sex differences in remembering the locations of objects in an array: Location-shifts versus location-exchanges. *Evol. and Human Behav.* 18: 155–163.

Johansson, S., and Ola Nygren. 1991. The missing girls of China: A new demographic account. *Pop. and Devel. Rev.* 17(1): 35–51.

Johnson, G. 1986. Kin selection, socialization, and patriotism: An integrating theory. *Politics and Life Sci.* 4:128–139.

Johnson, G., S. H. Ratwik, and T. J. Sawyer. 1987. The evocative significance of kin terms in patriotic speech. In V. Reynolds, V. Falger, and I. Vine (eds.), *The Sociobiology of Ethnocentrism: Evolutionary Dimensions of Xenophobia, Discrimination, Racism, and Nationalism,* 137–174. London: Croom Helm.

Johnson, N. E., and S. Lean. 1985. Relative income, race and fertility. *Population Studies* 39: 99–112.

Johnson, R. L., and E. Kapsalis. 1998. Menopause in free-ranging rhesus macaques: Estimated incidence, relation to body condition, and adaptive significance. *Intl. J. Primatol.* 19: 751–765.

Johnson, S. B., and R. C. Johnson. 1991. Support and conflict of kinsmen in Norse earldoms, Icelandic families, and the English royalty. *Ethol. and Sociobiol.* 12: 211–220.

Jolly, A. 1966. Lemur social behavior and primate intelligence. *Science* 153: 501–506.

Jolly, A. 1985. *The Evolution of Primate Behavior.* 2d ed. New York: Macmillan.

Jones, D. 1995. Sexual selection, physical attractiveness, and facial neoteny. *Current Anthropology* 36: 723–748 (including commentary).

Jones, D. 1996. An evolutionary perspective on physical attractiveness. *Evolutionary Anthropology* 5: 97–109.

Jones, D., and K. Hill. 1993. Criteria of facial attractiveness in five populations. *Human Nature* 4: 271–296.

Jones, E. C. 1975. The post-reproductive phase in mammals. In P. van Keep and C. Lauritzen (eds.), *Frontiers of Hormone Research*, vol. 3, 1–20. Basel: Karger.

Jörberg, L. 1972. *A History of Prices in Sweden: 1732–1914.* 2 vols. Lund, Sweden: CWK Gleerup.

Jörberg, L. 1975. Structural change and economic growth in nineteenth-century Sweden. In Steven Koblik, ed., *Sweden's Development from Poverty to Affluence, 1750–1970*, 92–135. Minneapolis: University of Minnesota Press.

Jorgensen, W. 1980. *Western Indians.* San Francisco: Freeman.

Josephson, S. C. 1993. Status, reproductive success, and marrying polygynously. *Ethol. and Sociobiol.* 14: 391–396.

Judge, D. S. 1995. American legacies and the variable life histories of women and men. *Human Nature* 6: 291–323.

Judge, D. S., and S. B. Hrdy. 1992. Allocation of accumulated resources among close kin: Inheritance in Sacramento, California, 1890–1984. *Ethol. and Sociobiol.* 13: 495–522.

Just, W. 1970. *Military Men.* New York: Knopf.

Kadlec, J. A., and W. Drury. 1968. Structure of the New England Herring Gull population. *Ecology* 49: 644–676.

Kagan, J. 1981. Universals in human development. In R. H. Monroe, R. L. Monroe, and B. B. Whiting (eds.), *Handbook of Cross-cultural Human Development*, 53–62. New York: Garland Press.

Kahneman, D., and A. Tversky. 1982. On the study of statistical intuitions. In D. Kahneman, P. Slovic, and A. Tversky (eds.), *Judgements under Uncertainty: Heuristics and Biases*, 493–508. Cambridge, U.K.: Cambridge University Press.

Kano, T. 1996. Male rank order and copulation rate in a unit-group of bonobos at Wamba, Zaïre. In W. C. McGrew, L. Marchant, and T. Nishida (eds.), *Great Ape Societies*, 135–145. Cambridge, U.K.: Cambridge University Press.

Kaplan, H. 1994. Evolutionary and wealth flows theories of fertility: Empirical tests and new models. *Pop. and Devel. Rev.* 20: 753–791.

Kaplan, H. 1997. The evolution of the human life course. In K. Wachter and C. Finch (eds.), *Between Zeus and Salmon: The Biodemography of Longevity.* Washington, D.C.: National Academy Press.

Kaplan, H., and K. Hill. 1985. Hunting ability and reproductive success among male Ache foragers: Preliminary results. *Current Anthropology* 26: 131–133.

Kaplan, H., and J. B. Lancaster. 1999. The life histories of men in Albuquerque: An evolutionary-economic analysis of parental investment and fertility in modern society. In L. Cronk, N. Chagnon, and W. Irons (eds.), *Adaptation and Human Behavior: An Anthropological Perspective.* Hawthorne, N.Y.: Aldine de Gruyter.

Kaplan, H., J. B. Lancaster, J. A. Bock, and S. E. Johnson. 1995. Fertility and fitness among Albuquerque men: A competitive labour market theory. In R.I.M. Dunbar (ed.), *Human Reproductive Decisions,* 96–136. London: St. Martin's Press, in association with the Galton Institute.

Karlsen, K. F. 1987. *The Devil in the Shape of a Woman: Witchcraft in Colonial New England.* New York: Norton.

Karsten, R. 1923. *Blood Revenge, War, and Victory Feasts among the Jibaro Indians of Eastern Ecuador.* Smithsonian Institution, Bureau of American Ethnology, Bulletin 79.

Kasarda, J. D., J.O.G. Billy, and K. West. 1986. *Status Enhancement and Fertility: Reproductive Responses to Social Mobility and Educational Opportunity.* New York: Academic Press.

Keegan, J. 1987. *The Mask of Command.* London: Jonathan Cape.

Keegan, J., and A. Wheatcroft. 1986. *Zones of Conflict: An Atlas of Future Wars.* New York: Simon and Schuster.

Keeley, L. H. 1996. *War before Civilization: The Myth of the Peaceful Savage.* Oxford: Oxford University Press.

Kenrick, D., M. Trost, and V. Sheets. 1996. Power, harassment, and trophy mates: The feminist advantages of an evolutionary perspective. In D. Buss and N. Malamuth (eds.), *Sex, Power, Conflict: Evolutionary and Feminist Perspectives,* 29–53. New York: Oxford University Press.

Keohane, R. O. 1984. *After Hegemony: Cooperation and Discord in the World Political Economy.* Princeton, N.J.: Princeton University Press.

Keohane, R. O., and E. Ostrom (eds.). 1995. *Local Commons and Global Interdependence: Heterogeneity and Cooperation in Two Domains.* London: Sage.

Keverne, E. B., R. Fundele, M. Narashima, S. C. Barton, and M. A. Surani. 1996. Genomic imprinting and the differential roles of parental genomes in brain development. *Devel. Brain Res.* 92: 91–100.

Keyfitz, N. 1985. *Applied Mathematical Demography.* 2d ed. New York: Springer-Verlag.

Kirkpatrick, M. 1996. Genes and adaptation: A pocket guide to the theory. In Michael Rose and George Lauder (eds.), *Adaptation*, 125–146. New York: Academic Press.

Klare, M. T., and P. Kornbluh (eds.). 1988. *Low-Intensity Warfare: Counterinsurgency, Proinsurgency, and Anti-Terrorism in the Eighties.* New York: Pantheon.

Knodel, J. 1981. Remarriage and marital fertility in Germany during the eighteenth and nineteenth centuries: An exploratory analysis based on German village genealogies. In J. Dupaquier, E. Helin, P. Laslett, and M. Livi-Bacci (eds.), *Marriage and Remarriage in Past Populations.* New York: Academic Press.

Knodel, J. 1988. *Demographic Behavior in the Past: German Village Populations in the 18th and 19th centuries.* Cambridge, U.K.: Cambridge University Press.

Knodel J., and K. A. Lynch. 1985. The decline of remarriage: Evidence from German village populations in the eighteenth and nineteenth centuries. *J. Family Hist.* 10(1): 34–59.

Knodel, J., and M. Wongsith. 1991. Family size and children's education in Thailand: Evidence from a national sample. *Demography* 28: 119–131.

Knodel, J., N. Havanon, and W. Sittitrai. 1990. Family size and the education of children in the context of rapid fertility decline. *Pop. and Devel. Rev.* 16(1): 31–62.

Knodel, J., C. Saengtienchai, B. S. Low, and R. Lucas. 1997. An evolutionary perspective on Thai sexual attitudes. *J. Sex Research* 34(3): 292–303.

Kohlberg, L. L. 1981. *Essays on Moral Development*, vol. 1: *The Philosophy of Moral Development.* San Francisco: Harper and Row.

Kohlberg, L. L. 1984. *Essays on Moral Development*, vol. 2: *The Psychology of Moral Development.* San Francisco: Harper and Row.

Konner, M. J. 1981. Evolution of human behavior development. In R. H. Monroe, R. L. Monroe, and B. B. Whiting (eds.), *Handbook of Cross-Cultural Development*, 3–51. New York: Garland Press.

Kottak, C. 1990. Culture and "economic development." *American Anthropologist* 92: 723–731.

Kottak, C., and A. C. G. Costa. 1993. Ecological awareness, environmentalist action, and international conservation strategy. *Human Organization* 52: 335–343.

Krebs, C. J. 1985. *Ecology: The Experimental Analysis of Distribution and Abundance.* New York: Harper and Row.

Krebs, J. R., and N. B. Davies. 1993. *An Introduction to Behavioral Ecology.* 3d ed. Sunderland, Mass.: Sinauer Associates.

Krebs, J. R., and N. B. Davies (eds.). 1991. *Behavioral Ecology: An Evolutionary Approach.* 3d ed. London: Blackwell Scientific.

Krebs, J. R., and N. B. Davies (eds.). 1997. *Behavioral Ecology: An Evolutionary Approach.* 4th ed. London: Blackwell Scientific.

Kruuk, H. 1972. *The Spotted Hyena: A Study of Predation and Social Behavior.* Chicago: University of Chicago Press.

Kumm, J., and M. W. Feldman. 1997. Gene-culture coevolution and sex ratios: II. Sex-chromosomal distorters and cultural preferences for offspring sex. *Theor. Pop. Biol.* 52: 1–15.

Kurland, J. 1979. Matrilines: The primate sisterhood and the human avunculate. In N. Chagnon and W. Irons (eds.), *Evolutionary Biology and Human Social Behavior,* 145–180. North Scituate, Mass.: Duxbury.

Kurland, J. A., and S.J.C. Gaulin. 1984. The evolution of male parental investment: Effects of genetic relatedness and feeding ecology on the allocation of reproductive effort. In D. M. Taub (ed.), *Primate Paternalism,* 259–306. New York: Van Nostrand Reinhold.

Lack, D. 1947. The significance of clutch size. *Ibis* 89: 302–352.

Lack, D. 1966. *Population Studies of Birds.* Oxford: Oxford University Press.

Laland, K. N., J. Kumm, and M. W. Feldman. 1994. Gene-culture coevolutionary theory: A test case. *Current Anthropology* 36(1): 131–156 (including commentary and response).

Lancaster, J. B. 1986. Human adolescence and reproduction: An evolutionary perspective. In J. B. Lancaster and B. A. Hamburg (eds.), *School-Age Pregnancy and Parenthood,* 17–38. New York: Aldine de Gruyter.

Lancaster, J. B. 1991. A feminist and evolutionary biologist looks at women. *Yearbook Phys. Anthropol.* 34: 1–11.

Lancaster, J. B., and B. J. King. 1992. An evolutionary perspective on menopause. In V. Kerns and J. K. Brown (eds.), *In Her Prime: New Views of Middle-Aged Women,* 7–15. 2d ed. Urbana: University of Illinois Press.

Langston, N., S. Rohwer, and D. Gori. 1997. Experimental analyses of intra- and intersexual competition in red-winged blackbirds. *Behavioral Ecology* 8: 524–533.

Larrick, R. P., R. E. Nisbett, and J. N. Morgan. 1993. Who uses the cost-benefit rules of choice? Implications for the normative status of microeconomic theory. *Organiz. Behav. and Human Decision Processes* 56: 331–347.

Lavely, W., and R. B. Wong. 1998. Revising the Malthusian narrative: The comparative study of population dynamics in late Imperial China. *J. Asian Stud.* 57: 714–748.

Leacock, E. 1955. Matrilocality in a simple hunting economy. *Southwestern J. Anthropol.* 11: 31–47.

Leacock, E., and R. Lee. 1982. *Politics and History in Band Societies.* Cambridge, U.K.: Cambridge University Press.

Le Boeuf, B., and J. Reiter. 1988. Lifetime reproductive success in Northern elephant seals. In T. H. Clutton-Brock (ed.), *Reproductive Success: Studies of Individual Variation in Contrasting Breeding Systems,* 344–383. Chicago: University of Chicago Press.

Ledyard, J. O. 1995. Public goods: A survey of experimental research. In J. H. Kagel and A. E. Roth (eds.), *The Handbook of Experimental Economics*, 111–194. Princeton, N.J.: Princeton University Press.

Lee, J., and C. Campbell. 1997. *Fate and Fortune in Rural China: Social Organization and Population Behavior in Liaoning, 1774–1873*. Cambridge, U.K.: Cambridge University Press.

Lee, J., and S. Guo (eds.). 1994. *Qingdai huanhzu renkou xingwei yu shehui huanjing* (Population behavior and social setting of the Qing imperial lineage). Beijing: Beijing University Press.

Lee, J., C. Campbell, and F. Wang. 1993. The last emperors: An introduction to the demography of the Qing imperial lineage. In D. Rehler and R. Schofield (eds.), *New and Old Methods in Historial Demography*, 361–382. Oxford: Oxford University Press.

Lee, Richard. 1979a. Politics, sexual and non-sexual, in an egalitarian society. In Eleanor Leacock and Richard Lee (eds.), *Politics and History in Band Societies*, 37–59. Cambridge, U.K.: Cambridge University Press.

Lee, Richard. 1979b. *The !Kung San*. London: Cambridge University Press.

Lee, Ronald D. 1990. Comment: The second tragedy of the commons. *Pop. Devel. Rev.* 16 (suppl.): 315–322.

Leimar, O. 1996. Life-history analysis of the Trivers and Willard sex-ratio problem. *Behavioral Ecology* 7: 316–325.

Lenski, R. E., and M. Travisano. 1994. Dynamics of adaptation and diversification: A 10,000-generation experiment with bacterial populations. *Proc. Natl. Acad. Sci.* 91: 6808–6814.

Lessells, C. M. 1991. The evolution of life history strategies. In J. R. Krebs and N. B. Davies (eds.), *Behavioural Ecology*, chap. 2. 3d ed. Oxford: Blackwell Scientific.

Lesthaeghe, R., and C. Wilson. 1986. Modes of production, secularization and the pace of the fertility decline in western Europe, 1870–1930. In A. J. Coale and S. C. Watkins (eds.), *The Decline of Fertility in Europe*. Princeton, N.J.: Princeton University Press.

Leutenegger, W. 1979. Evolution of litter size in primates. *American Naturalist* 114: 881–903.

Lever, J. 1976. Sex differences in the games children play. *Social Problems* 23: 478–487.

Lever, J. 1978. Sex differences in the complexity of children's play and games. *Amer. Sociol. Rev.* 43: 471–483.

Lewellen, T. C. 1983. *Political Anthropology*. South Hadley, Mass.: Bergen and Garvey.

Lewis, K. 1998. Pathogen resistance as the origin of kin altruism. *J. Theoret. Biol.* 193: 359–363.

Lewis, O. 1959. *Five Families: Mexican Case Studies in the Culture of Poverty*. New York: Basic Books.

Lewis, O. 1970. *Anthropological Essays.* New York: Random House.

Lewontin, R. C. 1970. The units of selection. *Ann. Rev. Ecol. and Syst.* 1: 1–18.

Li, L., and M. K. Choe. 1997. A mixture model for duration data: Analysis of second births in China. *Demography* 34: 189–197.

Lindert , P. H. 1978. *Fertility and Scarcity in America.* Princeton, N.J.: Princeton University Press.

Linton, R. 1939. Marquesan culture and analysis of Marquesan culture. In Abram Kardiner (ed.), *The Individual and His Society,* 137–250. New York: Columbia University Press.

Lips, J. E. 1947. Naskapi law. *Transact. Amer. Philos. Soc.* 37: 379–491.

Logan, M., and H. N. Qirko. 1996. An evolutionary perspective on maladaptive traits and cultural conformity. *Amer. J. Human Biol.* 8: 615–629.

Lo-Johannsson, F.A.I. 1981. *Sveriges Rikes Lag: Gillard och Antagen Parisdagen Ar, 1734, faksimilutgava.* Malmo, Sweden: Giulunds.

Lorenz, K. 1966. *On Aggression.* New York: Harcourt Brace Jovanovich.

Lovelock, J. 1988. *The Ages of Gaia.* New York: Norton.

Lovelock, J., and L. Margulis. 1974. Biological modulation of the earth's atmosphere. *Icarus* 21: 471–489.

Low, Bobbi S. 1978. Environmental uncertainty and the parental strategies of marsupials and placentals. *American Naturalist* 112: 197–213.

Low, Bobbi S. 1979. Sexual selection and human ornamentation. In Napoleon Chagnon and Bill Irons (eds.), *Evolutionary Biology and Human Social Behavior,* 462–486. North Scituate, Mass.: Duxbury Press.

Low, Bobbi S. 1988a. Measures of polygyny in humans. *Current Anthropology* 29(1): 189–194.

Low, Bobbi S. 1988b. Pathogen stress and polygyny in humans. In Laura Betzig, Monique Borgerhoff Mulder, and Paul Turke (eds.), *Human Reproductive Behaviour: A Darwinian Perspective,* 115–128. Cambridge, U.K.: Cambridge University Press.

Low, Bobbi S. 1989a. Occupational status and reproductive behavior in 19th century Sweden: Locknevi parish. *Social Biology* 36: 82–101.

Low, Bobbi S. 1989b. Cross-cultural patterns in the training of children: An evolutionary perspective. *J. Comp. Psychol.* 103: 311–319.

Low, Bobbi S. 1990a. Sex, power, and resources: Ecological and social correlates of sex differences. *J. Contemp. Sociol.* 27: 45–71.

Low, Bobbi S. 1990b. Human responses to environmental extremeness and uncertainty: A cross-cultural perspective. In E. Cashdan, (ed.), *Risk and Uncertainty in Tribal and Peasant Economies,* 229–255. Boulder, Colo.: Westview Press.

Low, Bobbi S. 1990c. Marriage systems and pathogen stress in human societies. *American Zoologist* 30: 325–339.

Low, Bobbi S. 1990d. Fat and deception. *Ethol. and Sociobiol.* 11: 67–74.

Low, Bobbi S. 1990e. Land ownership, occupational status, and reproductive behavior in 19th-century Sweden: Tuna parish. *American Anthropologist* 92(2): 457–468.

Low, Bobbi S. 1991. Reproductive life in nineteenth-century Sweden: An evolutionary perspective on demographic phenomena. *Ethol. and Sociobiol.* 12: 411–448.

Low, Bobbi S. 1992. Sex, coalitions, and politics in preindustrial societies. *Pol. and Life Sci.* 11(1): 63–80.

Low, Bobbi S. 1993a. Ecological demography: A synthetic focus in evolutionary anthropology. *Evolutionary Anthropology* 1993: 106–112.

Low, Bobbi S. 1993b. An evolutionary perspective on lethal conflict. In H. Jacobson and W. Zimmerman (eds.), *Behavior, Culture, and Conflict in World Politics.* Ann Arbor: University of Michigan Press.

Low, Bobbi S. 1994a. Human sex differences in behavioral ecological perspective. *Analyse & Kritik* 16: 38–67.

Low, Bobbi S. 1994b. Men in the demographic transition. *Human Nature* 5(3): 223–253.

Low, Bobbi S. 1996. The behavioral ecology of conservation in traditional societies. *Human Nature* 7: 353–379.

Low, Bobbi S. 1999. Sex, wealth, and fertility: Old rules, new environments. In L. Cronk, N. Chagnon, and W. Irons (eds.), *Adaptation and Human Behavior: An Anthropological Perspective.* Hawthorne, N.Y.: Aldine de Gruyter.

Low, Bobbi S., and A. L. Clarke. 1991. Occupational status, land ownership, migration, and family patterns in 19th century Sweden. *J. Family Hist.* 16(2): 117–138.

Low, Bobbi S., and A. L. Clarke. 1992. Resources and the life course: Patterns in the demographic transition. *Ethol. and Sociobiol.* 13: 463–494.

Low, Bobbi S., and A. L. Clarke. 1993. Historical perspectives on population and environment: Data from 19th-century Sweden. In G. Ness, W. Drake, and S. Brechen (eds.), *Population-Environment Dynamics: Ideas and Observations.* Ann Arbor: University of Michigan Press.

Low, Bobbi S., R. D. Alexander, and K. M. Noonan. 1987. Human hips, breasts and buttocks: Is fat deceptive? *Ethol. and Sociobiol.* 8: 249–257.

Low, Bobbi S., R. D. Alexander, and K. M. Noonan. 1988. Response to Judith Anderson's comments on Low, Alexander, and Noonan (1987). *Ethol. and Sociobiol.* 9: 325–328.

Low, Bobbi S., A. L. Clarke, and K. Lockridge. 1992. Toward an ecological demography. *Pop. Devel. Rev.* 18(1): 1–31.

Luker, K. 1996. *Dubious Conceptions: The Politics of Teenage Pregnancy.* Cambridge, Mass.: Harvard University Press.

Lumsden, C. J., and E. O. Wilson. 1981. *Genes, Mind, and Culture: The Co-evolutionary Process.* Cambridge, Mass.: Harvard University Press.

Lutz, W. 1994. *The Future Population of the Earth.* London: Earthscan Press.

MacArthur, R. H., and E. O. Wilson. 1967. *The Theory of Island Biogeography.* Princeton, N.J.: Princeton University Press.

Maccoby, E. E., and C. N. Jacklin. 1974. *The Psychology of Sex Differences.* Palo Alto, Calif.: Stanford University Press.

Mace, G. M. 1990. Birth, sex ratio, and infant mortality rates in captive western lowland gorillas. *Folia Primatologia* 55(3–4): 156–165.

Mace, R. 1996. Biased parental investment and reproductive success in Gabbra pastoralists. *Behav. Ecol. and Sociobiol.* 38: 75–81.

Mace, R. 1998. The coevolution of human fertility and wealth inheritnce strategies. *Phil. Trans. R. Soc. London* B 353: 389–397.

MacKellar, F. L. 1996. On human carrying capacity. *Pop. Devel. Rev.* 22: 145–156.

MacKellar, F. L. 1997. Population and fairness. *Pop. Devel. Rev.* 23: 359–376.

Macunovich, D. J. 1998. Fertility and the Easterlin hypothesis: An assessment of the literature. *J. Pop. Econ.* 11: 53–111.

Maechling, C., Jr. 1988. Counterinsurgency: The first ordeal by fire. In M. T. Klare, and P. Kornbluh (eds.), *Low-Intensity Warfare: Counterinsurgency, Proinsurgency, and Anti-Terrorism in the Eighties,* 21–48. New York: Pantheon.

Malamuth, N. 1996. The confluence model of sexual aggression: Feminist and evolutionary perspectives. In D. Buss and N. Malamuth (eds.), *Sex, Power, Conflict: Evolutionary and Feminist Perspectives,* 269–295. New York: Oxford University Press.

Mallett, M. 1987. *The Borgias.* Chicago: Academy Press.

Malmström, A. 1981. *Successionsratt II.* Uppsala, Sweden: Iustrus Forlag.

Manners, R. A. 1967. The Kipsigis of Kenya: Culture change in a "model" East African tribe. In R. Needham (ed.), *Contemporary Change in Traditional Societies,* 207–359. Urbana: University of Illinois Press.

Manson, J., and R. Wrangham. 1991. Intergroup aggression in chimpanzees and humans. *Current Anthropology* 32: 369–390.

Marsh, H., and T. Kasuya. 1986. Evidence for reproductive senescence in female cetaceans. *Report of the International Whaling Commission,* Special Issue 8: 83–95.

Martin, R. D. 1983. Human brain evolution in an ecological context. 52nd James Arthur Lecture on the Evolution of the Human Brain. New York: American Museum of Natural History.

Marrow, P., R. Johnstone, and L. Hurst. 1996. Riding the evolutionary streetcar: Where population genetics and game theory meet. *TREE* 11: 445–446.

Mason, K. O. 1997. Explaining fertility transitions. *Demography* 34: 443–454.

Maynard Smith, J. 1969. The status of neo-Darwinism. In C. H. Waddington (ed.). *Towards a Theoretical Biology, vol. 2: Sketches,* 82–89. Edinburgh: Edinburgh University Press.

Maynard Smith, J. 1974. The theory of games and the evolution of animal conflict. *J. Theor. Biol.* 47: 209–221.

Maynard Smith, J. 1978. *The Evolution of Sex.* Cambridge, U.K.: Cambridge University Press.

Maynard Smith, J. 1979. Game theory and the evolution of behaviour. *Proc. Roy. Soc. London* B 205: 475–488.

Maynard Smith, J. 1982. *Evolution and the Theory of Games.* Cambridge, U.K.: Cambridge University Press.

Maynard Smith, J., and G. A. Parker. 1976. The logic of asymmetrical contests. *Animal Behavior* 24: 159–175.

Maynard Smith, J., and G. R. Price. 1973. The logic of animal conflicts. *Nature* 246: 15–18.

Mazur, L. A. (ed.). 1994. *Beyond the Numbers: A Reader on Population, Consumption, and the Environment.* Washington, D.C.: Island Press.

McBurney, D. H., S.J.C. Gaulin, T. Devineni, and C. Adams. 1997. Superior spatial memory of women: Stronger evidence for the gathering hypothesis. *Evol. and Human Behav.* 18: 165–174.

McCabe, K., S. Rassenti, and V. L. Smith. 1998. Reciprocity, trust, and payoff privacy in extensive form bargaining. *Games and Econ. Behav.* 24: 10–24.

McCandless, B., A. Roberts, and T. Starnes. 1972. Teachers' marks, achievement test scores, and aptitude relations with respect to social class, race, and sex. *J. Educ. Psych.* 63: 153–159.

McCullough, J. M., and E. Y. Barton. 1991. Relatedness and mortality risk during a crisis year: Plymouth colony 1620–1621. *Ethol. and Sociobiol.* 12: 195–209.

McCullough, M. 1952. The Ovimbundu of Angola. In D. Forde, *Ethnographic Survey of Africa,* part 2, West Central Africa. London: International African Institute.

McGrew, W. C., L. F. Marchant, and T. Nishida (eds.). 1996. *Great Ape Societies.* Cambridge, U.K.: Cambridge University Press.

McInnis, R. M. 1977. Childbearing and land availability: Some evidence from individual household data. In Ronald Lee (ed.), *Population Patterns in the Past,* 201–227. New York: Academic Press.

McNeill, William H. 1963. *The Rise of the West: A History of the Human Community.* Chicago: University of Chicago Press.

McNeill, William H. 1982. *The Pursuit of Power: Technology, Armed Force, and Society Since A.D. 1000.* Chicago: University of Chicago Press.

Mead, M. 1935. *Sex and Temperament in Three Primitive Societies.* New York: William Murrow.

Mealey, L. 1985. The relationship between social status and biological success: A case study of the Mormon religious hierarchy. *Ethol. and Sociobiol.* 6: 249–257.

Mealey, L., and W. Mackey. 1990. Variation in offspring sex ratio in women of differing social status. *Ethol. and Sociobiol.* 11: 83–95.

Mech, D. 1977. Productivity, mortality, and population trends of wolves in northeastern Minnesota. *J. Mammalogy* 58: 559–574.

Mendels, F. F. 1981. *Industrialization and Population Pressure in Eighteenth-Century Flanders.* New York: Arno Press.

Merrigan, P., and Y. St.-Pierre. 1998. An econometric and neoclassical analysis of the timing and spacing of births in Canada from 1950 to 1990. *J. Popul. Econ.* 29–51.

Mesquida, C. G., and N. I. Wiener. 1996. Human collective aggression: A behavioral ecology perspective. *Ethol. and Sociobiol.* 17: 247–262.

Mesterton-Gibbons, M., and L. Dugatkin. 1992. Cooperation among unrelated individuals: Evolutionary factors. *Quart. Rev. Biol.* 67: 267–281.

Michod, R. E., and B. R. Levin (eds.). 1988. *The Evolution of Sex.* Sunderland, Mass.: Sinauer.

Mildvan, A. S., and B. L. Strehler. 1960. A critique of theories of mortality. In B. L. Strehler (ed.), *The Biology of Aging,* 45–78. New York: American Institute of Biological Sciences.

Miles, D. R., and G. Carey. 1997. Genetic and environmental architecture of human aggression. *J. Pers. Soc. Psych.* 72: 207–217.

Miller, E. 1975. War, taxation, and the English economy in the late 13th and early 14th centuries. In J. M. Winter (ed.), *War and Economics in Development.* Cambridge, U.K.: Cambridge University Press.

Mitani, J. 1984. The behavioral regulation of monogamy in gibbons (*Hylobates muelleri*). *Behav. Ecol. and Sociobiol.* 15: 225–229.

Mitterauer, M., and R. Sieder. 1982. *The European Family.* Oxford: Blackwell.

Moir, A., and D. Jessel. 1991. *Brain Sex: The Real Difference between Men and Women.* New York: Lyle Stewart.

Møller, A. P. 1987. Variation in badge size in male house sparrows *Passer domesticus*—evidence for status signalling. *Animal Behaviour* 35: 1637–1644.

Møller, A. P. 1989. Viability costs of male tail ornaments in a swallow. *Nature* 339: 132–135.

Møller, A. P. 1994. *Sexual Selection and the Barn Swallow.* Oxford: Oxford University Press.

Monroe, R. L., and R. H. Monroe. 1975. *Cross-cultural Human Development.* Monterey, Calif.: Brooks/Cole.

Moore, J. H. 1990. The reproductive success of Cheyenne war chiefs: A counter example to Chagnon. *Current Anthropology* 31: 169–173.

Moore, T., and D. Haig. 1991. Genomic imprinting in mammalian development: A parental tug-of-war. *Trends in Genetics* 7: 45–49.

Morell, V. 1998. A new look at monogamy. *Science* 281: 1982–1983.

Morin, P., J. Wallis, J. Moore, and D. Woodruff. 1994. Paternity exclusion in a community of chimpanzees using hypervariable simple sequence repeats. *Mol. Ecol.* 3: 469–478.

Mosk, C. 1983. *Patriarchy and Fertility: Japan and Sweden 1880–1960.* New York: Academic Press.

Mueller, U. 1991. Social and reproductive success: Theoretical considerations and a case study of the West Point Class of 1950. *ZUMA: Zentrum für Umfragen, Methoden und Analysen.*

Mueller, U. 1993. Social status and sex. *Nature* 363: 490.

Mueller, U., and A. Mazur. 1996. Facial dominance in West Point cadets predicts military rank 20+ years later. *Social Forces* 74: 823–850.

Mueller, U., and A. Mazur. 1998. Reproductive constraints on dominance competition in male *Homo sapiens. Evol. and Human Behav.* 19: 387–396.

Murdock, G. P. 1967. *Ethnographic Atlas.* Pittsburgh: University of Pittsburgh Press.

Murdock, G. P. 1981. *Atlas of World Cultures.* Pittsburgh: University of Pittsburgh Press.

Murdock, G. P., and C. Provost. 1973. Factors in the division of labor by sex. *Ethnology* 12: 203–225.

Murdock, G. P., and D. White. 1969. Standard cross-cultural sample. *Ethnology* 8: 329–369.

Neese, R. M. 1990. Evolutionary explanations of emotions. *Human Nature* 1: 261–289.

Nesse, R. M., and K. C. Berridge, 1997. Psychoactive drug use in evolutionary perspective. *Science* 278: 63–66.

Nice, M. M. 1962. Development of behavior in precocial birds. *Trans. Linnaean Soc. New York* 8: 1–211.

Niefert, M. R., J. M. Seacat, and W. E. Jobe. 1985. Lactation failure due to insufficient glandular development of the breast. *Pediatrics* 76: 823–828.

Nisbett, R. E. 1980. *Human Inference: Strategies and Shortcomings of Social Judgement.* Englewood Cliffs, N.J.: Prentice Hall.

Nisbett, R. E. 1996. *Culture of Honor: The Psychology of Violence in the South.* Boulder, Colo.: Westbiew Press.

Nisbett, R. E., G. T. Fong, D. R. Lehman, and P. W. Cheng. 1987. Teaching reasoning. *Science* 238: 625–631.

Nishida, T. 1979. The social structure of chimpanzees of the Mahale Mountains. In D. A. Hamburg and E. R. McCown (eds.), *The Great Apes,* 73–122. Menlo Park, Calif.: Benjamin Cummings.

Nishida, T. 1983. Alpha status and agonistic alliance in wild chimpanzees (*Pan troglodytes schweinfurthii*). *Primates* 24: 318–336.

Nishida, T., and M. Hiraiwa-Hasegawa. 1986. Chimpanzees and bonobos: Cooperative relationships among males. In B. B. Smuts, D. L. Cheney, R. M. Seyfarth, R. W. Wrangham, and T. T. Struhsaker (eds.), *Primate Societies,* 165–178. Chicago: University of Chicago Press.

Nishida, T., and K. Hosaka. 1996. Coalition strategies among adult chimpanzees of the Mahale Mountains, Tanzania. In W. C. McGrew, L. Marchant, and T.

Nishida (eds.), *Great Ape Societies*, 114–134. Cambridge, U.K.: Cambridge University Press.

Nishida, T., M. Hiraiwa-Hasegawa, and Y. Takahata. 1985. Group extinction and female transfer in wild chimpanzees in the Mahale Mountains. *Z. Tierpsych.* 67: 284–301.

Nishida, T., H. Takasaki, and Y. Takasaki. 1990. Demography and reproductive profiles. In T. Nishida (ed.), *The Chimpanzees of Mahale Mountains.* Tokyo: University of Tokyo Press.

Nkrumah, N. In press. The effect of land and tree tenure rights on deforestation patterns in common pool resource areas in the forest savannah transition zone, Ghana, West Africa.

Noë, R., F.B.M. de Waal, and J.A.R.A.M. van Hoof. 1980. Types of dominance in a chimpanzee colony. *Folia Primatol.* 34: 90–110.

Norberg, A., and M. Rolén. 1979. Migration and marriage: Some empirical results from Tuna parish, 1865–1894. In J. Sundin and E. Söderlund (eds.), *Time, Space, and Man: Essays on Microdemography.* Stockholm: Almqvist and Wiksell.

Nordenskiold, E. 1949. The Cuna. In J. Steward (ed.), *Handbook of South American Indians,* vol. 4. Washington, D.C.: U.S. Government Printing Office.

Nowak, M., and K. Sigmund. 1994. The alternating Prisoner's Dilemma. *J. Theor. Biol.* 168: 219–226.

Nowak, M., R. May, and K. Sigmund. 1995. The arithmetics of mutual help. *American Scientist,* June 1995, 76–81.

O'Brien, D. 1977. Female husbands in southern Bantu society. In Alice Schlegel (ed.), *Sexual Stratification: A Cross-Cultural View,* 109–126. New York: Columbia University Press.

Oelschlanger, M. 1991. *The Idea of Wilderness: From Prehistory to the Age of Ecology.* New Haven: Yale University Press.

Omark, D. R., and M. S. Edelman. 1975. A comparison status hierarchy in young children: An ethological approach. *Soc. Sci. Info.* 14: 87–107.

Orians, G. H. 1969. On the evolution of mating systems in birds and mammals. *American Naturalist* 103: 589–603.

Ostergren, R. 1990. Patterns of seasonal industrial labor recruitment in a nineteenth-century Swedish parish: The case of Matfors and Tuna, 1846–1873. Demographic Data Base Report No. 5, 1–100. Umeå, Sweden.

Ostrom, E. 1990. *Managing the Commons.* Cambridge, U.K.: Cambridge University Press.

Ostrom, E. 1998a. A behavioral approach to the rational choice theory of collective action. *Amer. Pol. Sci. Rev.* 92: 1–22.

Ostrom, E. 1998b. Coping with the tragedies of the commons. Paper given at the 1998 meeting of the Society for Politics and the Life Sciences/American Political Science Association, Boston, 1998.

Ostrom, E., J. Burger, C. B. Field, R. B. Norgaard, and D. Policansky. 1999. Revisiting the commons: Local lessons, global challenges. *Science* 284: 278–282.

Ostrom, E., R. Gardner, and J. Walker. 1994. *Rules, Games, and Common-Pool Resources.* Ann Arbor: University of Michigan Press.

Otterbein, K. F. 1970. *The Evolution of War: A Cross-Cultural Study.* Cambridge, Mass.: HRAF Press.

Oye, K. A. 1986. *Cooperation under Anarchy.* Princeton, N.J.: Princeton University Press.

Packer, C. 1986. The ecology of sociality in felids. In D. I. Rubenstein and R. W. Wrangham (eds.), *Ecological Aspects of Social Evolution,* 429–451. Princeton, N.J.: Princeton University Press.

Packer, C., and A. E. Pusey. 1983. Adaptations of female lions to infanticide by incoming males. *American Naturalist* 121: 716–28.

Packer, C., and A. E. Pusey. 1984. Infanticide in carnivores. In G. Hausfater, S. B. Hrdy (eds.), *Infanticide: Comparative and Evolutionary Perspectives.* New York: Aldine de Gruyter.

Packer, C., D. A. Collins, A. Sindimwe, and J. Goodall. 1995. Reproductive constraints on aggressive competition in female baboons. *Nature* 373: 60–63.

Pagel, M. 1999. Mother and father in surprise genetic agreement. *Nature* 397: 19–20.

Palmer, C. T., B. E. Fredrickson, and C. F. Tilley. 1997. Categories and gatherings: Group selection and the mythology of cultural anthropology. *Evol. and Human Behav.* 18: 291–308.

Parish, A. R. 1996. Female relationships in bonobos *(Pan paniscus):* Evidence for bonding, cooperation, and female dominance in a male-philopatric species. *Human Nature* 7: 61–96.

Parker, G. A. 1974. Assessment strategy and the evolution of fighting behaviour. *J. Theor. Biol.* 47: 223–243.

Parker, G. A. 1984. Evolutionarily stable strategies. In J. R. Krebs and N. B. Davies (eds.), *Behavioral ecology,* 30–61. Oxford: Blackwell Scientific.

Parker, G. A., R. R. Baker, and V.G.F. Smith. 1972. The origin and evolution of gamete dimorphism and the male-female phenomenon. *J. Theor. Biol.* 36: 529–553.

Parry, J. P. 1979. *Caste and Kinship in Kangara.* London: Routledge and Paul.

Partridge, L., and L. Hurst. 1998. Sex and conflict. *Science* 281: 2003–2008.

Paul, A., J. Kuester, A. Timme, and J. Arnemann. 1993. The association between rank, mating effort, and reproductive success in male Barbary macaques. *Primates* 34: 491–502.

Peacock, N. 1990. Comparative and cross-cultural approaches to the study of human female reproductive failure. In C. J. Rousseau (ed.), *Primate Life History and Evolution,* 195–220. New York: Wiley-Liss.

Peacock, N. 1991. An evolutionary perspective on the patterning of maternal investment in pregnancy. *Human Nature* 2(4): 351–385.

Pebley, A., and W. Mbugua. 1989. Polygyny and fertility in sub-Saharan Africa.

In R. Lesthaeghe (ed.), *Reproduction and Social Organization in sub-Saharan Africa*. Berkeley: University of California Press.

Pellison, M. 1897. *Roman Life in Pliny's Time*. Philadelphia: George W. Jacobs & Co.

Peng, X. 1991. *Demographic Transition in China*. Oxford: Clarendon Press.

Pennington, R., and H. Harpending. 1988. Fitness and fertility among Kalahari !Kung. *Amer. J. Phys. Anthropol.* 77: 303–319.

Pennisi, E. 1998. A genomic battle of the sexes. *Science* 281: 1984–1985.

Pérusse, D. 1993. Cultural and reproductive success in industrial societies: Testing the relationship at proximate and ultimate levels. *Behav. and Brain Sci.* 16: 267–322.

Pérusse, D. 1994. Mate choice in modern societies: Testing evolutionary hypotheses with behavioral data. *Human Nature* 5(3): 255–278.

Peter, Prince of Greece and Denmark. 1963. *A Study of Polyandry*. The Haguye: Mouton.

Pfister, U. 1989a. Proto-industrialization and demographic change: The Canton of Zurich revisited. *J. Econ. Hist.* 18: 629–62.

Pfister, U. 1989b. Work roles and family structure in proto-industrial Zurich *J. Interdisciplinary Hist.* 20: 83–105.

Piaget, J. 1932. *The Moral Judgement of the Child*. New York: Free Press.

Plomin, R., J. C. DeFries, and G. E. McClearn. 1990. *Behavioral Genetics: A Primer*. 2d ed. New York: Freeman.

Plouse, S. 1993. *The Psychology of Judgement and Decision Making*. New York: McGraw-Hill.

Podolsky, R. H. 1985. Colony formation and attraction of the Laysan Albatross and the Leach's Storm-Petrel. Ph.D. diss., University of Michigan, Ann Arbor.

Pomiankowski, A. 1987a. The 'handicap principle' does work—sometimes. *Proc. Roy. Soc. London* B 127: 123–145.

Pomiankowski, A. 1987b. The costs of choice in sexual selection. *J. Theor. Biol.* 128: 195–218.

Pope, T. 1990. The reproductive consequences of male cooperation in the red howler monkey: Paternity exclusion in multi-male and single-male troops using genetic markers. *Behav. Ecol. and Sociobiol.* 27: 439–567.

Promislow, D., and P. Harvey. 1990. Living fast and dying young. A comparative analysis of life history traits among mammals. *J. Zoology* 220: 417–437.

Pulliam, H. R., and C. Dunford. 1980. *Programmed to Learn: An Essay on the Evolution of Culture*. New York: Columbia University Press.

Pusey, A. E., and C. Packer. 1987. Dispersal and philopatry. In B. Smuts, D. L. Cheney, R. M. Seyfarth, R. W. Wrangham, and T. T. Struhsaker (eds.), *Primate Societies*, 250–266. Chicago: University of Chicago Press.

Pusey, A. E., and C. Packer. 1997. The ecology of relationships. In J. R. Krebs and N. B. Davies (eds.), *Behavioural Ecology,* chap. 11. 4th ed. Oxford: Blackwell.

Pusey, A. E., J. Williams, and J. Goodall. 1997. The influence of dominance rank on the reproductive success of female chimpanzees. *Science* 277: 828–831.

Putnam, R. D. 1993a. *Making Democracy Work: Civic Tradition in Modern Italy.* Princeton, N.J.: Princeton University Press.

Putnam, R. D. 1993b. *Double-edged Diplomacy: International Bargaining and Domestic Politics.* Berkeley: University of California Press.

Qirko, H. N. 1999. The maintenance and reinforcement of celibacy in institutionalized settings. In E. J. Sobo and S. Bell (eds.), *Anthropological Perspectives on Sexual Abstinence: Celibacy Examined.* Madison: University of Wisconsin Press.

Rank, M. A. 1989. Fertility among women on welfare: Incidence and determinants. *Amer. Sociol. Rev.* 54: 296–304.

Ransel, D. 1988. *Mothers in Misery: Child Abandonment in Russia.* Princeton, N.J.: Princeton University Press.

Rao, V. 1993a. Dowry "inflation" in rural India: A statistical investigation. *Population Studies* 47: 283–293.

Rao, V. 1993b. The rising price of husbands: A hedonic analysis of dowry increases in rural India. *J. Pol. Econ.* 101: 666–677.

Rao, V. 1997. Wife-beating in rural south India: A qualitative and econometric analysis. *Soc. Sci. Med.* 44: 1169–1180.

Rappaport, R. A. 1968. *Pigs for the Ancestors: Ritual in the Ecology of a New Guinea People.* New Haven: Yale University Press.

Rattray, R. S. 1923. *Ashanti.* New York: Negro Universities Press.

Ravenholt, R. T., and J. Chao. 1984. *World Fertility Trends.* Population Reports, series J, no. 2, George Washington University Department of Medical and Public Affairs, Washington, D.C.

Redford, K. 1991. The ecologically noble savage. *Cultural Survival Quart.* 15(1): 46–48.

Reeve, H. K. 1989. The evolution of conspecific acceptance thresholds. *American Naturalist* 133: 407–435.

Reeve, H. K. 1998. Acting for the good of others: Kinship and reciprocity with some new twists. In C. Crawford and D. Krebs (eds.), *Handbook of Evolutionary Psychology: Ideas, Issues, and Applications,* 43–85. Mahwah, N.J.: Erlbaum.

Reeve, H. K., and P. Nonacs. 1992. Social contracts in wasp societies. *Nature* 359: 823–825.

Reik, W., and A. Surani (eds.). 1997. *Frontiers in Molecular Biology: Cenomic Imprinting.* Oxford: Oxford University Press.

Rice, D. 1990. *Shattered Vows: Priests Who Leave.* New York: Triumph Books.

Richards, A. I. 1951. The Bemba of north-eastern Rhodesia. In E. Colson and M. Gluckman (eds.), *Seven Tribes of Central Africa,* 164–193. Oxford: Oxford University Press.

Richardson, L. 1960. *Statistics of Deadly Quarrels.* Pittsburgh: Boxwood Press.

Richerson, P. J., and R. Boyd. 1992. Cultural inheritance and evolutionary ecology. In E. A. Smith and B. Winterhalder (eds.), *Evolutionary Ecology and Human Behavior,* 61–92. New York: Aldine de Gruyter.

Ricklefs, R. E. 1983. Avian postnatal development. In D. S. Farmer, J. R. King, and K. C. Parkes (eds.), *Avian Biology,* vol. 7, 1–83. New York: Academic Press.

Ridgeway, C., E. Boyle, K. Kuipers, and D. Robinson. 1998. How do status beliefs develop? The role of resources and interactional experience. *Amer. Sociol. Rev.* 63: 331–350.

Ridley, M. 1993. *The Red Queen: Sex and the Evolution of Human Nature.* New York: Viking.

Ridley, M. 1996. *The Origins of Virtue.* New York: Viking.

Ritchie, M. 1990. Optimal foraging and fitness in Colombian ground squirrels. *Oecologia* 82: 56–67.

Ritchie, M. 1991. Inheritance of optimal foraging behavior in Colombian ground squirrels. *Evolutionary Ecology* 5:146–159.

Robey, B., S. O. Rutstein, and L. Morris. 1993. The fertility decline in developing countries. *Scientific American,* December 1993, 60–67.

Rodseth, L., R. W. Wrangham, A. M. Harrigan, and B. B. Smuts. 1991. The human community as a primate society. *Current Anthropology* 32: 221–241 (discussion 241–254).

Roff, D. A. 1992. *The Evolution of Life Histories: Theory and Analysis.* New York: Chapman and Hall.

Rogers, A. R. 1990. The evolutionary economics of human reproduction. *Ethol. and Sociobiol.* 11: 479–495.

Rogers, A. R. 1991. Conserving resources for children. *Human Nature* 2(1): 73–82.

Rogers, A. R. 1993. Why menopause? *Evolutionary Ecology* 7: 406–420.

Rogers, A. R. 1995. For love or money: The evolution of reproductive and material motivations. In R. I. M. Dunbar (ed.), *Human Reproductive Decisions,* 76–95. London: St. Martin's Press, in association with the Galton Institute.

Roscoe, J. 1923. *The Bakitara or Banyoro; The First Part of the Report to the Mackie Ethnological Expedition to Central Africa.* Cambridge, U.K.: Cambridge University Press.

Rose, M. R., and G. V. Lauder (eds.). 1996. *Adaptation.* San Diego: Academic Press.

Rose, S. 1997. *Lifelines: Biology beyond Determinism.* Oxford: Oxford University Press.

Rosenblatt, P., and M. R. Cunningham. 1976. Sex differences in cross-cultural perspective. In B. B. Lloyd and J. Archer (eds.), *Explorations in Sex Differences.* New York: Academic Press.

Røskaft, E., A. Wara, and Å. Viken. 1992. Reproductive success in relation to resource access and parental age in a small Norwegian farming parish during the period 1700–1900. *Ethol. and Sociobiol.* 13: 443–461.

Ross, C. 1998. Primate life histories. *Evolutionary Anthropology* 7: 54–63.

Ross, M. H. 1983. Political decision making and conflict: Additional cross-cultural codes and scales. *Ethnology* 22: 169–192.

Roth, H. L. (ed. note of Brooke Low). 1892. The Natives of Borneo. *J. Anthropol. Inst. Great Britain and Ireland* 22.

Rowell, T. E. 1988. Beyond the one-male group. *Behaviour* 104: 189–210.

Ruse, M. 1982. *Darwinism Defended: A Guide to the Evolution Controversies.* Reading, Mass.: Addison-Wesley.

Ruse, M. 1986. *Taking Darwin Seriously.* New York: Holt.

Ryan, M. 1998. Sexual selection, receiver biases, and the evolution of sex differences. *Science* 281: 1999–2003.

Sacher, G. A. 1959. Relationship of lifespan to brain weight and body weight in mammals. In G.E.W. Wolstenholme and M. O'Connor (eds.), *CIBA Foundation Symposium on the Lifespan of Animals.* Boston: Little, Brown.

Sade, D. S. 1967. Determinants of dominance in a group of free ranging rhesus monkeys. In S. A. Altmann (ed.), *Social Communication among Primates,* 99–114. Chicago: University of Chicago Press.

Schapera, I. 1930. *The Khoisan Peoples of South Africa.* London: Routledge and Kegan Paul.

Schein, E. 1957. Distinguishing characteristics of collaborators and resisters among American POWs. *J. Abnormal and Soc. Psych.* 55: 197–201.

Schlegel, A., and H. Barry. 1991. *Adolescence: An Athropological Inquiry.* New York: Free Press.

Schoen, R., Y. J. Kim, C. A. Nathanson, J. Fields, and N. M. Astone. 1997. Why do Americans want children? *Pop. Devel. Rev.* 23: 333–358.

Schofield, R. S., and D. Coleman. 1986. Introduction: The state of population theory. In D. Coleman and R. Schofield (eds.), *The State of Population Theory: Forward from Malthus,* 1–13. London: Blackwell.

Schultz, T. P. 1982. Family composition and income inequality. Yale University Economic Growth Center Paper 25.

Schultz, T. P. 1985. Changing world prices, women's wages, and the fertility transition: Sweden, 1860–1910. *J. Pol. Econ.* 93(6): 1126–1154.

Seeley, T. D. 1985. *Honeybee Ecology: A Study of Adaptation in Social Life.* Princeton, N.J.: Princeton University Press.

Segal, Nancy L. 1999. *Entwined Lives: Twins and What They Tell Us About Human Nature.* New York: Dutton.

Segal, N. L., and S. L. Hershberger. 1999. Cooperation and competition between twins: Findings from a Prisoner's Dilemma game. *Evol. and Human Behav.* 20: 29–51.

Seger, J., and J. W. Stubblefield. 1996. Optimization and adaptation. In Michael Rose and George Lauder (eds.), *Adaptation,* 93–123. New York: Academic Press.

Sharpe, P. 1990. The total reconstitution method: A tool for class-specific study? *Local Pop. Stud.* 44: 41–51.

Shaw, R. P., and Y. Wong. 1989. *The Genetic Seeds of Warfare: Evolution, Nationalism, and Patriotism.* Boston: Unwin Hyman.

Sherman, P. W. 1977. Nepotism and the evolution of alarm calls. *Science* 197: 1246–1253.

Sherman, P. W. 1981. Kinship, demography, and Belding's ground squirrel nepotism. *Behav. Ecol. and Sociobiol.* 8: 251–259.

Sherman, P. W., and H. K. Reeve. 1997. Forward and backward: Alternative approaches to studying human social evolution. In L. Betzig (ed.), *Human Nature: A Critical Reader,* 147–158. Oxford: Oxford University Press.

Sherwood, J. 1988. *Poverty in Eighteenth-Century Spain: Women and Children of the Inclusa.* Toronto: University of Toronto Press.

Shively, C., and D. G. Smith. 1985. Social status and reproductive success of male *Macaca fascicularis. Amer. J. Primatol.* 9: 129–135.

Sibley, R. M., and P. Calow. 1986. *Physiological Ecology of Animals: An Evolutionary Approach.* Oxford and Boston: Blackwell Scientific.

Sichona, F. J. 1993. The polygyny-fertility hypothesis revisited: The situation in Ghana. *J. Biosoc. Sci.* 25: 473–482.

Sigmund, K. 1993. *Games of Life.* Oxford: Oxford University Press.

Silk, J. B. 1987. Social behavior in evolutionary perspective. In B. B. Smuts, D. L. Cheney, R. M. Seyfarth, R. W. Wrangham, and T. T. Strusaker (eds.), *Primate Societies,* 318–329. Chicago: University of Chicago Press.

Silk, J. B., and R. Boyd. 1983. Cooperation, competition, and mate choice in matrilineal macaque groups. In S. Wasser (ed.), *Social Behavior of Female Vertebrates.* New York: Academic Press.

Silverman, I., and M. Eals. 1992. Sex differences in spatial abilities: Evolutionary theory and data. In J. H. Barkow, L. Cosmides, and J. Tooby (eds.), *The Adapted Mind,* 523–549. New York: Oxford University Press.

Silverman, I., and M. Eals. 1998. The evolutionary psychology of spatial sex differences. In C. Crawford and D. L. Krebs (eds.), *Handbook of Evolutionary Psychology,* 595–612. Mahwah, N.J.: Erlbaum.

Silverman, I., and K. Phillips. 1998. The evolutionary psychology of spatial sex differences. In C. Crawford and D. L. Krebs (eds.), *Handbook of Evolutionary Psychology,* 595–612. Mahwah, N.J.: Erlbaum.

Simon, J. 1974. *The Effects of Income on Fertility.* Carolina Population Center Monograph 19. Chapel Hill: University of North Carolina Press.

Simon, J. 1996. *The Ultimate Resource 2.* Princeton, N.J.: Princeton University Press.

Sing, C. F., M. Haviland, A. R. Templeton, and S. Reilly. 1995. Alternative strategies for predicting risk of atherosclerosis. In F. P. Woodford, J. Davignon, and A. Sniderman (eds.), *Atherosclerosis X,* 638–644. Amsterdam: Elsevier.

Sing, C. F., M. Haviland, A. R. Templeton, K. Zerba, and S. Reilly. 1992. Biological complexity and strategies for finding DNA variations responsible for inter-individual variation in risk of a common chronic disease, coronary artery disease. *Annals of Medicine* 24: 539–547.

Singer, J. David. 1980. Accounting for international war. *Ann. Rev. Sociol.* (1980): 349–376.

Singer, J. David. 1989. The political origins of international war. In J. Groebel and R. A. Hinde (eds.), *Aggresssion and War: Their Biological and Social Bases*, 202–220. Cambridge, U.K.: Cambridge Univ. Press.

Singh, D. 1993a. Adaptive significance of female attractivness: Role of waist-to-hip ratio. *J. Personality and Soc. Psych.* 65: 293–307.

Singh, D. 1993b. Body shape and women's attractiveness: The critical role of waist-to-hip ratio. *Human Nature* 4: 297–321.

Singh, D. 1994. Body fat distribution and perception of desirable female body shape by young Black men and women. *Int. J. Eating Disorders* 16: 289–294.

Singh, D., and S. Luis. 1995. Ethnic and gender concensus for the effect of waist-to-hip ratio on judgement of women' attractiveness. *Human Nature* 6: 51–65.

Sipe, A.W.R. 1990. *A Secret World: Sexuality and the Search for Celibacy.* New York: Brunner Mazel.

Skuse, D. H., R. S. James, D.V.M. Bishop, B. Coppin, P. Dalton, G. Aamodt-Leeper, M. Bacarese-Hamilton, C. Creswell, R. McGurk, P. A. Jacobs. 1997. Evidence from Turner's syndrome of an imprinted X-linked locus affecting cognitive function. *Nature* 387: 705–708.

Smith, B. H. 1990. The cost of a large brain. *Behav. and Brain Sci.* 13(2): 365–366.

Smith, B. H. 1991. Dental development and the evolution of life history in Hominidae. *Amer. J. Phys. Anthropol.* 86: 157–174.

Smith, B. H. 1992. Life history and the evolution of human maturation. *Evolutionary Anthropology* 1(4): 134–142.

Smith, E. A. 1992. Human behavioral ecology. I. *Evolutionary Anthropology* 1(1): 20–25. II, *Evolutionary Anthropology* 1(2): 50–56.

Smith, E. A. 1995. Inuit sex ratio: A correction and an addendum. *Current Anthropology* 36: 658–659.

Smith, E. A. 1998a. Is Tibetan polyandry adaptive? Methodological and metatheoretical analyses. *Human Nature* 9: 225–261.

Smith, E. A. 1998b. Commentary on D. S. Wilson's "Hunting, sharing and multilevel selection." *Current Anthropology* 39: 90–91.

Smith, E. A. 1999. Three styles in the evolutionary analysis of human behavior. In L. Cronk, N. Chagnon, and W. Irons (eds.), *Evolutionary Biology and Human Social Behavior 20 Years Later.* Hawthorne, N.Y.: Aldine de Gruyter.

Smith, E. A., and R. Boyd. 1990. Risk and reciprocity: Hunter-gatherer socioecology and the problem of collective action. In E. Cashdan (ed.), *Risk and Uncertainty in Tribal and Peasant Economies*, 167–191. Boulder, Colo.: Westview Press.

Smith, E. A., and S. A. Smith. 1994. Inuit sex-ratio variation. *Current Anthropology* 35: 595–624.

Smith, E. A., and B. Winterhalder (eds.). 1992. *Evolutionary Ecology and Human Behavior.* New York: Aldine de Gruyter.

Smith, J. E., and P. R. Kunz. 1976. Polygyny and fertility in nineteenth-century America. *Population Studies* 30: 465–480.

Smith, M. S., B. J. Kish, and C. B. Crawford. 1986. Inheritance of wealth as human kin investment. *Ethol. and Sociobiol.* 8(3): 171–182.

Smith, T., and T. Polacheck. 1981. Reexamination of the life table for northern fur seals with implications about population regulatory mechanisms. In C. W. Fowler and T. Smith (eds.), *Dynamics of Large Mammal Populations,* 99–120. New York: Wiley.

Smith, V. L. 1982. Microeconomic systems as an experimental science. *Amer. Econ. Rev.* 72: 923–955.

Smith, V. L. 1997. The two faces of Adam Smith. Southern Economic Association, Distinguished Guest Lecture, Atlanta, November 21, 1997. Online at http://www.econlab.arizona.edu.

Smuts, B. B. 1985. *Sex and Friendship in Baboons.* New York: Aldine de Gruyter.

Smuts, B. B. 1987a. Sexual competition and mate choice. In B. B. Smuts, D. L. Cheney, R. M. Seyfarth, R. W. Wrangham, and T. T. Strusaker (eds.), *Primate Societies,* 385–399. Chicago: University of Chicago Press.

Smuts, B. B. 1987b. Gender, aggression, and influence. In B. B. Smuts, D. L. Cheney, R. M. Seyfarth, R. W. Wrangham, and T. T. Strusaker (eds.), *Primate Societies,* 400–412. Chicago: University of Chicago Press.

Smuts, B. B. 1996. Male aggression against women: An evolutionary perspective. In D. Buss and N. Malamuth (eds.), *Sex, Power, Conflict: Evolutionary and Feminist Perspectives,* 231–268. New York: Oxford University Press.

Smuts, B. B., and D. J. Gubernick. 1992. Male-infant relationships in non-human primates: Paternal investment or mating effort: In B. Hewlett (ed.), *Father-Child Relations: Cultural and Biosocial Contexts,* 1–29. New York: Aldine de Gruyter.

Smuts, B. B., and R. Smuts. 1993. Male aggression and sexual coercion of females in nonhuman primates and other mammals: Evidence and theoretical implications. In P. Slater, J. Rosenblatt, M. Milinski, and C. Snowden (eds.), *Advances in the Study of Behavior* 1–63. New York: Academic Press.

Smuts, Robert. 1992. Fat, sex, class, adaptive flexibility, and cultural change. *Ethol. and Sociobiol.* 13: 521–542.

Snowden, D. A., R. L. Kane, W. Beeson, G. Burke, J. Sprafka, J. Potter, D. Jacobs, and R. Phillips. 1989. Is early natural menopause a biologic marker of health and aging? *Amer. J. Public Health* 79: 709–714.

Sober, E., and D. S. Wilson. 1998. *Do Unto Others: The Evolution and Psychology of Unselfish Behavior.* Cambridge, Mass.: Harvard University Press.

Speck, F. G. 1931. Montagnais-Naskapi bands and early Eskimo distribution in the Labrador Peninsula. *American Anthropologist* 33: 557–600.

Sperber, D., F. Cara, V. Girotto. 1995. Relevance theory explains the selection task. *Cognition* 57: 31–95.

Stearns, S. C., 1992. *The Evolution of Life Histories.* Oxford: Oxford University Press.

Stebbins, G. L. 1977. In defense of evolution: Tautology or theory? *American Naturalist* 111: 386–390.

Stephens, D. W., and J. R. Krebs. 1986. *Foraging Theory.* Princeton, N.J.: Princeton University Press.

Stewart, K. J., and A. H. Harcourt. 1986. Gorillas: variation in female relationships. In B. B. Smuts, D. L. Cheney, R. M. Seyfarth, R. W. Wrangham, and T. T. Strusaker, (eds.), *Primate Societies,* 155–164. Chicago: University of Chicago Press.

Stiller, L. F. 1973. *The Rise of the House of Gorkha: A Study in the Unification of Nepal, 1768–1816.* New Delhi: Manjusri Publishing House.

Stoessinger, J. G. 1982. *Why Nations Go to War.* New York: St. Martin's Press.

Strassmann, B. I. 1991. Sexual selection, paternal care, and concealed ovulation in humans. *Ethol. and Sociobiol.* 2: 31–40.

Strassmann, B. I. 1997. Polygyny as a risk factor for child mortality among the Dogon. *Current Anthropology* 38: 688–695.

Strassmann, B. I. 1999. Polygyny, family structure, and child mortality: A prospective study among the dogon of Mali. In L. Cronk, N. Changon, and W. Irons (eds.), *Evolution and Social Behavior.* New York: Aldine de Gruyter.

Strate, J. 1982. Warfare and political evolution: A cross-cultural test. Ph.D. diss., University of Michigan, Ann Arbor.

Strier, K. 1996. Male reproductive strategies in New World primates. *Human Nature* 7: 105–123.

Strong, W. D. 1929. Cross-cousin marriage and the culture of the northeastern Algonkian. *American Anthropologist* 31: 277.

Struhsaker, T. T., and L. Leland. 1987. Colobines: Infanticide by adult males. In B. B. Smuts, D. L. Cheney, R. M. Seyfarth, R. W. Wrangham, and T. T. Struhsaker (eds.), *Primate Societies,* 83–98. Chicago: University of Chicago Press.

Sugiyama, Y., S. Kawamoto, O. Takenaka, K. Kumazaki, and N. Miwa. 1993. Paternity discrimination and inter-group relationships of chimpanzees at Bossou. *Primates* 34: 545–552.

Sundin, J. 1976. Theft and penury in Sweden, 1830–1920: A comparative study at the county level. *Scandinavian J. Hist.* 1: 265–292.

Sundin, J. and L.-G. Tedebrand. 1981. Mortality and Morbidity in Swedish Iron Foundries 1750–1875. In Anders Brändström and Jan Sundin (eds.), *Tradition and Tansition: Studies in Microdemography and Social Change,* 105–159. Demographic Database, University of Umeå, Report no. 2.

Sundin, J., and L-G. Tedebrand. 1984. Migratory workers and settled workers: A study of demographic behaviour at Swedish 19th-century iron foundries. Paper delivered at the Ninth Annual Meeting, Social Science History Association, October 25–28, 1984, Ontario, Canada.

Swanton, J. R. 1911. *Indian Tribes of the Lower Mississippi Valley.* Smithsonian Institution, Bureau of American Ethnology, Bulletin 43.

Swanton, J. R. 1928a. *Social Organization and Social Usages of the Indians of the Creek Confederacy.* Forty-Second Annual Report of the Bureau of American Ethnology, 23–473.

Swanton, J. R. 1928b. Aboriginal culture of the Southeast. Forty-Second Annual Report of the Bureau of American Ethnology, 673–727.

Swartz, M., V. Turner, and A. Tuden (eds.). 1966. *Political Anthropology.* Chicago: Aldine de Gruyter.

Sweeney, T. A. 1992. *A Church Divided: The Vatican versus American Catholics.* New York: Prometheus Books.

Szal, J. A. 1972. Sex differences in the cooperative and competitive behaviors of nursery school children. M.S. thesis, Stanford University, Stanford, Calif.

Takahata, Y., H. Ihobe, and G. Idani. 1996. Comparing copulations of chimpanzees and bonobos: Do females exhibit proceptivity or receptivity? In W. C. McGrew, L. Marchant, and T. Nishida (eds.), *Great Ape Societies,* 146–155. Cambridge U.K.: Cambridge University Press.

Tambiah, S. J. 1966. Polyandry in Ceylon. In C. von Fuhrer-Haimendorf (ed.), *Caste and Kin in Nepal, India, and Ceylon.* London: Asia Publishing House.

Taşiran, A. C. 1995. *Fertility Dynamics: Spacing and Timing of Births in Sweden and the United States.* Amsterdam: Elsevier.

Taylor, P. D., and G. C. Williams. 1982. The lek paradox is not resolved. *Theor. Pop. Biol.* 22: 392–409.

Templeton, A. R. 1995. A cladistic analysis of phenotypic associations with haplotypes inferred from restriction endonuclease mapping or DNA sequencing. 5. Analysis of case/control sampling designs: Alzheimer's disease and the apoprotein E locus. *Genetics* 140: 403–409.

Terborgh, J., and A. W. Goldizen. 1985. On the mating system of the cooperatively breeding saddle-backed tamarin (*Saguinus fuscicollis*). *Behav. Ecol. Sociobiol.* 16: 293–299.

Thomas, D. S. 1941. *Social and Economic Aspects of Swedish Population Movements: 1750–1933.* New York: Macmillan.

Thompson, J., and M. Britton. 1980. Some socioeconomic differentials in fertility in England and Wales. In R. W. Hiorus. *Demographic Patterns in Developed Societies,* 1–10. London: Taylor and Franci.

Thornhill, R., and J. Alcock. 1983. *The Evolution of Insect Mating Systems.* Cambridge, Mass.: Harvard University Press.

Thornhill, A. R., and P. S. Burgoyne. 1993. A paternally imprinted X-chromosome retards the development of the early mouse embryo. *Development* 118: 171–174.

Thornhill, R., and S. W. Gangestad. 1994. Fluctuating asymmetry and human sexual behavior. *Psych. Science* 5: 297–302.

Thornhill, R., S. W. Gangestad, and R. Comer. 1995. Human female orgasm and mate fluctuating asymmetry. *Animal Behaviour* 50: 1601–1615.

Tiger, L., and R. Fox. 1971. *The Imperial Animal.* New York: Holt, Reinhart, and Winston.

Tilly, C. 1978. The historical study of vital processes. In C. Tilly (ed.), *Historical Studies of Changing Fertility*, 1–55. Princeton, N.J.: Princeton University Press.

Timaeus, I. M., and A. Reynar. 1998. Polygynists and their wives in sub-Saharan Africa: An analysis of five demographic and health surveys. *Population Studies* 52: 145–162.

Tooby, J., and L. Cosmides. 1988. The evolution of war and its cognitive foundations. Institute for Evolutionary Studies Technical Report 88–1.

Torres, A., and J. D. Forrest. 1988. Why do women have abortions? *Family Planning Persp.* 20(4): 169–176.

Trail, P. W. 1985. The intensity of sexual selection: Intersexual and interspecific comparisons require consistent measures. *American Naturalist* 126: 434–439.

Travisano, M., and R. E. Lenski. 1996. Long-term experimental evolution in Escherichia coli. IV. Targets of selection and the specificity of adaptation. *Genetics* 143: 15–26.

Travisano, M., J. A. Mongold, A. F. Bennett, and R. E. Lenski. 1995. Experimental tests of the roles of adaptation, chance, and history in evolution. *Science* 267: 87–90.

Trivers, R. L. 1971. The evolution of reciprocal altruism. *Quart. Rev. Biol.* 46: 35–57.

Trivers, R. L. 1972. Parental investment and sexual selection. In B. Campbell (ed.), *Sexual Selection and the Descent of Man*. Chicago: Aldine de Gruyter.

Trivers, R. L. 1974. Parent-offspring conflict. *American Zoologist* 14: 249–264.

Trivers, R. L. 1985. *Social Evolution*. Menlo Park, Calif.: Benjamin Cummings.

Trivers, R. L., and D. E. Willard. 1973. Natural selection of parental ability to vary the sex ratio. *Science* 179: 90–92.

Tronick, E. Z., R. B. Thomas, and M. Daltabuit. 1994. The Quechua manta pouch: A caretaking practice for buffering the multiple stressors of high altitude. *Child Development* 65: 1005–1013.

Turke, P. W. 1988. Helpers at the nest: Childcare networks in Ifaluk. In L. Betzig, M. Borgerhoff Mulder, and P. Turke (eds.), *Human Reproductive Behaviour: A Darwinian Perspective*. Cambridge, U.K.: Cambridge University Press.

Turke, P. W. 1989. Evolution and the demand for children. *Pop. and Devel. Rev.* 15(1): 61–90.

Turke, P. 1990. Which humans behave adaptively, and why does it matter? *Ethol. and Sociobiol.* 11: 305–339.

Turke, P. W. 1992. Theory and evidence on wealth flows and old-age security: A reply to Fricke. *Pop. and Devel. Rev.* 17(4): 687–702.

Turke, P. W., and L. Betzig. 1985. Those who can do: Wealth, status, and reproductive success on Ifaluk. *Ethol. and Sociobiol.* 6: 79–87.

Tutin, C.E.G. 1979. Mating patterns and reproductive strategies in a community of wild chimpanzees (*Pan troglodytes schweinfurtii*). *Behav. Ecol. and Sociobiol.* 6: 29–38.

United Nations. 1996. *U.N. World Population Prospects*, 1996 revision.

Vale, M. 1981. *War and Chivalry: Warfare and Aristocratic Culture in England, France, and Burgundy at the End of the Middle Ages*. Athens: University of Georgia Press.

van Schaik, C. P. 1989. The ecology of social relationships among female primates. In V. Standen and R. Foley (eds.), *Comparative Socioecology: The Behavioural Ecology of Humans and Other Mammals*, 195–218. Oxford: Blackwell.

van Schaik, C. P. 1996. Social evolution in primates: The role of ecological factors and male behaviour. *Proc. British Acad.* 88: 9–31.

van Schaik, C. P., and R. I. Dunbar. 1990. The evolution of monogamy in large primates: A new hypothesis and some crucial tests. *Behaviour* 115: 30–61.

Van Valen, L. 1973. A new evolutionary law. *Evolutionary Theory* 1: 1–30.

Vayda, A. P. 1960. *Maori Warfare*. Polynesian Society Maori Monographs No. 2.

Viazzo, P. P. 1990. *Upland Communities: Environment, Population, and Social Structure in the Alps since the Sixteenth Century*. Cambridge, U.K.: Cambridge University Press.

Vining, D. R. 1986. Social versus reproductive success: The central theoretical problem of human sociobiology. *Behav. Brain Sci.* 9: 167–187.

Vinovskis, M. A. 1972. Mortality rates and trends in Massachusetts before 1860. *J. Econ. Hist.* 32: 184–218.

Voland, E. 1984. Human sex-ratio manipulation: Historical data from a German parish. *J. Human Evol.* 13: 99–107.

Voland, E. 1989. Differential parental investment: Some ideas on the contact area of European social history and evolutionary biology. In V. Standen and R. A. Foley (eds.), *Comparative Socioecology: The Behavioural Ecology of Humans and Other Mammals*, 391–402. Special Publication No. 8, British Ecological Society.

Voland, E. 1990. Differential reproductive success within the Krummh"rn population (Germany, 18th and 19th centuries). *Behav. Ecol. and Sociobiol.* 26: 65–72.

Voland, E., and R.I.M. Dunbar. 1995. Resource competition and reproduction: The relationship between economic and parental strategies in the Krummh"rn population (1720–1874). *Human Nature* 6: 33–49.

Voland, E., E. Siegelkow, and C. Engle. 1990. Cost/benefit oriented parental investment by high status families: The Krummhörn case. *Ethol. and Sociobiol.* 12: 105–118.

Wade, M. J. 1979. Sexual selection and variance in reproductive success. *American Naturalist* 114: 742–747.

Wade, M. J. and S. J. Arnold. 1980. The intensity of sexual selection in relation to male sexual behavior, female choice, and sperm precedence. *Animal Behaviour* 28: 446–461.

Wall, R. 1984. Real property, marriage, and children: The evidence from four preindustrial communities. In R. M. Smith (ed.), *Land, Kinship, and the Life-Cycle*, 443–479. Cambridge, U.K.: Cambridge University Press.

Wang, F., J. Lee, and C. Campbell. 1995. Marital fertility control among the Qing nobility: Implications for two types of preventive checks. *Population Studies* 49: 383–400.

Warner, R. R., R. K. Harlan, and E. G. Leigh. 1975. Sex change and sexual selection. *Science* 190: 633–638.

Wasser, S. 1983. Reproductive competition and cooperation among female yellow baboons. In S. Wasser (ed.), *Social Behavior of Female Vertebrates*, New York: Academic Press.

Watkins, S. C. 1989. The fertility transition: Europe and the third world compared. In J. M. Stycos (ed.), *Demography as an Interdiscipline*, 27–55. New Brunswick, N.J.: Transaction Publishers.

Watson, P. 1978. *War on the Mind.* New York: Basic Books.

Watts, D. P. 1989. Infanticide in mountain gorillas: New cases and a reconsideration of the evidence. *Ethology* 81: 1–18.

Watts, D. P. 1996. Comparative socio-ecology of gorillas. In W. C. McGrew, L. F. Marchant, and T. Nishida (eds.), *Great Ape Societies*, 16–28. Cambridge, U.K.: Cambridge University Press.

Webster, D. 1975. Warfare and the evolution of the state: A reconsideration. *American Antiquity* 40(4): 464–470.

Webster, M., Jr., and S. Hyson. 1998. Creating status characteristics. *Amer. Sociol. Rev.* 63: 351–378.

Wen, X. 1993. Effect of son preference and population policy on sex ratios at birth in two provinces of China. *J. Biosoc. Sci.* 25: 509–521.

West-Eberhard, M. J. 1975. The evolution of social behavior by kin selection. *Quart. Rev. Biol.* 50: 1–33.

West-Eberhard, M. J. 1978. Temporary queens in Metapolybia wasps: Non-reproductive helpers without altruism? *Science* 200: 441–443.

West-Eberhard, M. J. 1989. Phenotypic plasticity and the origins of diversity. *Ann. Rev. Ecol. Syst.* 20: 249–278.

White, D. R. 1988. Rethinking polygyny: Co-wives, codes and cultural systems. *Current Anthropology* 29(4): 529–558.

White, D. R., and M. L. Burton. 1988. Causes of polygyny: Ecology, economy, kinship, and warfare. *Amer. Anthropol.* 90(4): 871–887.

Whiting, B., and C. P. Edwards. 1973. A cross-cultural analysis of sex differences in the behavior of children aged three through eleven. *J. Soc. Psych.* 91: 171–188.

Whiting, B., and C. P. Edwards. 1988. *Children of Different Worlds: The Formation of Social Behavior.* Cambridge, Mass.: Harvard University Press.

Whiting, B., and J. Whiting. 1975. *Children of Six Cultures: A Psychocultural Analysis.* Cambridge, Mass.: Harvard University Press.

Whitten, P. L. 1987. Infants and adult males. In B. B. Smuts, D. L. Cheney, R. M. Seyfarth, R. W. Wrangham, and T. T. Struhsaker (eds.), *Primate Societies*, 343–357. Chicago: University of Chicago Press.

Whittington, L. A., J. Alm, and E. Peters. 1990. Fertility and the personal exemption: Implicit pronatalist policy in the United States. *Amer. Econ. Rev.* 80: 545–556.

Whyte, M. K. 1978. Cross-cultural codes dealing with the relative status of women. *Etnology* 17: 211–237.

Whyte, M. K. 1979. *The Status of Women in Pre-industrial Society.* Princteon, N.J.: Princeton University Press.

Wickler, W. and U. Seibt. 1981. Monogamy in Crustacea and man. *Zofür Tierpsychologie* 57: 215–234.

Wilkinson, G. S. 1984. Reciprocal food sharing in the vampire bat. *Nature* 308: 181–184.

Wilkinson, P. 1986. Terrorism: international dimensions. In W. Gutteridge (ed.), *Contemporary Terrorism,* 29–56. New York: Facts on File.

Williams, C. L., and W. H. Meck. 1991. The organizational effects of gonadal steroids on sexually dimorphic spatial ability. *Psychoneuroendocrinology* 16: 155–176.

Williams, G. C. 1957. Pleiotropy, natural selection, and the evolution of senescence. *Evolution* 11: 398–411.

Williams, G. C. 1966. *Adaptation and Natural Selection.* Princeton, N.J.: Princeton University Press.

Williams, G. C. 1975. *Sex and Evolution.* Princeton, N.J.: Princeton University Press.

Williams, G. C. 1989. A sociobiological explanation of *Evolution and Ethics.* In J. Paradis and G. C. Williams, *Evolution and Ethics: T. H. Huxley's Evolution and Ethics with New Essays on Its Victorian and Sociobiological Context,* 179–214. Princeton, N.J.: Princeton University Press.

Williams, G. C. 1992a. *Natural Selection: Domains, Levels, and Challenges.* Oxford: Oxford University Press.

Williams, G. C. 1992b. Gaia, nature worship, and biocentric fallacies. *Quart. Rev. Biol.* 67: 479–486.

Wilson, D. S. 1975. New model for group selection. *Science* 189: 8701.

Wilson, D. S. 1980. *The Natural Selection of Populations and Communities.* Menlo Park, Calif.: Banjamin Cummings.

Wilson, D. S. 1998a. Game theory and human behavior. In L. Dugatkin and K. Reeves (eds.), *Game Theory and Animal Behavior,* 261–282. Oxford: Oxford University Press.

Wilson, D. S. 1998b. Hunting, sharing and multilevel selection. *Current Anthropology* 39: 73–86 (comments on pp. 86–97).

Wilson, D. S., and E. Sober. 1994. Reintroducing group selection to the human behavioral sciences. *Behav. and Brain Sci.* 17: 585–608.

Wilson, D. S., D. Near, and R. R. Miller. 1996. Machiavellianism: A synthesis of the evolutionary and psychological literatures. *Psych. Bull.* 119: 285–299.

Wilson, E. O. 1975. *Sociobiology: The New Synthesis.* Cambridge, Mass.: Harvard University Press.

Wilson, E. O. 1978. *On Human Nature.* Cambridge, Mass.: Harvard University Press.

Winter, J. M. 1989. Causes of war. In J. Groebel and R. A. Hinde (eds.), *Aggression and War: Their Biological and Social Bases,* 194–201. Cambridge, U.K.: Cambridge University Press.

Winterhalder, B. 1996a. A marginal model of tolerated theft. *Ethol. and Sociobiol.* 17: 37–53.

Winterhalder, B. 1996b. Social foraging and the behavioral ecology of intragroup transfers. *Evolutionary Anthropology* 5: 46–57.

Winterhalder, B. 1996c. Gifts given, gifts taken: The behavioral ecology of non-market, intragroup exchange. *J. Archaeological Research* 5: 121–168.

Winterhalder, B. 1997. Delayed reciprocity and tolerated theft. *Current Anthropology* 38: 74–75.

Winterhalder, B. 1999. Intragroup resource transfers. In C. B. Stanford and H. T. Bunn (eds.), *Meat Eating and Human Evolution.* Oxford: Oxford University Press.

Wolf, A. P. 1981. Women, widowhood and fertility in pre-modern China. In J. Dupâquier, E. Hélin, P. Laslett, M. Livi-Bacci, S. Sogner (eds.), *Marriage and Remarriage in Populations of the Past,* 139–147. New York: Academic Press.

Wood, J. W. 1990. Fertility in anthropological populations. *Ann. Rev. Anthropol.* 19: 211–242.

Wood, J. W. 1998. A theory of preindustrial population dynamics. *Current Anthropology* 39: 99–135.

Woolfenden, G. E., and J. W. Fitzpatrick. 1984. *The Florida Scrub Jay: Demography of a Cooperative Breeding Bird.* Princeton, N.J.: Princeton University Press.

Wrangham, R. 1979. Sex differences in chimpanzee dispersion. In D. Hamburg and E. McCown (eds.), *The Great Apes,* 481–490. Menlo Park, Calif.: Benjamin Cummings.

Wrangham, R. 1980. An ecological model of female-bonded primate groups. *Behaviour* 75: 262–300.

Wrangham, R. 1987. Evolution of social structure. In B. B. Smuts, D. L. Cheney, R. M. Seyfarth, R. W. Wrangham, and T. T. Struhsaker (eds.), *Primate Societies,* 282–296. Chicago: University of Chicago Press.

Wrangham, R., and D. Peterson. 1996. *Demonic Males: Apes and the Origins of Human Violence.* Boston: Mariner Books.

Wright, L. 1997. *Twins and What They Tell Us about Who We Are.* New York: Wiley.

Wright, P. 1984. Biparental care in Aotus trivirgatus and *Callicebus moloch.* In M. Small (ed.), *Female Primates,* 59–75. New York: Alan Liss.

Wright, S. 1945. Tempo and mode in evolution: A critical review. *Ecology* 26: 415–419.

Wright, S. 1949. Adaptation and selection. In G. L. Jepson, E. Mayr, and G. G. Simpson (eds.), *Genetics, Paleontology, and Evolution,* 365–386. Princeton, N.J.: Princeton University Press.

Wright, S. 1977. *Evolution and the Genetics of Populations,* vol. 3: *Experimental Results and Evolutionary Deductions.* Chicago: University of Chicago Press.

Wrigley, E. A. 1983a. The growth of population in eighteenth-century England: A conundrum resolved. *Past and Present* 98: 121–150.

Wrigley, E. A. 1983b. Malthus's model of a pre-industrial economy. In J. Dupâquier, A. Fauve-Chamoux, and E. Grebenik (eds.), *Malthus Past and Present,* 111–124. New York: Academic Press.

Wrigley, E. A., and R. Schofield. 1981. *The Population History of England, 1541–1871.* Cambridge, Mass.: Harvard University Press.

Wrong, D. 1980. *Class Fertility Trends in Western Nations.* New York: Arno Press.

Wrong, D. 1985. Trends in class fertility in Western nations. *Can. J. Econ. and Pol. Sci.* 24: 216–219.

Wu, J., and R. Axelrod. 1995. How to cope with noise in the iterated Prisoner's Dilemma. *J. Conflict Resol.* 39: 183–189.

Wuethrich, B. 1998. Why sex? Putting theory to the test. *Science* 281: 1980–1982.

Wynne-Edwards, V. C. 1962. *Animal Dispersion in Relation to Social Behaviour.* Edinburgh: Oliver and Boyd.

Yalman. N. 1967. *Under the Bo Tree.* Berkeley: University of California Press.

Yi, Z., T. Ping, G. Baochang, X. Yi, L. Bohua, and L. Yongping. 1993. Causes and implications of the recent increase in the reported sex ratio at birth in China. *Pop. and Devel. Rev.* 19(2): 283–302.

Yu, D. W., and G. H. Shepard, Jr. 1998. Is beauty in the eye of the beholder? *Nature* 396: 321–322.

Zahavi, A. 1975. Mate selection—a selection for the handicap principle. *J. Theor. Biol.* 53: 205–214.

Zahavi, A., and A. Zahavi. 1997. *The Handicap Principle: A Missing Piece of Darwin's Puzzle.* Oxford: Oxford University Press.

Zhang, J., J. Quan, and P. Van Meerbergen. 1994. The effect of tax-transfer policies on fertility in Canada, 1921–88. *J. Human Resources* 29: 181–202.

Author Index

Subject Index

Taxonomic Index

Society/Social Group Index